Fermilab

Fermilab

Physics, the Frontier, and Megascience

LILLIAN HODDESON,
ADRIENNE W. KOLB,
AND CATHERINE WESTFALL

The University of Chicago Press   Chicago and London

Illustration on pages xii–xiii: Birds in flight over the Fermilab prairie. (Courtesy of Angela Gonzales.) Illustration on closing pages: Reflections of the Fermilab frontier. (Courtesy of Angela Gonzales.)

The University of Chicago Press, Chicago 60637
The University of Chicago Press, Ltd., London

Paperback edition 2011
Printed in the United States of America

20 19 18 17 16 15 14 13 12 11   3 4 5 6 7

ISBN-13: 978-0-226-34623-6 (cloth)
ISBN-13: 978-0-226-34624-3 (paper)
ISBN-10: 0-226-34623-4 (cloth)
ISBN-10: 0-226-34624-2 (paper)

Library of Congress Cataloging-in-Publication Data

Hoddeson, Lillian.
Fermilab : physics, the frontier, and megascience / Lillian Hoddeson, Adrienne W. Kolb, and Catherine Westfall.
p. cm.
Includes bibliographical references and index.
ISBN-13: 978-0-226-34623-6 (cloth : alk. paper)
ISBN-10: 0-226-34623-4 (cloth : alk. paper)
1. Fermi National Accelerator Laboratory—History. 2. Particle accelerators—Research—United States. 3. Particles (Nuclear physics)—Research—United States. I. Kolb, Adrienne W. II. Westfall, Catherine. III. Title.
QC789.2.U62.F474 2008
539.7′30973—dc22   2008006254

∞ The paper used in this publication meets the minimum requirements of the American National Standard for Information Sciences—Permanence of Paper for Printed Library Materials, ANSI Z39.48–1992.

Dedicated to

CAROL AND MICHAEL,

CHRISTINE, JEFFREY, AND KAREN,

AND

FORREST

in their quests for new frontiers

Contents

Acronyms and Abbreviations

ACP	Advanced Computer Program
AEC	Atomic Energy Commission
AGS	Alternating Gradient Synchrotron (Brookhaven National Laboratory [Brookhaven, NY])
AIP	American Institute of Physics
ANL	Argonne National Laboratory (Argonne, IL)
APS	American Physical Society
AUI	Associated Universities, Inc.
BeV	Billion Electron Volts
BOB	Bureau of the Budget
BOO	Board of Overseers
BNL	Brookhaven National Laboratory (Brookhaven, NY)
CBA	Colliding Beams Accelerator
CDF	Colliding Detector at Fermilab (Fermi National Accelerator Laboratory [Batavia, IL])
CDG	Central Design Group (Lawrence Berkeley Laboratory [Berkeley, CA])
CEA	Cambridge Electron Accelerator
CEBAF	Colliding Electron Beams Accelerator Facility (Newport News, VA)
CERN	European Center for Nuclear Research (Geneva)
CESR	Cornell Electron Storage Ring (Ithaca, NY)
DESY	Deutsches Elektronen-Synchrotron (Hamburg)
DOE	Department of Energy
DPF	Division of Particles and Fields (American Physical Society)
DZero	Detector at DØ (Fermi National Accelerator Laboratory [Batavia, IL])

ECFA European Committee for Future Accelerators
ERDA Energy Research and Development Administration
FFAG Fixed-Field Alternating Gradient
FNAL Fermi National Accelerator Laboratory (Batavia, IL)
GAC General Advisory Committee
GAO General Accounting Office
GeV giga electron volts (= billion electron volts [BeV])
HEPAP High Energy Physics Advisory Panel
HERA Hadron Elektron Ring Anlage (Hamburg)
IAEA International Atomic Energy Agency (Vienna)
ICFA International Committee for Future Accelerators
IEEE Institute of Electrical and Electronics Engineers
IHEP Institute for High Energy Physics (Serpukhov, Russia)
ISA Intersecting Storage Accelerator (ISABELLE, Brookhaven National Laboratory [Brookhaven, NY])
ISR Intersecting Storage Rings (CERN)
IUPAP International Union of Pure and Applied Physics
JCAE Joint Committee on Atomic Energy (U.S. Congress)
JINR Joint Institute for Nuclear Research (Dubna, Russia)
KEK Ko Enerugii Butsurigaku Kenkyusho (Tsukuba, Japan)
KeV Thousand Electron Volts
LBJ Library Lyndon Baines Johnson Presidential Library and Museum (Austin, TX)
LBL/LBNL/LRL Lawrence Berkeley Laboratory (Berkeley, CA)
LEP Large Electron-Positron (CERN)
LHC Large Hadron Collider (CERN)
MeV million electron volts
MOU memorandum of understanding
MURA Midwestern Universities Research Association (Madison, WI)
NAL National Accelerator Laboratory (Fermi National Accelerator Laboratory [Batavia, IL])
NAS National Academy of Sciences
NASA National Aeronautics and Space Administration
NRC National Research Council
NSF National Science Foundation
OER Office of Energy Research
OMB Office of Management and Budget
PAC Physics Advisory Committee
PEP Positron-Electron Project (Stanford Linear Accelerator Center [Menlo Park, CA])
PMG Project Management Group
PS Proton Synchrotron (CERN)
PSAC President's Science Advisory Committee
RDS Reference Designs Study (Lawrence Berkeley Laboratory [Berkeley, CA])

RF	radio frequency
RFP	request for proposal
SLAC	Stanford Linear Accelerator Center (Menlo Park, CA)
SLC	Stanford Linear Collider (Stanford Linear Accelerator Center [Menlo Park, CA])
SLIC	segmented liquid ionization calorimeter
SMD	silicon microstrip detector
SPEAR	Stanford Positron-Electron Accelerator Rings
S$p\bar{p}$S	Super Proton-Antiproton Synchrotron (CERN)
SPS	Super Proton Synchrotron (CERN)
SSC	Superconducting Super Collider (Waxahachie, TX)
SURA	Southeastern Universities Research Association
TAC	Texas Accelerator Center (The Woodlands, TX)
TeV	trillion electron volts
TNL	truly national laboratory
TPL	Tagged Photon Laboratory (Fermi National Accelerator Laboratory [Batavia, IL])
TPMS	Tagged Photon Magnetic Spectrometer (Fermi National Accelerator Laboratory [Batavia, IL])
UA1	Underground Area 1 (Super Proton Synchrotron, CERN)
UA2	Underground Area 2 (Super Proton Synchrotron, CERN)
UPC	Underground Parameters Committee
URA	Universities Research Association, Inc.
VBA	Very Big Accelerator
WAG	Western Accelerator Group
ZGS	Zero Gradient Synchrotron (Argonne, IL)

Introduction

On a clear summer afternoon in 2007, as the authors of this book complete their manuscript, a passenger peers through the window of an airplane. As his plane flies into Chicago's O'Hare Field from the west, he notices a large ring on the ground below (see fig. I.1). Near it he sees a towering white structure, a group of colorful smaller buildings, an expanse of forest, open fields, and lakes.

"What is that ring?" he asks his neighbor.

"Fermilab," she replies. "It's a physics laboratory. The government supports research there into what the universe is made of."

"Why the ring?"

"It's the four-mile-round main ring of a machine called the Tevatron. It turns protons into tools for looking inside the atomic nucleus. Huge magnets steer the protons around the ring, while high voltages accelerate them. When they reach their peak energy they collide with targets, sometimes other particles.Teams of physicists study the collisions as they explore the scientific frontier. Strangely, Fermilab's site is maintained to resemble the American West as it was a century ago. There's even a herd of buffalo!"

* * *

To this hypothetical airplane conversation, intended to introduce Fermilab and its frontierlike site to a member of the public, this book adds a historical dimension and asks why and how Fermilab saw the birth of an approach we call "megascience." We draw on the powerful frontier imagery,

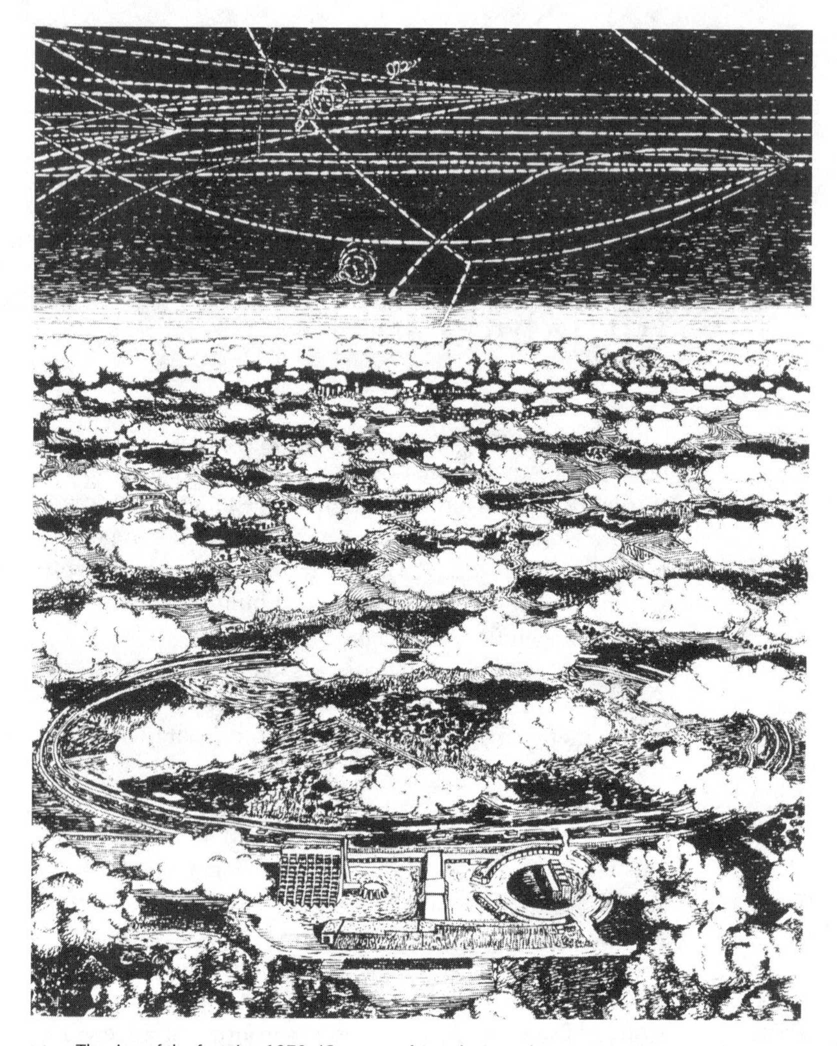

I.1 The ring of the frontier, 1972. (Courtesy of Angela Gonzales.)

which the physicists who work at Fermilab themselves use, and form it into a conceptual tool to help us tell Fermilab's story to a broad audience that includes historians and general readers, as well as physicists.

Many writers have used the term "megascience" interchangeably with "big science,"[1] to refer loosely to research supported by large sums of money, often in the form of government grants, and conducted over long time spans by large groups using large pieces of equipment.[2] In this book, however, the term "megascience" has a more specific meaning distinct from what is typically called big science. In most writings "big science" refers to the large-scale research conducted in the decades following World War II, when the funding for science in the United States was widely experienced as unlimited, allowing the research to grow very rapidly, with many parameters (e.g., size, cost, numbers of collaborators, time scales) increasing exponentially. Megascience, however, as we use the term, evolved in a time when the government's funding for science was slowing down. While many parameters still grew, they did so more slowly than in the first two postwar decades, causing dramatic changes in the nature of the research.

The scholarly focus of this book is the tension intrinsic to any pioneering exploration whether it occurs in a scientific field or as part of a country's history. Pulling the explorers forward is the call of the frontier, a force that drives them to pursue unknown territory. In experimental particle physcis, if funding allows, this call to exploration encourages experiments to grow larger, causing some of them to become long-lasting institutions in their own right, existing within their parent institutions. Resisting the call of the frontier are numerous constraints, many of them social or economic. The tension causes the exploration to take a particular form. One resisting trend is competition for limited resources at a laboratory, a counterforce that encourages the scientists to hold on to their precious investments, especially the larger ones, such as electronic detectors. Thus, the most striking feature of megascience is that its experiments seem no longer to "end," in the sense described by the historian of science Peter Galison.[3]

By-products of these changes and features of megascience are paradoxes, conflicts, and ironies which, while interesting for historians to explore, are often troubling to the scientists because they threaten to undermine fundamental aspects of their research. One paradox in this story of Fermilab is that every advance toward the frontier worked to limit the focus of physicists to fewer basic problems, resulting in a narrowing of the research frontier. A prominent conflict in the Fermilab story derived from the opposition between founding director Robert R. Wilson's ideal that science should be pursued by lone independent explorers and the reality that research into the heart of matter requires large, typically more bureaucratic, team efforts. Among many ironies in the story of Fermilab is that megascience evolved there directly out of Wilson's vision of small-scale research.

Out of such conflicting factors emerged the Fermilab of today, home to some two thousand employees (including staff physicists) and an additional two thousand five hundred "users," physicists from many countries who come to Illinois to work with Fermilab's facilities. In 1967, when the laboratory was established under the United States Atomic Energy Commission (AEC), it was called the National Accelerator Laboratory (NAL). It was renamed Fermi National Accelerator Laboratory (FNAL), or Fermilab, in 1974.

The site of this laboratory is unusual, for the 6,800 acres of government land on which the laboratory lies was designed to evoke the western frontier of the United States. The land appears largely unaffected by recent suburban sprawl around the laboratory. Such development was anticipated in 1966 when the federal and Illinois governments ruled that NAL's large open site would be preserved for scientific research. But while in 1967 the site consisted almost entirely of farmland, it now includes a vast restored prairie rich with native wildlife coexisting with state-of-the-art research facilities, often housed in brightly painted buildings. One finds nineteenth-century farmhouses and barns amid imaginative sculpture and modern architecture on the site. The laboratory's imposing seventeen-story central laboratory, Robert Rathbun Wilson Hall, the towering white structure that our imaginary airplane travelers noted, vaguely resembles a medieval cathedral in the French countryside.

The four-mile circular tunnel at the heart of the site, the "ring" central to our story, is the track for the racing protons, which are given their high energy by the laboratory's accelerator, the proton synchrotron called the Tevatron. Particle accelerators have been described as powerful microscopes, because the high-energy particles they produce can probe deeply into the nucleus, revealing its inner structure. Accelerators have also been compared to time machines, because they re-create conditions that existed in the early universe. The energized protons in the ring are indeed, as one of our travelers explained, among the physicists' powerful tools for probing frontiers of nature. They perform a role similar to that of the spyglass, the ax, or the plow used to explore the American West.

The story told in this book unfolds in three parts. Part 1 (chapters 2, 3, and 4) presents Fermilab's prehistory, beginning in 1960, before the founding of the laboratory in 1967. During this heady period of particle physics many scientific discoveries fueled a desire to break through that era's energy frontier of roughly 30 GeV and reach energies greater than were available then at Brookhaven National Laboratory (BNL) on Long Island, New York (33 GeV), or at the European nuclear research laboratory CERN in Geneva (28 GeV).[4] (A GeV is a unit of energy. In 1960 physicists used the two energy measures of BeV [billion electron volts] and GeV [giga electron volts] interchangeably in referring to the energy achieved by an electron energized to a billion volts. While most Europeans preferred the term GeV, many Americans, including Robert R. Wilson, used BeV. As GeV had become the more popular term by the time NAL was built, we use GeV throughout this book in an effort to avoid confusion, unless the term BeV appears in a direct quote.)[5]

In 1960, when American physicists still enjoyed the high level of support characteristic of the postwar period, a number of groups submitted proposals for high-energy accelerators in the 100–1000 GeV range to the Atomic Energy Commission (AEC), which had grown out of the wartime Manhattan Engineer District. To evaluate the flurry of requests, an important joint panel of President John F. Kennedy's Scientific Advisory Committee (PSAC) and the General Advisory Committee (GAC) of the AEC was assembled. This panel, chaired by Harvard physicist Norman F. Ramsey, would shape the course of particle physics for the next several decades by recommending, in May 1963, that a frontier proton accelerator in the energy range of several hundred GeV be designed by Ernest Lawrence's laboratory in Berkeley, then known as the Lawrence Radiation Laboratory (LRL) (later as the Lawrence Berkeley Laboratory [LBL] and now as the Lawrence Berkeley National Laboratory [LBNL]). Berkeley accelerator physicists designed a reliable 200 GeV synchrotron with a projected cost of $340 million. Berkeley's physicists expected to build the machine in California, but within only a year political forces and new ideas about how and where large-scale physics should be done dashed Berkeley's plan.

Wilson, who had been one of Lawrence's graduate students during the 1930s, and later became a professor at Cornell, made his dramatic entry into Fermilab's prehistory in September 1965, when he offered a well-publicized criticism of the Berkeley machine. Referring to the design as conservative, unimaginative, and far too expensive, he proposed instead an innovative and far less costly alternative based on principles he had employed in building accelerators at Harvard and Cornell.[6] The Berkeley physicists dismissed Wilson's proposal, but Congress appreciated his lower cost projection in this time of war in Vietnam. National politics, growing tensions between East Coast and West Coast physicists, and arguments by physicists from the Midwest claiming that they deserved a new accelerator in their region all affected the selection of the new accelerator's site. Both East and West Coast physicists were disappointed when in December 1966 a site in suburban Chicago was chosen for the new laboratory, contributing to the "postwar suburbanization of American physics" that historian David Kaiser has described.[7] Berkeley

I.2 The art, architecture, and buffalo of Robert R. Wilson's Fermilab, 1990. (Courtesy of FNAL Visual Media Services.)

physicists, who expected to lead the new laboratory, were shocked when Wilson was offered the position of director.

Part 2 of the book (chapters 5–8) tells how Wilson created the new laboratory his way. Since he had once aspired to be a cowboy, it was natural for him to apply the cultural idiom of the American frontier to this building process.[8] He used the familiar frontier rhetoric to help his staff accept the hardships of work under primitive conditions and to encourage them to take cost-saving risks in building the accelerator. Wilson embraced frugality, not only as a social need but also as part of his personal aesthetic and frontier ideology. He expressed his vision of research as performed ideally by lone, independent scientists working in frontierlike conditions in a limited funding environment. Hoping to redeem what the Manhattan Project had wrought upon the world, Wilson and his deputy director, University of Illinois physicist Edwin L. Goldwasser, planned a utopian laboratory intended as a place of beauty (fig. I.2) devoted to the peaceful pursuit of particle physics and contributing to cultural and social advancement.[9]

A crucial component of Wilson's laboratory was its new philosophy in which access by outside users would be determined solely on the merits of their research, not according to the prestige or location of their home institution. The concept of the "truly national laboratory" (TNL)—that a national physics facility should be governed by a democratic user policy—had been proposed in 1963 by Columbia University experimental physicist Leon M. Lederman. It was to signify that the new laboratory would adhere to TNL principles that Wilson named it the National Accelerator Laboratory (NAL).

Slashing costs in every way, while using frontier rhetoric to encourage creativity and motivate his staff, Wilson argued that any technology that worked the first time was overdesigned and thus overpriced. Despite numerous crises, including a particularly troubling one in 1971 when a large fraction of the accelerator's magnets shorted out, Wilson's new laboratory and its accelerator were completed on time, at a lower cost than budgeted, with almost twice their design energy, and with more experimental areas than had originally been planned. The experimental research under Wilson suffered because of Wilson's imposed frugality, but as we demonstrate, for example, in our account of the discovery of the "bottom quark" (chapter 7), much important research was nevertheless conducted at Wilson's laboratory.

Wilson turned promptly to the next frontier as soon as he completed the laboratory's original accelerator (the so-called Main Ring). He planned to double the energy of the accelerator by building a second ring of magnets in the Main Ring tunnel. The novel feature of this second ring was its use of superconducting magnets, then a new and untested technology for large accelerators. The plan proved viable; Wilson's Energy Doubler (also called the Energy Saver, the Doubler/Saver, and later the Tevatron) would also save power and money. But in the later years of Wilson's tenure he struggled unsuccessfully to gain funding for the Energy Doubler. Severe budgetary pressures brought him to resign and then to step down in 1978.

Part 3 (chapters 9–13) explains how Lederman, who became Fermilab's second director, revitalized the laboratory and succeeded, at least initially, in extending its horizons. Lederman brought Wilson's Energy Doubler to life, opening the trillion electron volt (TeV) energy domain. The research under Lederman also carried Fermilab into the era of megascience. This venture proceeded along two tracks simultaneously, one in the existing fixed-target program (discussed in chapter 11) and the other based on the use of colliding beams to reach higher energies (discussed in chapter 12). Long-lived "strings" of experiments and their follow-up efforts allowed the largest experimental programs to continue to operate for decades or more. Drawing on the historical analysis of philosopher of science Mark Bodnarczuk, we interpret these strings as part of the physicists' effort to cope with the growing investments needed for survival in the 1980s. We illustrate with a brief analysis of the birth of a particular experiment string, one that aimed to study charmed particles.

In approaching megascience along the colliding-beams track, Fermilab built two mammoth detectors to analyze the collisions, first CDF, the Colliding Detector at Fermilab, and then DZero, named for its location on the Main Ring. Growing up in the shadow of CDF, DZero was given fewer resources, but it in time grew comparable in scale to CDF. Larger than many earlier laboratories, the experiments at CDF and DZero bore fruit in their 1995 codiscovery of the "top quark," arguably the outstanding achievement of American megascience.

In the second half of his tenure as director, Lederman attempted to extend Fermilab's horizon beyond the range of the Tevatron, proposing the 20 TeV on 20 TeV Superconducting Super Collider (SSC). Just as the call of the frontier had motivated Wilson to propose the Energy Doubler, a project that carried him into territory he could not navigate, Lederman's ambition to build the SSC at Fermilab brought him into stormy seas. The storms did not subside during the tenure of Fermilab's subsequent directors, as we briefly outline in the epilogue.

As of this writing, Fermilab has offered high-energy physicists almost four decades of passage into the scientific hinterland of subnuclear physics. During this time countless secrets of nature have been uncovered. But it was only a matter of time before this adventure would end. Serious political and economic circumstances presently threaten the survival of the Tevatron, and perhaps of Fermilab. "The whole point of the frontier had been to vanish," wrote historian William Cronon in an interpretation of the historian Frederick Jackson Turner's classic work on the frontier. The purpose, Cronon claimed, "was to prepare the way for the civilization that would eventually replace it."[10]

ONE

The Call of the Frontier

From the conditions of frontier life came intellectual traits of profound importance. . . . That coarseness and strength combined with acuteness and inquisitiveness; that practical, inventive turn of mind, quick to find expedients; that masterful grasp of material things, lacking in the artistic but powerful to effect great ends; that restless nervous energy; that dominant individualism, working for good and for evil, and withal that buoyancy and exuberance which comes with freedom—these are traits of the frontier, or traits called out elsewhere because of the existence of the frontier. FREDERICK JACKSON TURNER[1]

Every culture has its stories about those who were attracted to exploring frontiers and seeking new worlds. In science, from antiquity onward, naturalists and natural philosophers have employed the imagery of frontier exploration to help express their identity as pioneers who pursue the limits of man's understanding of nature. For experimental physicists, their laboratory is an exploration vessel, like the sailing ships or covered wagons of earlier explorers, or the spacecraft of today's astronauts. In high-energy physics, Fermilab has functioned as a flagship for almost four decades.

Voyages of exploration need their patrons. Many of the classic stories about exploring frontiers (e.g., of the pursuits of Meriwether Lewis and William Clark) pivot on the tension among those who explore, those who stay behind, and those concerned with economic gain. The United States government's support of weapons research during World War II created the large national institutions that have been funding much of American science since then. Today's physicists routinely use the word "frontier" in their research proposals because government funding agencies still respond well to frontier rhetoric.[2] In return for the support they offer, these patrons shape the explorations, at least to some extent. Control is an ongoing theme in frontier narratives. In physics the wish to build costly accelerators for research in particle physics has had to mesh with the government's interest in supporting a group of technically trained researchers for reasons of national security.[3]

The historical notion of the frontier received one of its most articulate and influential treatments in an address delivered in July 1893 in Chicago by the historian Frederick Jackson Turner (fig. 1.1). Speaking at the World's Columbian Exposition, the world's fair held to celebrate the four hundredth anniversary of the landing of Columbus in the Western Hemisphere, Turner argued that when American settlers confronted the western frontier during the eighteenth and nineteenth centuries, they engaged in a process that cultivated social and cultural traits, such as individualism, democracy, and political equality.[4] To the frontier itself—to the "hither edge of an area of free land"—Turner ascribed the power to foster a national character. Turner claimed: "The frontier is the line of most rapid and effective Americanization. The wilderness masters the colonist. It finds him a European in dress, industries, tools, modes of travel, and thought. It takes him from the railroad car and puts him in the birch canoe. It strips off the garments of civilization and arrays him in the hunting shirt and the moccasin. . . . At the frontier the environment is at first too strong for the man. He must accept the conditions which it furnishes, or perish."[5]

Turner went so far as to suggest that the act of confronting the frontier generated positive attitudes, such as optimism, individualism, confidence, dynamism, and the courage to venture beyond the safe and familiar, producing an "antipathy to control."[6] The members of frontier communities thus arrived, in Turner's view, at a distinctly American outlook which was (at least in principle) individualistic, progressive, practical, and democratic.[7] Subsequent historians were quick to criticize Turner's sweeping thesis, pointing out many overlooked aspects of frontier life, for example, the brutality committed against Native Americans, animals, or the land itself. More recently historians have noticed that Turner's "American" outlook is male gendered and white. While Turner's thesis fell into some scholarly disfavor, the rhetoric and imagery of its grand narrative, as well as its provocative description of America's national growth, remained alive and passed into American popular culture.[8] Frontier stories became part of the American mythology, and they continue to be dramatized, not only in histories read by generations of Americans but also in productions by the entertainment industry or in

1.1 Historian Frederick Jackson Turner, author of the 1893 "Frontier Thesis," presented at the World's Columbian Exposition in Chicago. (Courtesy of The Wisconsin Historical Society, image no. WHi-28376.)

actual re-creations, like Frontierland in Disneyland and countless "frontier towns."[9]

Ideas about the frontier have also permeated the rhetoric of American politics. Even before Turner's address, President Theodore Roosevelt published two articles on his own frontier experience ("Frontier Types" [1888] and "In Cowboy Land" [1893], in the *Century Magazine*). Subsequently, President and Professor Woodrow Wilson glorified the American frontier in his popular articles. Politicians on both sides of President Franklin D. Roosevelt's New Deal found the notion useful. Proponents of the policy argued that "the end of the frontier" meant that government must offer the security and opportunities formerly provided by the West, while opponents claimed that government meddling in the marketplace would undermine American individualism and self-reliance.[10]

Roosevelt connected the frontier with science when, in November 1944, he sought advice from Vannevar Bush on how the government should proceed with science after the Second World War. Roosevelt was optimistic about the benefits of science, for he wrote: "New frontiers of the mind are before us, and if they are pioneered with the same vision, boldness, and drive with which we have waged this war we can create a fuller and more fruitful employment and a fuller and more fruitful life."[11] An MIT professor of electrical engineering, Bush was then serving as the director of the White House Office of Scientific Research and Development. He responded positively to Roosevelt's frontier imagery, employing more frontier imagery when he replied in July 1945 with rhetoric intended to galvanize government support of basic research: "The pioneer spirit is still vigorous within this Nation. Science offers a largely unexplored hinterland for the pioneer who has the tools for his task. The rewards of such exploration both for the Nation and for the individual are great. Scientific progress is one essential key to security as a nation, to our better health, to more jobs, to a higher standard of living, and to our cultural progress." Thus, in 1945 Bush painted an image of scientists confronting an endless scientific frontier in the nation's interest. He was, however, worried about the damaging consequences of closing such frontiers. In his memoirs he confessed that he became anxious years earlier, "when I was told [as a child in 1919] that the frontier had been occupied, that all of man's wants had been met, that science had come to the end of a trail."[12]

Bush then used this frontier imagery effectively in his 1945 pamphlet *Science, the Endless Frontier,* which ushered in the golden age of American science funding. As other historians have shown, the major American national funding agencies of today (such as the National Science Foundation, established in 1950) grew out of the discussions that Bush stimulated in 1945, and this period of well-supported government-funded research, which lasted through the 1960s, resulted in much scientific progress and many Nobel prizes.[13] Science grew so rapidly that scholars of science, noting its apparently exponential growth, used the term "big science." Alvin Weinberg, the director of Oak Ridge National Laboratory, popularized the term in the mid-1960s.[14]

During the 1960s, heroic frontier imagery illuminated the rhetoric of many politicians who spoke about science and technology. When John F. Kennedy accepted his party's presidential nomination in July 1960, he spoke in his "New Frontier" address of "the frontier of the 1960s, a frontier of unknown opportunities and perils, a frontier of unfulfilled hopes and threats." He referred to "uncharted areas of science and space, unsolved problems of peace and war, unconquered pockets of ignorance and prejudice, unanswered questions of poverty and surplus." And as he inspired the nation to undertake space exploration he asked Americans "to be new pioneers on that new frontier."[15]

Scientists took advantage of the imagery's popularity and used the rhetoric even when communicating among themselves, sometimes in code. No physicist misunderstood when Arthur Holly Compton, the director of the Chicago Metallurgical Lab, reported to James Conant, the president of Harvard, on December 2, 1942, "The Italian navigator has just landed in the new world." They knew it meant that Enrico Fermi had achieved the first self-sustaining nuclear chain reaction.[16] Three decades later, Fermi would be a model for his Los Alamos colleague Robert R. Wilson, when Wilson created the National Accelerator Laboratory (fig. 1.2).

References to the frontier appeared often in the writings and speeches of Fermilab's founding fathers. They readily tapped into the deeply rooted spirit evoked by frontier imagery. Wilson was familiar with the imagery, having been born in Frontier, Wyoming, and having spent some of his early years as a cowboy. While he did not, as far as we know, ever refer to Frederick Jackson Turner, the imagery Wilson used was remarkably similar to Turner's. Wilson's rhetoric, however, reflected his interest in designing the new laboratory as a workplace for rugged, self-reliant, determined researchers with a zeal for conquering the unknown with their hands-on efforts. His imagery also differed from Turner's in that his ideal included equality for minorities, women, and outside users.

When Wilson brought his perspective to the final congressional authorization hearings of the JCAE in 1969, Senator John Pastore asked a witness, Paul McDaniel of the AEC, what the new laboratory would offer for national defense. As Edwin Goldwasser, Wilson's deputy director, remembered, "the witness on the stand was dumbfounded and speechless." Wilson, who was sitting beside the science statesman Norman Ramsey, then asked whether he might answer instead. "Norman, although he was unaware of any precedent for audience participation in a Congressional hearing, told Bob to go ahead and try to be recognized," said Goldwasser.[17] The message that Wilson offered became legend: "[It] has only to do with the respect with which we regard one another, the dignity of

1.2 Robert Rathbun Wilson in his Cornell University office, 1966. (Cornell University. Photograph by Sol Goldberg. Division of Rare and Manuscript Collections, Cornell University Library.)

men, our love of culture. . . . It has to do with, are we good painters, good sculptors, great poets? I mean all the things we really venerate and honor in our country and are patriotic about. It has nothing to do directly with defending our country except to make it worth defending."[18] Against the background of the Vietnam War, escalating urban violence, declining prospects for scientific employment, and eroding research funding for the first time since World War II, this message was heartening.

Wilson's version of the frontier rhetoric forged an identity at Fermilab that unified his workforce.[19] NAL's frontier theme was well articulated by Wilson's first associate director, M. Stanley Livingston, who wrote in 1968, "There is in mankind a driving urge to explore the unknown."[20] Linking the frontier with the research about to begin at the new laboratory, he wrote, "In past ages much of this exploration was geographical—the search for new continents and new seas." He then explained, "In our generation the most challenging frontiers lie in the search for new knowledge about nature and about man, and the most dramatic progress has been made on the frontiers of science." One search was to identify the ultimate constituents of matter. "The frontier of high energy and the infinitesimally small is a challenge to the mind of man. If we can reach and cross this frontier, our generation will have furnished a significant milestone in human history."[21] Livingston was accustomed to pursuing frontiers in his work using particle accelerators. He had built the first cyclotron four decades earlier while in graduate school at the University of California in Berkeley. This instrument for particle acceleration had been

1.3 M. Stanley Livingston (*left*) and Ernest O. Lawrence (*right*), standing by the 27-inch cyclotron magnet at the Lawrence Berkeley Laboratory, 1939. (Courtesy of American Institute of Physics.)

conceived two years earlier by Livingston's mentor Ernest O. Lawrence (fig. 1.3), who had also embraced the imagery of the frontier—for example, in his Nobel Prize acceptance speech in December 1939, titled "The New Frontiers in the Atom."[22]

To some, Wilson's heroic tales of pioneers struggling on the frontier offered relief or even hope during the turbulence of the late 1960s. But the application of his frontier imagery to Fermilab's site had obvious limits. The physicists clearly did not face the physical dangers of the Old West or of the atomic West. And even if the landscape contained vestiges of the prairie, everyone knew the Fermilab site had once contained a housing development and farms, and that the city of Chicago was not far away.

Fermilab was not the only research laboratory to use the frontier motif. For example, in the state of Washington, Hanford Laboratory and its company town Richland, as John W. Findlay notes in his history of the plutonium-producing community, "fashioned an identity by conjuring up images of an Old West full of untamed nature, harsh conditions, and individual sacrifice."[23] Like the workers at Fermilab, those at Hanford saw themselves as pioneers, although at Fermilab the territory being explored was particle physics, while at Hanford it was the "industrial frontier."[24]

At the turn of the twentieth century, Turner suggested that historians pay "attention to the frontier as a fertile field for investigation,"[25] but only a few historians of science have drawn on this imagery as a conceptual tool.[26] As the following chapters illustrate, historians of physics have an especially fertile opportunity to heed Turner's suggestion, given the broad application that physicists, and especially particle physicists, have made of the frontier both as image and metaphor. By unveiling the layers of meaning within the physicists' own use of the term "frontier," historians of science can forge a powerful tool to help them integrate within general history such major developments in science as the rise of large high-energy physics research laboratories.

PART ONE

An American Dream

TWO

The Several Hundred GeV Accelerator, 1959–1963

The scientists of the United States, native and foreign-born, have led the world in high energy physics. Over the last decade, most of the major inventions and discoveries in high energy physics have been made in U.S. laboratories. Several of these have been recognized by the award of the Nobel Prize.

NORMAN F. RAMSEY[1]

In late summer of 1960, the community of high-energy physics was buzzing with surprising news. The proton beams emerging from the first two large alternating-gradient accelerators were of much higher intensity than expected. Both machines—CERN's Proton Synchrotron (PS), operating in November 1959, and Brookhaven's Alternating Gradient Synchrotron (AGS), operating in July 1960—were in the range of 30 GeV.

The PS and AGS followed a line of important accelerator technologies pioneered by Ernest Lawrence's cyclotron, which he invented in 1929. But while in the first circular particle accelerator, the cyclotron, light positive ions spiraled outward in a constant magnetic field, gradually increasing their radius by the action of a constant-frequency alternating electric field synchronized with the circular movement of the particles, in the subsequent generation of circular accerators, known as synchrotrons, which had been invented by the Russian physicist Vladimir Veksler in 1944 as well as by Edwin McMillan in 1945, the circular radius of the particles was held constant by varying either the frequency of the accelerating voltage or the strength of the magnetic field. With particles having a constant, rather than a spiraling-outward path, the synchrotron was a vast improvement on the cyclotron, because the accelerator could take the shape of a doughnut; the reduction in material of a doughnut-shaped accerator drastically reduced its cost.

The alternating-gradient (or "strong-focusing") principle was the next major advance in circular accelerators after the synchroton. The principle behind it was developed collaboratively in 1952 by three Brookhaven physicists: Ernest Courant, Stanley Livingston, and Hartland Snyder. (It was learned later that the Greek electrical engineer Nicholas Christofilos had already invented strong focusing two years earlier.) The strong-focusing principle allowed much larger accelerators to be built at lower cost, because focusing the particle beam could be performed relatively easily using an alternating series of converging and diverging magnetic lenses.[2] The unexpected demonstration that the two first large strong-focusing machines had high intensity as well as higher energy assured physicists that the tradition of building ever larger and more powerful accelerators would continue for a long time.

Proposals from MURA, Caltech, and Berkeley, 1959–1960

Soon the lure of new data at higher energies inspired the planning of accelerators having much higher energy. Even with America's increasing military involvement in Southeast Asia, the myriad of social problems that erupted in America during the 1960s, the continuing appreciation of the scientific advances made by physicists in World War II caused the United States government to be willing to fund larger accelerators.[3] One of the first viable proposals emerged during a summer 1959 workshop held in Madison, Wisconsin, and sponsored by the Midwestern Universities Research Association (MURA). MURA's innovative Fixed-Field Alternating Gradient Accelerator (FFAG), then in design form, was to produce an intense proton beam by colliding two 10–15 GeV beams.[4] MURA hoped to build the machine in the Midwest. The primary aim of MURA's summer meeting was to design and generate support for the FFAG, which was also studied at the meeting in relation to other existing schemes for achieving high energy or high intensity. As this workshop took place before the PS or AGS began operating, many of the physicists attending the meeting thought that the only practical route to higher energies was by colliding particle beams. High-energy fixed-target machines

(machines whose beams were directly used in experiments), they believed, would be prohibitively expensive, and perhaps not even useful, because above 5 GeV the particles would be confined to a narrow cone and would therefore be difficult to distinguish.[5]

Alvin Tollestrup, an experimental physicist from Caltech who would later play a major role in Fermilab's history, was a participant in this workshop. He recalled that MURA's physicists were so confident in their FFAG design that they issued a challenge to everyone present to design a technically and economically feasible fixed-target machine that could produce the same energy at the point of collision that the FFAG would produce (i.e., 300 GeV).[6] Tollestrup's Caltech colleague Matthew Sands took this challenge up. An iconoclast by nature, Sands reinvented a scheme (which, in fact, Wilson and also Argonne physicist Lee Teng had suggested but not implemented several years earlier) based on forming a cascade of several accelerators. In this plan, an accelerated beam would be injected from one machine into the next higher-energy one.[7] Sands hoped to cut costs by accelerating the particles to high energy in a "booster" synchrotron and then, at a fairly high injection field, feeding the accelerated beam emerging from the booster into the main synchrotron, where it would reach several hundred GeV. Money would be saved because beams of particles have a smaller cross section after acceleration and thus the largest synchrotron would be able to function with magnets of smaller apertures.[8]

Most of the participants in the MURA workshop did not take Sands's proposal seriously, for it was thought that one could not control such a large system or work with accelerator magnets as small as Sands specified. But in computing the details with a subgroup of participants in the MURA study, including Tollestrup as well as Courant and M. Hildred Blewett from Brookhaven, Sands estimated that the cost of his 100 and 300 GeV cascade designs would be about the same order of magnitude as an FFAG accelerator of 10 GeV. Optimizing the parameters, Sands realized that he could achieve 300 GeV most efficiently by injecting beam into the main ring from a 10 GeV–range "rapid-cycling" booster synchrotron. The high repetition rate of such a booster would make it possible to produce a high-intensity beam. He calculated that the magnet aperture in the main accelerator ring could be as small as a few square centimeters.[9]

After the workshop, Sands and Tollestrup worked further on achieving 300 GeV. They were joined by their Caltech colleague, Robert Walker. To the booster synchrotron and the main ring in their proposed accelerator system, they added a linear accelerator to inject protons into the booster.[10] Next, Sands, Tollestrup, and Walker invited Snyder, Courant, and Hildred Blewett from Brookhaven, Kenneth Robinson from the Harvard-MIT Cambridge Electron Accelerator (CEA), and Robert Hulsizer from the University of Illinois to join their effort. Their report, "A Proton Synchrotron for 300 GeV," estimated the cost of the proposed machine, including preparation of the site, salaries, and accelerator components, at $77 million.[11] In May 1960, Sands, Tollestrup, and Walker proposed that Caltech formally support the small group that was working on the design and "initiate a project" for designing the next high-energy accelerator.[12]

By that time, several physicists who were part of a planning group at Lawrence's laboratory in Berkeley had also conceived a proton synchrotron in the several hundred GeV range.[13] The Berkeley laboratory was now headed by Edwin McMillan, whom Lawrence had picked as his successor. McMillan's interest in building a larger accelerator had been sparked in March 1960 when CERN director John Adams visited Berkeley bringing news about the operation of the PS. On August 12 that year, McMillan convened an accelerator planning committee to consider building a strong-focusing synchrotron similar to the CERN PS and the Brookhaven AGS but of higher energy.[14] Along with Luis Alvarez, Glenn Seaborg, and Robert Wilson, McMillan had been one of "Lawrence's boys" (fig. 2.1), the group of young men Lawrence gathered around him in the 1930s to help him bring the Berkeley laboratory to prominence.[15]

When Sands learned about McMillan's new accelerator planning group, he wondered whether the Caltech and Berkeley groups should join forces. Two months later, Berkeley physicists Lloyd Smith, McMillan, and David Judd attended a meeting at Caltech at which the Caltech and Berkeley groups "agreed that as much interchange of ideas should take place as possible without setting up a formal connection, and that we should try to reach agreement on a single proposal for the West Coast."[16] Caltech had, in any case, judged Sands's accelerator project too expensive to support on its own. A sponsoring group named the Western Accelerator Group (WAG) was then formed. WAG included, besides Caltech, the University of California at Los Angeles, the University of California at San Diego, and the University of Southern California. But the University of California at Berkeley did not join WAG. In March 1961, Sands explained to John Blewett, a leading accelerator physicist at Brookhaven, that the WAG effort was "still working independently but in close communication with Berkeley."[17]

2.1 Lawrence and his "boys," 1939. Lawrence is in the center of the front row; McMillan is fourth from the left in the middle row; Wilson is third from the right and Oppenheimer is sixth from the right in the top row. (Courtesy of Emilio Segre Visual Archives, American Institute of Physics.)

Meeting on Ultra-high-energy Accelerators

Discussions about building larger accelerators were also taking place in international meetings sponsored by the High-Energy Physics Commission of the International Union of Pure and Applied Physics (IUPAP). In 1959, Soviet and American physicists explored a proposal for a joint U.S.-USSR accelerator project as one of several cooperative ventures in unclassified atomic energy research. Meetings between Soviet premier Nikita Khrushchev and U.S. president Dwight D. Eisenhower had encouraged scientific exchange between their countries. One result was a U.S.-USSR information exchange agreement signed on November 24, 1959, by AEC chairman John A. McCone and his counterpart, chairman of the USSR Administration of Atomic Energy Vasily S. Emelyanov. When McCone asked Emelyanov whether Soviet physicists would be interested in exploring an international accelerator collaboration, Emelyanov proposed to McCone in July 1960 that it might be "convenient to take advantage of the presence of our high energy physics specialists in the USA for the Rochester Conference to conduct a preliminary exchange of views."[18]

On learning of this Soviet interest during a visit to the USSR, Robert R. Wilson of Cornell took the initiative to organize an informal meeting at the 1960 Rochester Conference to discuss the new "ultra-high-energy" accelerators. Held on August 28, 1960, this meeting provided a convenient forum for exploring the pros and cons, as well as the methodology, of experimenting at much higher energy.[19] While the Soviets hardly spoke at the meeting (perhaps because of the period's mercurial political climate), the Americans and Europeans hotly debated the proposed ultra-high-energy accelerators. American physicists tended to be confident about the possibility. J. Robert Oppenheimer, one of the American enthusiasts, felt that higher energies might reveal new unstable heavy particles. Pointing out that "we do not know what we shall find," Oppenheimer considered it "likely that essential novelty will appear" at higher energy. In any case, he added, "a knowledge of what does in fact occur in this domain will take us a long way" toward understanding subatomic matter. The Germans tended to be more skeptical. Werner Heisenberg remarked that since cosmic ray data had revealed little at higher energies, they might "find nothing of great interest" at higher energies. Others considered the disappointing cosmic ray evidence "inconclusive," because of the limitations of cosmic ray detection techniques and the small number of events. In his conference summary, Wilson judged that "the range of opinion between optimism and pessimism is fairly uniformly populated by physicists—but shaded a bit toward those who are optimistically minded."[20]

A variety of accelerator designs in the range of 100–1000 GeV were discussed at Wilson's meeting—linear accelerators, colliding-beam machines, and proton synchrotrons. MURA members supported their FFAG, but most of those present considered colliding beams to be a technology "of the future," because they required detectors having submicrosecond time resolution, a requirement not yet met by that period's primary detection apparatus, the bubble chamber.[21] Also discussed were the new proposals from California in which protons would be accelerated to hundreds of GeV in a circular fixed-target synchrotron. According to Wilson, the group agreed that for about $100 million—or at most $200 million—it would be feasible to push the design of a conventional alternating-gradient proton synchrotron to 100 GeV, or even higher, and that this sum might even cover the first round of experiments. By pushing the tolerances, they could hope to attain 1000 GeV at a cost of less than $1 billion, "really a bargain of course," concluded Wilson.[22]

Proposals by Brookhaven, Berkeley, and Caltech—August 1960–December 1961

The issue of building an accelerator in the 100–1000 GeV range was then taken up by many physics advisory panels. In December 1960, a joint panel was assembled under Emanuel Piore, IBM's director of research, consisting of the President's Science Advisory Committee (PSAC) and the General Advisory Committee (GAC) of the AEC. Piore issued the panel's report, which recommended approval of the electron linear accelerator at the Stanford Linear Accelerator Center (SLAC), intensive studies of strong-focusing proton synchrotrons, including studies oriented toward much higher energy "within the next few years," and continued collaboration with the Soviets.[23] The Piore Panel was lukewarm about MURA's FFAG, because the unexpectedly high intensity of the Brookhaven AGS and CERN PS had eroded the FFAG's relevance. While the AGS and PS beams were still roughly a hundred times less intense than those projected for the FFAG, the panel pointed out that by increasing energy instead of intensity physicists gained "the additional advantage of extending the range of study of all known primary and secondary phenomena and of adding the potential of new phenomena such as undiscovered particles." The Piore Panel also noted that to accommodate the next generation of accelerators, the annual high-energy physics budget might need to be increased by as much as $200 million by 1970.[24]

Brookhaven's plan to build an accelerator in the 300–1000 GeV range was discussed at a meeting arranged by the AEC in response to the Soviet and American plans to collaborate on an accelerator project. It was held on September 16, 1960, at the American Institute of Physics (AIP) in New York City. This meeting was attended by two delegations of physicists: a Soviet one led by Vladimir Veksler from the Joint Institute for Nuclear Research (JINR) at Dubna and an American delegation led by Brookhaven director Leland Haworth. The delegations agreed that their nations should sponsor national study groups that could later cooperate. A "comprehensive intergroup discussion" was planned for the upcoming International Accelerator Conference at Brookhaven during the following summer, in 1961. In an effort to avoid competition with ongoing national efforts, the participants agreed that their international accelerator should have an energy greater than 300 GeV.[25] John Blewett assembled a group at Brookhaven to prepare a preliminary design study for the American contingent of this international collaboration.

Meanwhile, the national effort to build a 200–300 GeV accelerator continued in California. In April 1961, WAG and Berkeley submitted their separate proposals to the AEC. WAG requested $593,000 to support a fifteen-month study of the design for a new proton synchrotron project to be directed jointly by Sands and Robert Bacher, a former Los Alamos division leader and former AEC commissioner who was then chairman of Caltech's physics department. The proposal included Sands's earlier September 1960 report on the 300 GeV cascade synchrotron, as well as other reports expanding Sands's ideas.[26] Among those who appreciated Sands's design was Wilson, who wrote to Sands in April 1961: "I have been watching your efforts with the 300 GeV machine with open-mouthed admiration. It seems to me that you are working on the right problem at the right time, and I am sure that something will come of it all."[27]

The Berkeley proposal, which McMillan submitted in late April 1961, described a somewhat more ambitious machine than did WAG's, with a higher price tag. The request was for $3 million for a two-year period to design an expanded alternating-gradient synchrotron in the 100–300 GeV range. The research included extensive injector studies.[28] The WAG proposal was in many ways more attractive than Berkeley's, for its design, which had been studied longer and more extensively, stressed economy and introduced a more innovative injector scheme. Also WAG planned to augment its nucleus of designers with outside experts who would obtain some support from their home institutions. On the other hand, Berkeley had an established reputation and far more experience in building large accelerators than any of the WAG universities. As Tollestrup and Walker recalled, the Berkeley physicists tended to discount WAG's proposal, or at best see it as part of the Berkeley project.[29] Whether Berkeley ever gave serious consideration to WAG's efforts is uncertain.

One reason Berkeley's budget topped WAG's was that Berkeley estimated that 105 man-years of effort were needed to create its design, while WAG assumed that only ten man-years would be needed to realize its design. Berkeley planned to construct prototypes of all the major components.[30] By April 1961, both Berkeley and Brookhaven were arguing that they were the only reasonable candidates for building the new machine and that if the project went to an organization other than Berkeley or Brookhaven, it would be delayed for several years because of the time spent gathering the necessary personnel and creating a smoothly functioning operation. Edward Lofgren wrote to McMillan on April 6 that a strong case could be made to the AEC for choosing Berkeley over Brookhaven "in terms of history, demonstrated competence and organization." He also argued to McMillan that Berkeley needed a new accelerator to maintain its front-rank position in high-energy physics, given that the Bevatron would soon no longer be a unique facility. As he explained, "governments of this and other countries have built up many new large laboratories more or less in the image of Berkeley at the demand of scientists all over the world."[31] By this time it was clear that Berkeley did not plan to join WAG. Thus, two independent California proposals for a several GeV accelerator were submitted to the AEC, one from WAG and the other from Berkeley.

Lofgren's memo to McMillan also raised an organizational theme then new to high-energy physics. He explained that the increased scale of the proposed accelerator demanded that it be based at a "national facility" open to users from all parts of the country. To make the facility national, Lofgren explained, one could organize an accelerator council of representatives having expertise in high-energy physics and appropriate geographic representation from top universities and laboratories. In Lofgren's scheme, this council would act as an advisory board and report to the director of the laboratory, but the laboratory would be built and operated by Berkeley on a nearby site.[32] WAG's proposal that month included a similar idea: "An accelerator in the hundred-BeV class would necessarily be a national facility serving the national scientific community."[33]

The AEC was sympathetic to both California proposals, as Paul McDaniel, the director of the AEC's Division of Research, explained in a letter to Oppenheimer. But funding more than one independent design study seemed "especially unwarranted in view of the close geographical proximity of the two groups in question."[34] The commission decided not to approve either design study until new procedures for handling this situation were in place. This decision led McMillan, who represented Berkeley's proposal, and Bacher, who represented WAG's, to work toward making their two studies more palatable to the AEC. In a joint letter to McDaniel, they pledged "to work toward the objective of a single proposal for the construction of a high-energy accelerator to be submitted to the Atomic Energy Commission," and they explained that a steering group had been established to coordinate the Berkeley and WAG efforts.[35] Still uncomfortable with the situation, the AEC asked twenty-three prominent physicists to review the two very different proposals. While the reviewers generally favored both studies, many felt that the country "can only afford one 300 BeV accelerator and indeed only one 300 BeV study group."[36]

The discussions continued during summer studies held at both Berkeley and Brookhaven involving physicists from many institutions in the United States and Western Europe. The Berkeley meeting, which ran from June to August in the summer of 1961 and involved about thirty physicists, considered experiments that could be performed in the new energy range. The group's thorough examination of experimental and theoretical issues helped build the scientific case for funding the project and bolstered support for the effort within the physics community.[37] The Brookhaven meeting, with roughly two dozen participants who met from late July through early September of 1961, focused on issues of accelerator design. The invitees from the Soviet Union did not attend, perhaps because of the Berlin crisis that summer. Nonetheless, as John Blewett reported, the mood at the meeting was positive. The participants enthusiastically discussed a number of the new technical ideas, including WAG's suggestions of cascade injection and building magnets of smaller aperture and thus of smaller size and cost. They also discussed beam injection and extraction, and the novel idea proposed by Thomas Collins of the CEA to leave long, empty, straight sections between the magnets in the main ring to allow room for radio frequency (RF) utility, beam extraction, and other accelerator functions.[38]

By August 1961, Brookhaven had revised its design report to reflect the summer study discussions. A series of papers, later to be edited by Luke C. L. Yuan into a collection titled *The Nature of Matter*, spoke of the research potential of the new machine. Part of the report, edited by Blewett, included a preliminary design for a 1000 GeV alternating-gradient proton synchrotron. Using rough estimates of the parameters for the injector and radio frequency systems, and reducing drastically the amount of steel used in magnets based on experience with the AGS, Blewett projected that the 1000 GeV machine would cost $675 million. He interpolated that similar 700 GeV and 300 GeV machines would cost approximately $500 million and $300 million, respectively, including accelerator components, buildings, site preparation, salaries, and a 10% contingency. In his conclusion, Blewett noted that "a real machine design will emerge only when a group of competent physicists and engineers is assembled for a full-scale attack on the problem."[39]

The high cost of the proposals and the consensus that the United States could afford only one large accelerator and one study group put the three proposals in clear competition. The WAG proposal was more complete, more innovative, and less expensive than Berkeley's, but Berkeley could boast having an experienced team of accelerator builders and a reputation for constructing reliable accelerators. Brookhaven also had a solid accelerator-building tradition, but its having only recently completed the AGS (in 1960) was a disadvantage. Since the 1950s Berkeley and Brookhaven had alternated in building the next largest machines, and McMillan insisted that Berkeley should have the next turn.[40]

No major technical obstacles emerged during the work of designing, but the last four months of 1961 were a tense time for all the groups, especially as support for any of them was far from assured, despite backing from the 1960 Piore Panel. SLAC had recently experienced trouble gaining congressional approval, even with the project's support from the 1958 Piore Panel. The situation was murky because the AEC had not yet defined its decision-making process. And while each group's individual input needed recognition, it was important that the physicists present a united front.[41]

WAG and Berkeley continued to discuss the possibility of collaborating. They convened a steering committee, which met in September, October, and November 1961. The September meeting outlined technical assignments: Berkeley would focus on designing Linac injectors and the RF system; WAG would concentrate on synchrotron injectors. Both groups would work on site issues and magnet design. The October and November meetings considered various possible Berkeley-WAG collaboration schemes, agreeing that the working design group should "have a central location with broad provision for national participation." McMillan and Bacher guessed that "some sort of national committee and/or national study effort" would be proposed.[42]

In late December 1961, representatives from Brookhaven, Berkeley, MURA, and WAG met in Los Angeles to make further plans for the new accelerator. Bacher and McMillan emphasized the importance of campaigning for the accelerator within both the government and the physics community. Blewett presented an organizational plan. Without too much debate the group also agreed that there should be a central study group at an existing laboratory authorized to prepare a design and cost estimate. The leader of this group would be advised by a committee of experts with nationwide representation. But in contrast to Lofgren's plan, this group agreed that the advisory committee would report to the AEC, rather than to the laboratory director.[43]

As for where the work should be done, representatives from MURA, Brookhaven, and Berkeley each presented arguments for why their home laboratories should be chosen. The members of WAG made no bid, for they realized that they stood no chance against the politically powerful groups at Berkeley and Brookhaven.[44] Brookhaven physicists argued that they were the nation's experts in strong-focusing synchrotrons and could participate only if Brookhaven was the central site, for otherwise the effort would detract from AGS development. Berkeley representatives contended that only their laboratory possessed the necessary resources to host the study because Brookhaven was involved in AGS development. The meeting ended without a resolution.[45] In February 1962, both Berkeley and Brookhaven asked to be the base of the design study. With a technical steering committee added to its proposal, Berkeley's estimate for the study, which would be completed in two years, was $3 million.[46] Brookhaven estimated that its design study would cost $4 million with completion in three years, incorporating an electron synchrotron into its design and planning to explore schemes for accelerating protons to energies as high as 1000 GeV. Fifteen years younger than Berkeley, Brookhaven did not stress its history but made the argument that with its progressive style of consortium management Brookhaven had "developed the broad views and cooperative attitude expected of a National Laboratory."[47]

The AEC's Dilemma, 1962

The AEC worried about the "bad feelings" brewing at the "large laboratories" in response to the competing proposals from Berkeley and Brookhaven. The presidential science adviser, MIT's Jerome Wiesner, warned that steps needed to be taken "cautiously, but fast" before things got "out of control." Discussing the situation with the AEC's chairman, Glenn Seaborg, who had earlier served as chancellor of the University of California at Berkeley, Wiesner explained in February 1962 that commissioner Leland Haworth, the former director of Brookhaven, had been handling the matter for the commission.[48] Seaborg raised a ticklish point. As a former member of Berkeley's staff, he had many close Berkeley associations; he had shared with McMillan not only the Nobel Prize but residence at the Berkeley Faculty Club. Similarly, Haworth had many close associations at Brookhaven. To avoid possible conflicts of interest, Paul McDaniel of the AEC, a trained physicist and skilled bureaucrat, assumed responsibility for the synchrotron project until mid-1963.[49]

On April 19, in letters sent to McMillan, Maurice Goldhaber, who in 1961 became the director of Brookhaven, and twenty-eight other scientists, McDaniel made clear that the federal government was not committed to "the location or mode of operation of such a facility if it were authorized," and that the commission concurred in the belief that the management of the new laboratory "necessarily should be separate from the management of any other laboratory." He asked for comments on organizational issues, such as establishing a committee for policy advice, and reminded his correspondents that "a large new accelerator facility for energies of several hundred BeV will receive authorization only if the scientific community is in general agreement on the need for such a facility."[50]

In responding on May 8, McMillan spoke against making too radical a change. He insisted that the amount of outside participation should be settled later and that arrangements for managing the laboratory should be left open. He also noted that it might be most efficient for an existing laboratory to build and operate the accelerator, at least in its early stages, and commented that if Brookhaven were chosen as the center for the study, few Berkeley researchers were likely to go there and that he doubted much work would be done at Berkeley, since "the main part of the accelerator is so much of an integrated whole that parts of the design can not be split off without serious loss of efficiency."[51] This was early evidence for the conservative approach McMillan would consistently take in subsequent discussions of the management of the new laboratory. His tradition-bound view would soon clash with the new AEC ideas of what was appropriate for an expensive national facility.

Goldhaber, on the other hand, expressed willingness to align with the new AEC guidelines for siting and management. Not only did he agree with the approach outlined by McDaniel, but he also applauded having 30%–40% of the work performed by outside laboratories, proclaiming that Brookhaven "would strongly oppose any suggestions which would tend to reduce this degree of multi-participation." Moreover, despite earlier comments that Brookhaven should not be involved unless the study were located at BNL because that would detract from the AGS effort, Goldhaber stated that the laboratory expected "to be deeply involved in the study regardless of its location."[52] Unlike McMillan, Goldhaber never had a particular interest in machine design. He could be more open minded, having delegated most of the designing responsibility to Cosmotron and AGS veterans G. Kenneth Green and John Blewett. And, unlike Berkeley researchers, Brookhaven researchers had mixed feelings about whether the proposed new high-energy machine should be their top priority; some felt it was more important to promote an AGS upgrade. Finally, even though some Brookhaven users felt that the laboratory's management was not open enough, as a younger laboratory, Brookhaven was in principle oriented toward serving outside users.[53]

An August 1962 AEC summary of responses to the letters circulated by McDaniel reveals overwhelming support for the design study. While several respondents suggested that there be two focal points, and a few others favored neutral sites, most cast a vote either for Brookhaven or Berkeley. Between these two, "the choice was 6–2 in favor of LRL [Berkeley]," chiefly because the correspondents thought "BNL should continue to be heavily occupied with getting the AGS into full utilization, while Berkeley should have a greater strength of engineering talent available for the study program."[54] McDaniel noted that the commission's preference for establishing the laboratory at a new site under the direction of a geographically balanced management organization would be complicated by the difficulty of obtaining the required authorization from Congress. He remarked, however, that if a major laboratory like Berkeley were given central responsibility for the study, it would be "important to have made provisions which will insure that the study program have a national character." He also made clear that the location of the study would not necessarily dictate the site or management of the facility.[55] He then recommended that there be created "a special High Energy Accelerator Advisory Committee," with broad national representation, eventually to become known as HEPAP, the High Energy Physics Advisory Panel, which would "consider the broad implications of the study program both initially and throughout the course of the work."

In contrast to Lofgren's and McMillan's official pronouncements about the "national character" of the study group and the new facility, Lofgren's notes of an informal meeting that he and McMillan held with McDaniel and his staff on September 26, 1962, reveal that the AEC was in fact willing to tolerate a more old-fashioned arrangement. At meetings held to gather information for the AEC's presentation to the Bureau of the Budget (BOB, later to become OMB, the Office of Management and Budget), McMillan and Lofgren were told that the AEC planned to request funding for a "Western Regional" accelerator.[56] According to Lofgren, after some "sparring" over the definition of such a western accelerator, staff members "did not disagree with our idea that this meant a Berkeley designed, built, and operated accelerator with safeguards to insure proper use." McDaniel realized that the AEC needed formally to reserve the right to decide the site of the accelerator, as was standard practice for large construction projects. Wiesner had already indicated the political necessity of enforcing such a rule in this case. But in informal discussions with Berkeley representatives, McDaniel did not retract McMillan's traditional prerogatives. It would be up to Seaborg, as AEC chairman, to take such action, if he so chose.[57]

Brookhaven was mentioned in the AEC's request to BOB. In a two-track approach, a design study for a machine in the several hundred GeV range would be conducted by Berkeley, while Brookhaven worked on the design of a 1000 GeV machine slated for future construction. The FFAG was also included in the budget request, although the AEC admitted that it would not survive if only two of the three large accelerators were funded. The discussions were reminiscent of the earlier arrangements for constructing the Bevatron and Cosmotron, when it was decided that both laboratories could build machines of different energies.[58]

The Ramsey Panel, 1963

Officials in Washington DC were disturbed by the number and expense of the requests made for new particle accelerators and by tensions caused by the dark horse MURA.[59] Responding to this concern, in November 1962 Wiesner and the AEC convened another in the series of President's Science Advisory Committee/General Advisory Committee (PSAC/GAC) panels to offer advice on future needs in high-energy physics. Headed by Norman F. Ramsey of Harvard (fig. 2.2), this new panel would set the course of high-energy accelerator development for the next decade, and beyond.

Highly respected for his scientific capabilities, Ramsey had been a graduate student of Columbia University's Isidor I. Rabi, whose experiments with molecular beams had made it possible to measure the radio frequency spectrum of atomic nuclei. Ramsey's work with Rabi led to the discovery of the deuteron electric quadrupole moment. Ramsey had also worked on both the radar and atomic bomb projects during the Second World War, aided in the establishment of Brookhaven after the war, and served on a number of advisory boards before joining the GAC. He developed a reputation for being fair minded and diplomatic. These skills would prove useful as he and nine other eminent scientists appointed jointly to PSAC and GAC, and two ex officio members from other government agencies, worked on Ramsey's panel to investigate the issues of the several hundred GeV accelerator.[60]

In early December 1962, as the Ramsey Panel began its deliberations, Berkeley presented a revised proposal to the AEC, requesting a much higher budget of almost $5 million ($4,875,000) to support a two-year design study for an accelerator in the 100 GeV range. Much of the proposal focused on the use of the machine for research, providing a lengthy discussion of experiments to be done. The new proposal estimated that a 100 GeV machine would cost $152 million, while a 200 GeV machine would cost $263 million, including accelerator components, architect engineering costs, development costs and 20% technical contingency. The proposal also mentioned that Brookhaven was proceeding with studies for a 1000 GeV accelerator.[61]

2.2 Harvard professor of physics Norman F. Ramsey, chairman of the 1963 Joint PSAC-GAC Panel and longtime president of Universities Research Association, 1972. (Courtesy of Fermilab Visual Media Services.)

Unlike the cautious February 1962 version of the Berkeley proposal, the December proposal stated boldly that "LRL would assume full responsibility for design, construction, and operation of the accelerator and facilities." And in contrast with the February report, the December edition made only passing reference to outside collaboration, remarking that "if people and facilities are available some portions of the study could be carried on away from Berkeley."[62] Berkeley had thus revised its proposal along traditional lines, and the AEC seemed prepared to accept it, having obviously been in close contact with Berkeley during the document's preparation. The AEC transmitted Berkeley's revised proposal to the Ramsey Panel.

Meeting in fourteen sessions over a six-month period, the Ramsey Panel interviewed physicists from all over the nation, heard presentations from various government agencies, met with BOB, and engaged in constantly vigorous discussion. The panel had a number of other issues to consider besides the several hundred GeV accelerator, including Brookhaven's plans for a 1000 GeV machine and for storage rings for the AGS, MURA's FFAG, a 10 GeV electron synchrotron proposed by Robert Wilson at Cornell, various high-intensity meson accelerators, and funding for various detector developments and university user groups. The most controversial issue was the relative importance of MURA's FFAG proposal.[63]

After extensive discussion, the panel members decided that they could not "encourage new high-energy accelerator construction costing millions of dollars unless a useful extension of accelerator parameters is achieved, or a new technology advanced." Referring to the problem of explaining the strong interactions and working more on the weak interactions using neutrinos at higher energies, following the 1962 discovery of two types of neutrinos by Columbia physicists Leon Lederman, Melvin Schwartz, and Jack Steinberger, the panel judged proton energy—the parameter in question for the several GeV accelerator—to be "the single most important energy parameter to be extended." They also expressed hope of furthering research aimed at providing "a clue to the connections among the different kinds of basic forces." The panel members were aware that using high-intensity beams could allow detailed study of interactions of such low probability that present accelerators could not observe them, but they also recognized the difficult technological problems associated with high-intensity accelerators, including high radioactivity levels and the high cost of such machines.[64] For these reasons, the Ramsey Panel felt obliged to recommend only one high–intensity machine, the FFAG, but it was ranked lower than the high-energy machines.

The deliberations over Berkeley's proposal for its 100 GeV machine were heated. Edwin Goldwasser, a panel member from the University of Illinois, recalled that some resisted the proposal because of Berkeley's conservative stance on management or because the design energy of 100 GeV was seen as too low.[65] Paul McDaniel had noted in September 1962 that disapproval from the powerful Caltech physicist Robert

Bacher "could kill the project."[66] By January 1963 McMillan had agreed to raise the energy range to 150–300 GeV, thus pleasing both the panel and Bacher. As Bacher explained to Ramsey, no one in the Caltech group "had any real enthusiasm for a 100 BeV West Coast machine." In light of the increased energy range and further technical discussions, the Caltech group wished to add its "enthusiastic support of the study proposal contained in McMillan's letter."[67]

When the Ramsey Panel issued its report on April 26, 1963, it recommended "that the Federal Government—Authorize, at the earliest possible date, the construction by the Lawrence Radiation Laboratory, of a high energy proton accelerator at approximately 200 BeV energy."[68] In the vernacular of the time, the project became known as "the 200 BeV project." Berkeley was triumphant. Goldwasser later summarized the panel's rationale: "Berkeley had made enormous contributions to the field and had a large number of very talented people. If you want to optimize the success of such a large new venture, you have to get the best people to work on it. It could be hoped that differences in management philosophy could later be resolved."[69]

As the second priority, the Ramsey Panel recommended that Brookhaven construct the proposed AGS storage rings and begin "intensive design studies" for an accelerator in the 600–1000 GeV range. This recommendation eventually led to the Brookhaven project named ISABELLE, a superconducting colliding-beams accelerator (discussed in chapters 8, 10, and 13).[70] The panel also recommended that support be given to Cornell's 10 GeV machine (whose energy was later increased to 12 GeV) and to the construction of electron-positron storage rings at Stanford. As for MURA's FFAG, now estimated at $148 million, the panel recommended authorizing it for the energy of 12.5 GeV, instead of the 10 GeV originally proposed, but with the constraint that this project be authorized "without permitting this to delay the steps toward higher energy."[71] The panel spent hours debating its wording on MURA, aware that the recommendation would be controversial and perhaps decisive for the continued existence of the MURA group, which had been working for over ten years to fulfill its dream of bringing a first-rate accelerator to the Midwest. As Ramsey recalled, everyone realized that such an expensive "conditional" choice would not be kept in the current budget, and that the panel's recommendation was, in fact, the "kiss of death" for MURA.[72] The Ramsey Panel also recommended "an administrative structure with national representation to assure that all proposals for qualified scientists shall be considered on equal footing."[73] This recommendation marked a change which would be critical for the future of not only the new laboratory, but also of all large laboratories, including Berkeley and Brookhaven.

The goals driving the funding requests to support the recommendations of the Ramsey Panel were, more than any specific scientific discovery, based on continuing willingness by the U.S. government to support physicists in the aftermath of the Second World War. The subsequent Cold War, as historian Silvan Schweber wrote in 1987, then "cemented the wartime relationship." To the extent that the United States "saw itself in an international technological competition with the Soviet Union," the country felt compelled to remain ahead of the Soviets in its advanced technology. Science, and especially high-energy physics, with its large accelerators and state-of-the-art electronics, benefited from this conception, as "national security and national prestige became the major determinants for both the size and the pace of growth of the government budgets supporting research and development in general, and the physical sciences in particular." These sciences were mobilized "in the interests of national security."[74] In this context, Schweber argued, the support of scientists who had been active in research during World War II, especially those who had been associated with the Manhattan Project, including Robert Bacher, I. I. Rabi, J. Robert Oppenheimer, Ernest Lawrence, and Enrico Fermi, "were some of the most convincing advocates of high-energy physics, and the spectacular flowering of the field owes much to their effectiveness as proponents." Among the arguments put forth in the 1963 report of the Ramsey Panel was that "it is essential that the United States maintain its leading position in this area which ranks among our most prominent scientific undertakings."[75]

It was clear both to Congress and the White House that having the most powerful particle accelerators would yield a strategic advantage. Recently, the Cuban missile crisis of October 1962 had revealed how easily the delicate international balance could be upset. And it was known that the Soviet Union, even with its semblance of international cooperation in the 1959 Emelyanov-McCone agreements, was planning to build a 70 BeV proton synchrotron, UNK, in Serpukhov, a project that threatened to wrest the worldwide lead in high energy from the United States.

The BOB approved the Ramsey Panel's recommendations and supported the AEC's plan to circulate Berkeley's design study within the community, democratically inviting input. The panel thus opened the door to a more bureaucratic era in accelerator history, in which all proposals for a large project would be circulated beforehand, often in massive detail, before construction could be considered. Once again, the two accelerator laboratories with established reputations had attracted the funding to develop new accelerators. It appeared that tradition would hold and that Berkeley would again lead the way to higher energy.

The decision in Berkeley's favor was not universally applauded. Goldwasser remembers that when Rabi heard the news, there were "tears in his eyes" because he was concerned that under Berkeley's management the new accelerator would not offer a democratically equitable policy for all qualified users.[76] There were definite rumblings, however, about the possibility that the several hundred GeV machine might not be constructed by Berkeley. Ramsey, who had a number of long meetings with McMillan, recalled "warning him" that if Berkeley was too insistent in opposing the required committee with national representation, "there was a high probability that Berkeley would lose the project."[77] Such sentiments would play out in novel ways over the next several years, as the scientists and politicians who were concerned about democratic access to the accelerator or who were disappointed over the loss of MURA would figure critically in the decision to select a Midwestern site for the new laboratory.

THREE

The Berkeley Design, 1963–1965

The last centuries of science have been marked by an unabating struggle to describe and comprehend the nature of matter, its regularities, its laws, and the language that makes it intelligible. The successes in this struggle, from the Sixteenth Century until our own day, have inspired the whole scientific enterprise, and lighted the world of technology, and the whole of man's life. They have informed the education and the devotion of young people. They have played an ineluctable part in the growth, the health, the spirit, and the nature of science. We are now, despite tempting and brilliant topical successes, deep in the agony of this struggle. J. ROBERT OPPENHEIMER[1]

Soon many physicists challenged Berkeley's plan to build the several hundred GeV accelerator. Their sentiments arose partly from long-standing East Coast–West Coast or Coast-Midwest scientific rivalries, and partly from the users' dissatisfaction over limited access to the existing laboratories on both coasts.[2] Midwestern political forces during the mid-1960s strengthened these challenges with a move to democratize American high-energy physics.

The Truly National Laboratory

Not long after the Ramsey Panel offered its recommendations, the PSAC appointed Myron L. "Bud" Good of the University of Wisconsin to chair a committee of accelerator users authorized to review the Ramsey Panel's recommendations. Meeting in June of 1963, the Good Panel underscored "the need for emphasizing the national character of new facilities," stressing that "access should be available not only to the accelerator, but to on-site support activities, and that outside groups should have a voice and responsibility in certain aspects of laboratory management."[3] This facet of the Ramsey Panel's recommendations was a sticking point for McMillan, who, in adhering to Berkeley's traditional attitude, downplayed this particular issue. McMillan's continued resistance to granting such rights to outside users of the proposed several hundred GeV accelerator was an indication of the conflict that would soon erupt between McMillan and his advisory committee.

Berkeley's policy toward outside users dated back to the 1930s, when Lawrence built his laboratory in Berkeley with private funding. Although he accepted requests from outside researchers to conduct experiments using Berkeley's machines, he solicited these by his invitation. His control of the use of the machines continued even after federal funding began during World War II, to the annoyance of many physicists outside the Berkeley circle.[4] When Brookhaven started in 1947, it planned to give access to its Cosmotron to the physicists from the nine member institutions of Associated Universities, Inc. (AUI), all in the northeastern United States.[5] By the early sixties, the idea of a national laboratory offering open access and other rights to all qualified users, regardless of their home bases, began to gain momentum. In 1962 physicists Edwin Goldwasser from the University of Illinois, Roger Hildebrand from the University of Chicago, and Father Theodore Hesburgh, the president of Notre Dame, were among those who promoted the open user access concept in the course of creating the program for Argonne's new Zero Gradient Synchrotron (ZGS). MURA physicists also joined the campaign.[6]

The ideal of a laboratory offering democratic outside user rights gained much wider acceptance when Leon Lederman presented the rationale persuasively at a Brookhaven summer study held in June 1963.[7] The lively forty-two-year-old physicist from Columbia University, already well known for leading a number of important experiments, had served on both McMillan's advisory committee and the Good Panel. He objected strongly to Berkeley's and Brookhaven's limitations on outside users. At a Brookhaven users' meeting he presented the argument for an outside-user-friendly laboratory in a paper on the "truly national laboratory" (TNL). The acronym TNL was an intended pun with which Lederman criticized BNL (Brookhaven National Laboratory) for not functioning as a TNL. As Lederman noted, cooperation and enthusiasm for the new facility would be assured only "when it is clear that the new facilities are accessible as a right to any physicist bearing a competitively acceptable proposal." The ideal laboratory, he said, would have complete on-site facilities for outside users, resources for facilitating individual experiments, an accessible and pleasant site, scheduling and advisory committees to assure fairness in the allocation of beam time, free communication between management and users, and a strong laboratory director "responsible to a governing body of wide national representation." Unlike BNL, the TNL Lederman described would give young researchers from different parts of the United States equitable access to the highest energies at a facility organized to award beam time according to merit. It would offer good experimental facilities and be a place where users would feel "at home and loved."[8]

Progress of Berkeley's 200 GeV Design Study

After being authorized by Congress, Berkeley's design effort progressed rapidly, with Lofgren serving as director and Lloyd Smith as Lofgren's assistant. The staff included fifty-five full-time workers, including physicists, engineers, programmers, draftsmen, and other support personnel. Ultimately, more than 130 people contributed to creating the Berkeley design. They were aided by a series of meetings that brought together accelerator experts from Berkeley and Brookhaven, as well as from CERN, where plans were underway for the Super Proton Synchrotron (SPS), CERN's 300 GeV strong-focusing synchrotron, which would eventually reach 450 GeV.[9]

The Berkeley physicists paid special attention to designing the 600–700 magnets for the main ring of the machine, constructing prototype models and choosing the best injector accelerator. They felt the need to produce a reliable design, for as LBL physicist Denis Keefe later explained, their "feet were being held to the fire." For greater reliability the selection of a magnet design turned quickly away from the more economical H-shaped magnets to the conventional, but more costly, C-shaped magnets. As for the machine itself, Berkeley initially favored accelerating the beam first in a linear accelerator (Linac) that directly fed its beam into the main synchrotron. But as the studies progressed, the problems of Linac injection appeared increasingly formidable and Sands's idea of injecting from a booster synchrotron appeared more attractive.[10]

Despite McMillan's resistance to the idea of appointing a national advisory committee, he wasted no time fulfilling what he saw as his obligation to do so. On May 10, 1963, he sent Paul McDaniel a suggested list of participants, including candidates from various parts of the

United States: four from the East, four from the West, and two from the Midwest. He explained that all ten had "agreed to serve as such a committee, and have been informed that their appointment is subject to AEC approval." McDaniel assured McMillan that the AEC would probably approve the list.[11] Almost three months later, McMillan submitted the advisory committee charter to the AEC, proposing that the ten-member committee have national representation, be nominated by the Berkeley director, be approved by the AEC, and report to the Berkeley director. The committee would then "set its own agenda" and have the right to "consider all technical, scientific, and administrative matters connected with the high-energy accelerator design study."[12]

Back in January 1963, Lofgren had met with McDaniel and other AEC Division of Research staff to discuss guidelines for preparing Berkeley's final proposal, including cost estimates and the arrangements for obtaining architect-engineering services. On the question of siting, the staff of the AEC Division of Research expressed willingness to work with Berkeley to find a California site. Lofgren mentioned Camp Parks, a 3,636-acre site in the Livermore-Amador Valley, thirty-seven miles from San Francisco. AEC staff member Phillip McGee agreed to inquire whether the site was already earmarked for national defense purposes. The AEC staff advised Berkeley researchers to go through the proper channels but admitted that "some simplification of the procedures may be worked out." Lofgren thought it likely they would "get a very high degree of cooperation from the Research Division Staff."[13]

An Invitation from the Midwest—MURA

Meanwhile, after learning of the Ramsey Panel's low ranking of the FFAG, MURA physicists threw themselves into an eleventh-hour attempt to save their project. It was easy for MURA to elicit support from Midwestern university presidents, who had nothing to lose and much to gain if a prestigious facility was built in their region.[14] Particularly active in the debate were Elvis Stahr, the president of Indiana University, and Bernard Waldman, a physicist at Notre Dame as well as MURA's staff director. For a long time Waldman had "felt Midwesterners were not getting a fair shake." After all, the only Midwestern accelerator was Argonne's weak-focusing 12 GeV ZGS, a project born in the midst of troubles between MURA and Argonne and advanced primarily because of the AEC's desire to have a machine built to surpass a new 10 GeV Soviet machine. But when the ZGS came online in 1963, its operation was disappointing compared with the 30 GeV machines at Brookhaven and CERN which had come into operation three years earlier. Waldman "wondered if the Midwest would be a permanent step-child."[15]

Midwestern politicians realized the advantages of championing the FFAG. By winning the nearly $150 million project for his constituents, a congressman could promise not only new jobs, but also greater prestige and new industrial opportunities for his district. With the economic recession in 1962, these politicians expressed their long-standing concern about the inequality of geographic distribution of research funding. Minnesota's Democratic senator Hubert Humphrey was prepared to act on the lack of support for MURA.[16] Receiving less funding for research than East and West Coast institutions irked Midwesterners who pointed out that their region included many internationally ranked physicists, such as Enrico Fermi, who was based at Chicago until his death in 1954, and John Bardeen, who in 1951 moved from Bell Laboratories to the University of Illinois in Urbana-Champaign.

Thanks to Stahr's efforts during the summer of 1963, President Kennedy received a steady stream of form letters signed by Midwestern congressmen stating that MURA "represents only a crucial first step in restoring the Midwest to its rightful place—along with the East and West Coasts—as a major center for educational research." Similar letters were sent to the BOB and the AEC. MURA argued that the Midwest deserved a first-class machine and that the FFAG was an innovative design that inspired respect among accelerator specialists.[17] But in July 1963, MURA received another jolt when it learned that the majority of the members of the AEC's General Advisory Committee (GAC) "felt there is not adequate justification at this time for approval of the MURA accelerator construction." They pointed to the high cost of the machine and to the fact that it would not extend the high-energy frontier.[18] While many physicists approved of the FFAG, most felt that if they had to choose between a new high-energy or high-intensity machine, high energy was the choice.[19]

The GAC's damning decision placed the AEC in an awkward position. It was time for the AEC's chairman, Glenn Seaborg, to become more directly involved. Believing in the power of negotiation, Seaborg explored alternative schemes, such as incorporating MURA's design into future plans to upgrade the Brookhaven AGS, or building the MURA machine near Argonne, perhaps with the same management.[20] MURA's prospects grew dimmer when in mid-November 1963 a reconvened Ramsey Panel reaffirmed its earlier position.[21] Later that month, President Kennedy, who in the eyes of MURA researchers was on their side, was assassinated and the nation entered a period of mourning.

As the United States emerged from its bereavement a few weeks later, BOB director Kermit Gordon cast the bureau's vote on MURA. In a memorandum to the new president, Lyndon B. Johnson, Gordon recommended that MURA not be authorized for various reasons: the lack of "resounding scientific support" for the machine, potential adverse affects on funding for the 200 GeV machine, Argonne's newly funded ZGS, and "current fiscal stringency."[22] The AEC made its decision later in December, just a month before the budget was to be presented to Congress. In a memorandum to Johnson, Seaborg explained that while the commission still favored the authorization of MURA, the project was not to interfere with high-energy proposals and could not be placed "in as high a relative priority" as other items.[23]

The official rejection of MURA's project to build the FFAG occurred on December 20, 1963, in President Johnson's office. The president met there with Wiesner, Seaborg, Gordon and other BOB representatives, along with several Midwestern congressmen, including Senator Humphrey (who would in January 1965 become Johnson's vice president), and a number of MURA supporters, including Stahr, Waldman, and Goldwasser. After listening to the advantages of building the FFAG, Johnson read a prepared statement listing its disadvantages. Unknown to the MURA representatives, Wiesner, who would soon return to MIT, had prepared the statement ahead of time at the president's insistence. After making a vague remark about the difficulty of funding MURA, Johnson dismissed the group. Waldman remembered: "I felt dazed. I hadn't been sure we would win, but I never thought it would end like this."[24]

Johnson's decision to cancel the MURA machine sent waves of dismay through Congress and the Midwestern physics community. Seaborg recorded in his diary that Johnson called him several hours after the meeting. Acknowledging that they were "going to be in serious trouble in Minnesota," the president asked Seaborg to be prepared to offer something to the congressmen who, like Humphrey, were likely to be upset. The next day, Stahr wrote an anguished letter asking Humphrey to urge Johnson to reconsider the MURA proposal. The letter complained that the statement read by Johnson "omitted some highly relevant matters, and its prejudice showed on its face. . . . I don't know who wrote it, but it sounded exactly as if it might have been prepared by someone from New England or the West Coast." An angry Humphrey quickly wrote to Johnson, enclosing Stahr's letter.[25] On January 20, the AEC issued a press release officially announcing MURA's fate.[26] MURA representatives felt betrayed, for they had been reassured that both Wiesner and Seaborg supported their project. Johnson clearly did not, as he explained in his reply to Humphrey on January 16, 1964, and at a meeting the next day with Wiesner, Seaborg, and Frederick Seitz. The president's budget, submitted to Congress in late January 1964, did not even mention the FFAG. For the first time, an existing accelerator effort had been canceled.[27] After the death of the FFAG, the physics community mended its fences as best it could, but the bitterness that Midwestern physicists felt about MURA's defeat would fuel their lobbying for a large accelerator in their region.

Reality Sets In: Management and Siting Plans Evolve

The FFAG concept continued to play a role in the discussions of high-intensity machines. Some MURA physicists joined Brookhaven's effort to design its 1000 GeV machine. Many more physicists reconsidered the 200 GeV accelerator and turned their attention to the issue of deciding its site. Most Midwestern physicists felt that it would be unfair to build the next accelerator anywhere but in their region. Meanwhile, leaders of the physics community, including Seaborg, Ramsey, and Seitz, launched intense negotiations in an effort to quell the troubles in the Midwest. Seitz, a respected solid-state physicist then head of the Department of Physics at the University of Illinois, would soon become the first full-time president of the National Academy of Sciences (NAS).

For government leaders the arrangements for the 200 GeV accelerator were taking on a new complexity. The problem, as Seaborg told Goldwasser on January 23, was that "on the horizon" were "two accelerators and three sites." He warned that it might "wind up as one accelerator, with an order for everybody to get together, if the scientists fail to come to agreement." The vocal dissent of Midwestern scientists and politicians made it difficult for Seaborg to justify funding accelerators on either the East or West Coast. As the expensive accelerator needed the unanimous support of the entire physics community, the Midwestern physicists would have to be appeased. In his conversation with Goldwasser, Seaborg stated that "the site of the 200 GeV accelerator . . . is still undetermined."

A further complicating factor was that a $100 million linear accelerator was being built at SLAC. The AEC and the Stanford faculty had crossed swords over control of the research program, including the question of who should control user access. Although in the end Stanford refused to allow joint appointments with the new laboratory, SLAC researchers continued to resist attempts to give guaranteed access to outside users.[28] In a conversation with George Beadle, president of the University of Chicago, Seaborg remarked that "the Stanford accelerator is the last of its kind.

The new ones will be of the consortium management type." As Seaborg's notes indicate, Berkeley's prospects for managing and building the new accelerator at a California site were noticeably weaker in early 1964 than they had been only six months earlier.[29]

McMillan felt the shift in mood. On April 13, 1964, Milton G. White, director of the Princeton-Pennsylvania Accelerator, disseminated a letter addressed "To Friends of Fundamental Particle Physics." He proclaimed that "realistic proponents of high energy machines . . . must strive for a realistic goal which takes full cognizance of the political and economic realities of the present era," and he suggested that a "new, contracting, managing group . . . be set up to run the laboratory for the benefit of all scientists everywhere." Asking "that all regional, parochial interests be swept aside," he proposed that the new laboratory "be located in some central, but on the average convenient, area not previously associated with high energy physics."[30]

On April 19, McMillan responded to White, declaring that an accelerator was not "an industrial plant, to be located where economics dictate and operated by hired hands from the region and managers exported from the home office." He argued that those designing, constructing, and operating a large accelerator "should have a sense of identification and continuity that is hard to establish in a new group." Stressing that scientists work best in a multidisciplinary environment close to an established university, McMillan insisted that it would be time consuming and difficult to form a nationwide management group. On a more personal note, he said he resented "the implication that the interest of Berkeley is regional and parochial." He explained that Berkeley researchers were only "seeking to continue the line of development started many years ago by Professor Lawrence," a goal they felt was compatible with creating a national facility. McMillan was "distressed by the bald assertion" that Berkeley plans were "doomed to failure," fearing that this unfortunate opinion, which many had already expressed to him, would gain credibility from repetition. He concluded that "the best way for 'friends of fundamental particle physics' to promote the development of high energy facilities" was to "support fully the recommendations of the Ramsey Panel."[31]

Ironically, a few weeks after McMillan thus evoked the Ramsey Panel, Ramsey sent McMillan a warning based on other findings of the panel. In a May 13 letter, written at the request of Lederman, Ramsey said that he personally didn't "care as to the form of the laboratory administration provided it is fair and efficient, with favorable arrangements for visiting scientists from universities deprived of the opportunity of having their own accelerators." But, as he carefully explained, "a large number of physicists believe that various formal organizational requirements must be met before they can support a unique facility laboratory." In his opinion such views arose because of the troubled relations between Midwestern physicists and Argonne and because "a number of institutions and physicists are—rightly or wrongly—jealous of the University of California." In view of the current mood, Ramsey urged McMillan to "make a reasonable number of concessions to assure those who might become opponents of the accelerator that the laboratory will be fairly administered without too much of the advantage going to a single institution."[32] While such opinions had been implied by the Good Panel, by Lederman's paper on the TNL, and by White and others, McMillan had never before been confronted so directly.

McMillan discovered the magnitude of the concessions being requested in reading a May 26, 1964, letter from William "Jack" Fry, the chairman of McMillan's advisory committee. Noting the committee's "satisfaction with the technical progress" and its "high confidence in the capabilities of the LRL group to produce a successful design," Fry presented a management scheme reminiscent of the one discussed by the AEC in 1962 and by the Ramsey Panel in 1965, but certainly different from McMillan's plan. Speaking on behalf of the advisory committee, Fry proposed that McMillan form "a Corporation of a national group of major universities active in the field of high energy physics." The idea was that this corporation, led by a board of trustees with national representation, would sponsor the proposal to construct the accelerator "based on the results provided by the current Berkeley design study." But although "the talents and experience" of Berkeley scientists and engineers would be "utilized fully" during the design and construction of the accelerator, Fry and his committee suggested that as the facility approached its "operating and research phase," it "should become independent of LRL." And while LRL could hire senior scientific and technical staff members, permanent appointments would require approval of the corporation, which would eventually appoint a director "responsible to the Corporation for the program of the laboratory."[33]

As expected, McMillan opposed Fry's suggested scheme. After the meetings with McMillan, Fry wrote to committee member Lee Teng that it was now "clear" that Berkeley would "not accept a National Board for the construction stage." In an attempt to iron out a mutually acceptable plan, Fry planned a fall meeting between Berkeley and its advisory committee.[34]

The View from Washington

Washington's response to these discussions must be placed into the historical context of the mid-1960s, when some members of Congress began questioning both the social value of science and the role of scientists in policy making, given the huge increase that had taken place in the proportion of funding for basic research in relation to overall federal expenditures during the previous decade and a half.[35] Accelerator projects occupied an especially visible portion of the high-energy physics budget because of their high cost of construction. By 1963 this subfield appeared uncomfortably costly and faced criticism from those who questioned the merits of "big science," the term coined in this period by the director of Oak Ridge National Laboratory, Alvin Weinberg, and subsequently discussed at length by such historians as Derek de Solla Price.[36] For several years, Weinberg, a longtime proponent of nuclear reactor development, had been issuing warnings about the dire implications of the high cost of large-scale research. His criticism sharpened with the threat of severe budget cuts. In a series of publications he described a method for assessing the relative value of various scientific programs, remarking that if one considered "relevance to the sciences in which it is embedded, relevance to human affairs, and relevance to technology—high-energy physics rates poorly."[37]

At the same time, high-energy physics was becoming a cause célèbre in the fight for equitable geographic distribution of research funding. Although Johnson insisted in his January 1964 letter to Humphrey that the Midwest had already received its fair share of funding, the president also stated that he shared the "strong desire to support the development of centers of scientific strength in the Midwest."[38] The Midwestern politicians and physicists agreed that they deserved such scientific centers and considered MURA's well-publicized defeat an example of the unfair distribution of federal monies. Thus, Midwestern physicists and politicians were arguing for their fair allocation of resources to deserving researchers at a time when the number of first-class facilities was dwindling. Their combined interests formed a strong alliance for a Midwestern accelerator.

Funding considerations were not lost on the Joint Committee on Atomic Energy (JCAE), the powerful congressional overseer of the AEC and the traditional champion of energy programs, including those of particle accelerators. In the annual hearings on the AEC budget held in early March 1964, Chester Holifield, JCAE chairman and the representative from California, pointed out that the high-energy physics budget had jumped from $53 million in 1960 to $135 million in 1964. He noted that the AEC shouldered about 90% of the financial burden of the subfield but reminded the executive-branch officers that basic research appropriations were "sometimes hard to sustain in the Congress . . . much harder than it is for applied research and development." Holifield was concerned that "the overall budget of the AEC stays about the same, but the high-energy part of it is increasing." Conceding that it was "perfectly natural" for scientists to have a limitless supply of ideas and ambition for exploring them, he worried that in accommodating such researchers, decision makers were "squeezing to death many fields of science." "Very frankly," he concluded, "the Congress is becoming alarmed."[39]

After this warning, Holifield began to quiz Johnson's new science adviser, Princeton chemist Donald Hornig, about proposed accelerators. Pointing out that the MURA accelerator had been cut from the executive-branch budget because of tight constraints, and noting that millions were being spent on the design of the 200 GeV accelerator. Holifield queried: "What assurance does the executive branch provide that this planned 200-billion-electron-volt accelerator will not also be scrapped in time?" To champion AEC budget requests, Holifield insisted, the JCAE needed a defensible national policy for high-energy physics which had explicit presidential approval. The policy, which should take into account long-term projected costs for such expensive projects as the proposed 200 and 1000 GeV accelerators, should not act as "a rigid plan," but as "a general plan with a forecast of expenditures." He added that in light of the current high cost of accelerators it would be necessary periodically to reassess the scientific utility of proposed projects, "what has happened in science in the meantime, . . . and what the fiscal situation of the Nation is."[40]

At the March 1964 JCAE hearing, questions were again raised about the increasing cost of high-energy physics and its effect on other scientific fields. Holifield quoted statements by the nuclear physicist Eugene Wigner, who was anxious to promote nuclear reactors and had complained, as a member of the 1960 Piore Panel, that high-energy physics was draining an undue amount of scientific manpower. Wigner also pointed to the funding shortage in other areas, such as low-energy nuclear physics.[41] Another vocal critic of high-energy physics was the chemist Philip Abelson, codiscoverer with McMillan of the element neptunium, and later editor of *Science*. Although Abelson had served on the Ramsey Panel and endorsed its final recommendations, he later decried the level of physics funding for the 200 GeV accelerator. In a January 1966 article in the *Saturday Review*, Abelson proclaimed: "Never, in the history of science have so many fine minds been supported on such a grand scale, and worked so diligently, and returned so little to society for its patronage."[42] Goldwasser, a member of the Ramsey Panel, later remarked about Abelson: "We thought we'd educated him in the course of the meetings . . . but we were wrong."[43]

Hornig, on the other hand, believed that high-energy physics was making rapid progress. By the end of March 1964, he submitted a three-page policy statement to Senator John O. Pastore from Rhode Island, the JCAE vice-chairman, in which he noted that "it has been true historically that in the long run these understandings have had a very great impact on science and technology and on all mankind." Thus, he argued, it was "in the national interest to continue vigorous exploration" of the field. He urged supporting two accelerators, with the second eventually providing energies of 1000 GeV in about twenty years. Hornig also wrote to Johnson, enclosing his letter to Pastore and presenting the time schedules and expenditure estimates for the proposed machines. He judged that the 200 GeV project would be up for authorization in fiscal year 1967 (July 1966 to June 1967) and cost about $280 million, with completion by June 1973. Although the figures for the 800–1000 GeV machine were more speculative, he guessed that the larger accelerator would be up for authorization in fiscal year 1968 (July 1967 to June 1968), cost about $800 million, and be ready for initial operation by June 1976. To provide a more complete national policy statement on high-energy physics, the AEC subsequently made plans to compile a much longer report to be ready in time for the JCAE hearings to be held the next spring.[44]

To counter the opposition of Abelson, Weinberg, and Wigner, Milton White organized a special session at the April 1964 annual meeting of the American Physical Society (APS). At the roundtable discussion that concluded the session, Weinberg faced newly elected NAS president Seitz, who chaired the meeting, four high-energy physicists, and George E. Pake, a Washington University physicist who described himself as "sympathetic" to high-energy physics. While many panel members cited problems caused by the growing size and expense of high-energy physics, all but Weinberg proclaimed the field's superiority. C. N. Yang, for example, then at the Institute for Advanced Studies, spoke of the "nobility" of the aim of high-energy physics; Goldwasser called the field "the essence of pure science."[45]

In June, Weinberg encountered further public resistance to his opposition to high-energy physics when *Physics Today* published side-by-side letters from Weinberg and Victor Weisskopf, then the director general of CERN. Weisskopf endorsed Weinberg's assessment criteria but argued that high-energy physics fared well when judged by such a standard. Agreeing with Weinberg that the field held little "technological merit," he insisted that its value stemmed "from the fact that this field is basic and relevant for all sciences and therefore touches questions which all thinking human beings are deeply interested in."[46] Weinberg reiterated his contention, explaining at the APS meeting: "Science which commands great public support must be justified on grounds that originate outside the particular branch of science demanding support: it must rate high in social, technological, or scientific merit, preferably in all three."[47]

As Weinberg's words lingered, leaders of the physics community worried about the dissension arising from Midwestern high-energy physicists and from those agitating for outside user rights. And even though Wigner's and Weinberg's criticisms of high-energy physics remained a minority opinion, their dissension reinforced congressional skepticism about the value of basic research at a time when budgets were being cut. At the April APS meeting, Seitz sagely mused: "One is reminded of the way in which the competition among the nations of Western Europe in the last century has had the effect of decreasing the collective strength of all." He added: "Wisdom would seem to indicate that the family of high-energy physicists must somehow learn to resolve its differences and speak with a unified voice."[48]

The Formation of URA

The idea that a consortium might manage the 200 GeV accelerator facility was often suggested during the mid-1960s. McMillan and his advisory committee had been engaged in intense negotiations about the issue for over a month at the time when Goldwasser wrote to McMillan in September 1964, "I believe that the actual control must be in the hands of an organization similar, perhaps, to AUI but national in its scope." In the same letter, Goldwasser expressed his concern about the internal dissent in the physics community over the new accelerator: "We have now advanced one more generation to a point at which national motivation expressed essentially unanimously by the entire scientific community is required in order to obtain the government support that is needed."[49]

Later that month, McMillan's advisory committee gathered in Berkeley for a two-day meeting to devise what they considered the best management scheme, presenting this in draft form to McMillan on October 7. They distributed the draft to various Berkeley staff members. The scheme included a national management corporation, but unlike the earlier plan suggested in May, the regents of the University of California, with whom the corporation was to form a "joint venture," would supervise the design and construction of the accelerator. The University of California was allowed a vote in selecting the new laboratory director, but after construction of the machine the laboratory would "become the sole responsibility of the National Corporation."[50]

In late October 1964, thirteen members of the design team at Berkeley responded to this scheme in a memorandum addressed to Lofgren. While they supported the general idea of national access and national participation, they were horrified by the prospect of losing the traditional prerogatives of machine builders to use the machine they had built. Lofgren endorsed the memo and passed it on to his advisory committee.[51]

Tensions resulting from the disagreement about whether machine builders should have priority in using the accelerator they built were hardly unique to Berkeley; in the 1950s at Brookhaven, Cosmotron builders, including, ironically, Lederman, had argued for just this sort of priority. Berkeley was facing an unprecedented clash between its elitism and changing times. One participant remembered that the growing tension joined a more general feeling of anger. Coincidentally, tempers in Berkeley's physics community rose just as the first major demonstrations of the Free Speech Movement erupted on the Berkeley campus in late September 1964 to protest regulations restricting political activities on the campus.[52]

Meanwhile, on October 29, Fry forwarded the unanimous report of his committee to McMillan. McMillan was incensed to find this report but a slightly edited version of the October 7 document, which included the joint venture scheme. Three days later, McMillan complained to Wolgang K. H. "Pief" Panofsky, his longtime colleague, first director of SLAC, and an advisory committee member, that he was "deeply upset" by his advisory committee's refusal to consider his design team's complaints. In a November 3 letter to the advisory committee, Fry remarked that both McMillan and Lofgren "were very unhappy with the report," and that it was now "clear that neither of them want a National Committee to administer the laboratory, which clearly is in opposition to our views." After minor editing, a final report was delivered to McMillan on November 11, 1964.[53]

As the negotiations between the Berkeley design group and its advisory committee grew heated, yet another National Academy of Sciences panel released its conclusions on the future of high-energy physics. During the spring of 1964, the National Research Council (NRC) of the NAS had convened a Physics Survey Committee, headed by George Pake, "to identify the most significant research problem in various subfields of physics, as a basis for estimating levels of support necessary to assure the balanced development of the field as a whole over the next 5 to 10 years." Pake organized nine panels, including one on high-energy physics under the direction of Caltech's Robert Walker, a former WAG researcher.[54] Like the Good Panel, the Walker Panel endorsed the findings of the Ramsey Panel, but it went farther in defining recommendations for the next accelerator facility. Three of its four points did not seriously contradict Berkeley's plans: that the laboratory should have (1) a written charter to define its "national character," (2) a strong director independent from the control of those wishing to use the accelerator, and (3) a committee with national representation to approve experiments. The fourth, however, stated: "The body which decides the major policies of the laboratory and to which the director is responsible should be national in representation."[55]

McMillan's design team fully expected their leader to reject the national management scheme. It faced an even greater likelihood of opposition from Berkeley's star team of inside users, which was just then struggling to adjust to outside user representation in the Bevatron's scheduling committee. Indeed, McMillan himself saw the push for non-Berkeley management as an unworkable idea prompted by jealousy and opportunism.[56] On November 23, 1964, McMillan wrote to McDaniel, remarking diplomatically that the advisory committee report was in general "a very thoughtful document." He agreed that a national management corporation "offers a valid mechanism for assuring national representation, and in the present context was probably the best way." Despite this concession, he urged that the matter of turning over eventual operation of the machine to the corporation "be kept open for further discussion."[57]

At this point, Seitz made a surprising move. He considered the new laboratory vital for the national science program and he was worried that the bickering between McMillan and his advisory committee would doom the laboratory's funding prospects. Empowered by his position as president of the National Academy of Sciences, Seitz simply commandeered the management plans and settled them. He first asked AUI whether it might take on the project. When AUI declined, Seitz took the decisive step himself.[58] As he recalled, "We set up this organization, funded it through the Academy to begin with, and then began forming a board with the official view that the whole issue of the next machine should be reviewed. That caused some difficulty with California's delegation in Congress. . . . But when the whole high-energy physics community—with the exception, obviously, of the people at Berkeley—piled on it was very hard not to move forward." According to Seitz, Seaborg could "probably have stopped us." But, as Seitz assessed, "Glenn was a very rational scientist and realized that with the whole high-energy physics community, which he respected, in an uproar, we ought to move ahead." To help with the legalities of creating the new consortium Seitz brought in the NAS's lawyer, his friend Leonard Lee Bacon.[59]

In late November 1964, Seitz invited the presidents of twenty-five major research universities to a meeting sponsored by the NAS to draw up future plans for the new laboratory. This special meeting, held on January 17, 1965, marked a turning point in planning the new accelerator. Seitz, Hornig, and Seaborg addressed the assembled university presidents and explained why a national organization was needed to manage the 200 GeV facility. In the course of vigorous discussion, it was proposed that a new nationally based organization modeled on AUI be formed. Named the Universities Research Association, Inc. (URA), its initial goal would be to construct and administer the 200 GeV accelerator.[60] The university presidents would act as overseers of the new laboratory, while a group of administrators and senior scientists would formulate its operating policies. For the time being, Berkeley would continue its design work on the accelerator, but the new management organization would take charge of selecting its site through a national site contest.[61]

Marshaling Support

While awaiting the Berkeley group's preliminary design for the new accelerator, due in July 1965, the AEC sought to gain the financial support it needed from the White House and the BOB. In the previous round of authorization hearings in 1964, the JCAE had emphasized the difficulties of obtaining congressional approval because of the rising costs of high-energy physics research. Seaborg and Seitz knew it was time to mount an intensified campaign to marshal support for the 200 GeV accelerator, for the earlier criticisms from Weinberg, Wigner, and other physicists had hurt the prospects of funding such an expensive high-energy physics project.

In January 1965, Luke Yuan of Brookhaven published the essays he had been gathering since mid-1961, written by twenty-eight well-known theoretical physicists, in a collection titled *The Nature of Matter: Purposes of High Energy Physics*. Plans for this document had emerged when Brookhaven physicists sensed "misunderstanding of the objectives of high-energy physics" while they were working to design the 600–1000 GeV accelerator. They recognized that a clearer explanation was "urgently needed among high-energy physicists, the scientific community as a whole, and the general public."[62] The volume began with a foreword by Oppenheimer, who articulated traditional accelerator justifications, including the technological spin-offs and benefits to other sciences. He evoked some of the famous contributions of nuclear physics, noting that it was "not impossible" that physicists would produce "an unanticipated discovery of profound importance to technology and to human welfare." He put greatest emphasis on the cultural value of fundamental physics research, insisting that the struggle to understand the nature of matter had not only enriched science and technology, but shed light on "the whole of man's life."[63] This justification of basic research as enrichment for the human spirit was a natural argument for physicists who saw themselves as explorers at the frontiers of understanding the basic structure of matter.

This view of science as a beacon for culture was reinforced and amplified in other essays. For example, Julian Schwinger proclaimed: "The world view of the physicist sets the style of the technology and the culture of the society, and gives direction to future progress."[64] Several essays pointed to the success of the SU(3) classification scheme developed by Murray Gell-Mann and Yuval Ne'eman independently in 1961, which brought order to the confusion about particles by organizing strongly interacting particles into families of eight and ten. The scheme gained general acceptance after correctly predicting the mass and existence of the omega minus, a new particle found in 1964 by Nick Samios and his colleagues at the Brookhaven AGS. As Tsung Dao "T. D." Lee explained, an accelerator "in the range of 200 to 1000 BeV" would "certainly be crucial" in exploring the "detailed dynamics of this strong SU(3) symmetrical interaction."[65]

While these sentiments were being transmitted to a general audience, the AEC was sending a similar message to government leaders. On January 25, 1965, Seaborg sent President Johnson the "Policy for National Action in the Field of High Energy Physics," a full-length AEC report written to augment Hornig's spring 1964 policy statement. Like Yuan's collection, this report was optimistic about the future success of high-energy physics, characterizing the field as "an exciting and vigorous field of basic research which ranks high among our most prominent scientific undertakings." Describing recent advances, the research potential of extending the high-energy frontier, and the potential for technological spin-offs, the report also noted that the recent bounty of particle discoveries might "mark the beginning of one of those intensely productive and exhilarating eras in which apparently uncorrelated facts accumulated over many years are tied together, and science makes a 'quantum leap' forward."[66]

The report insisted that it was "in the national interest to support vigorous advancement of high-energy physics" and asserted that the nation's high-energy physics program currently required "increased financial support, especially for the provision of advanced accelerators and equipment." Specific plans included improvements to several existing accelerators, construction of the 200 GeV accelerator, and continued planning for the 600–1000 GeV machine. Favoring this ambitious plan, the commission ignored congressional concern about the rising proportion of basic research funding and projected a steady increase in the national high-energy physics program, estimating that funding in 1970 should be almost twice that for 1965.[67]

The following day, on January 26, Johnson sent Seaborg's bold assertion of the AEC's prerogatives to the JCAE, along with his vote of confidence, noting that it provided "a useful guideline for decision-making in the development of high-energy physics."[68] One factor that probably favored Johnson's support for increased funding for high-energy physics was his close relationship with Seaborg. Seaborg saw the promotion of basic research as an important part of his mission as AEC chairman. As the policy report demonstrated, Seaborg was willing to lobby strongly for high-energy physics funding, even though his own scientific specialty was transuranic chemistry.

The public hearings of a JCAE subcommittee, on March 2, 3, 4, and 5, 1965, offered an opportunity for the physics community and the AEC to express their opinions on the status of high-energy physics while also giving both groups the chance to gauge the mood of the powerful congressional committee. Numerous physicists made presentations on overall research objectives, recent accomplishments, and the status of research programs at various laboratories in the United States, Europe, and the USSR. Gell-Mann explained the utility of SU(3) and amused his audience with his explanation that the term "quark" had been taken from James Joyce's *Finnegan's Wake*. Lofgren spoke on the status of the 200 GeV design study, showing diagrams of a four-stage machine that included a Cockcroft-Walton preaccelerator, a linear accelerator, a booster synchrotron, and a larger synchrotron. Besides incorporating Sands's idea of a booster synchrotron, the design also drew on Collins's long, straight sections.[69] Lofgren did not present a detailed cost breakdown, which would be included in the upcoming design report, but when quizzed on the cost of the machine, he remarked that "construction cost will go a little higher and the operating cost a little lower" than the $280 million construction cost mentioned in the AEC's policy statement. The congressmen did not object to this admission and, in general, seemed ready to accept the broad outlines of the Berkeley design.[70]

Although the atmosphere of the rest of the hearings was similarly friendly, several issues sparked heated disagreement. Predictably, a round-table discussion on the relation of high-energy physics to other sciences, led by Seitz, and including both Wigner and Yuan, displayed sharply divergent views on the social value of the field. Although the members of the JCAE let the panel members do most of the talking, they sided several times with those arguing the scientific merits of high-energy physics. They took a more critical stand, however, when discussing management plans for the new accelerator laboratories, in particular, the NAS management organization then being planned by the university presidents gathered by Seitz.[71]

At the beginning of the hearing, the JCAE questioned whether the university presidents on the overseeing board were competent to make technical decisions, whether the physicists in the policy-making board would be able to tackle large-scale management issues, and whether the new organization would dilute the AEC's control of the project. Considering the expense of the 200 GeV accelerator, California representative Craig Hosmer, a member of the JCAE, said it was "incumbent" upon the JCAE "to insure that the management and organizational setup is of the type that is designed and calculated to produce the highest return for the taxpayers' dollar." He added that he did not "believe that a group of university presidents or a group of high-energy nuclear physicists are perhaps the best people in the world . . . to determine the final management setup."[72]

In a panel discussion on management requirements for future laboratories, JCAE members pressed Seaborg to explain how the 200 GeV facility would be managed differently from other large laboratories that were already considered national facilities. Seaborg reiterated that they planned a management scheme similar to Brookhaven. When questioned, he admitted that Berkeley was "less of a national laboratory than the other laboratories." This negative message about Berkeley was emphasized by a chart presented by McDaniel showing that inside users at Berkeley consumed 65% of accelerator time, compared with 20% and 35% at Argonne and Brookhaven, respectively.[73]

The JCAE was even more adamant about the new accelerator's siting than about its management plans. Although the broad outlines of Berkeley's 200 GeV design were accepted without complaint, a discordant note was hit in Lofgren's presentation after he remarked that many aspects of the 200 GeV design, such as foundations and water and power availability, were "very sensitive" to the specific site chosen, and that the Berkeley group had therefore used the Camp Parks site in California as an example "to accomplish definite design studies." Representative Hosmer immediately asked whether Lofgren was trying to "sell" them the California site. This was a sensitive issue because a Denver group led by physicist Edward U. Condon had prepared a statement in February advocating a Boulder-Denver-area site for the new accelerator and by early March the AEC had also received recommendations for sites in Hanford, Washington, and St. Louis, Missouri. Hosmer decried such attempts "to jump the gun . . . and start a big national fight about where this thing is going to be." He warned that if the offenders did not stop, he would "do everything possible . . . to block the possibility of their ever getting this machine." Since several contenders for the site had already emerged and tempers were already inflamed over the regional distribution of research funds, there was no easy way to avoid a site competition. But the JCAE clearly wanted the AEC to take firm charge of the politically volatile contest. Thus, when summarizing the hearings, Representative Melvin Price from Illinois remarked that in the near future he expected "the AEC will be prepared to tell us what they have already done in respect to site selection and what they plan to do."[74]

In light of the generally positive JCAE response to the AEC policy statement and the affirmative stance taken on the social value of the field, it was likely that the JCAE would champion the new accelerator project, assuming that its concerns were addressed. But the JCAE's lack of enthusiasm about the new management organization created an uncomfortable problem for Seaborg and Seitz. They were reluctant to disband URA, because it seemed a viable vehicle for proving to congressmen and physicists alike that the new accelerator facility would be truly national in character. In addition, such an organization was desperately needed to maintain support within the physics community. Although dissenting opinions on the value of high-energy physics had not eroded the support of the JCAE at the March hearings, physics leaders knew that without unified support in the physics community they had little chance of success.

The committee's stand on site selection also caused headaches. At the January 1965 meeting of URA's university presidents, the participants discussed the possibility that the new organization would both manage the laboratory and select its site. AEC commissioner Gerald Tape explained that the AEC wanted to avoid taking sole responsibility for site selection. The idea arose to have the NAS appoint a committee to help. By letting the NAS group assume the task of mediating the site selection, the AEC could avoid the sort of accusations they had suffered in the aftermath of MURA's defeat. Because of its prestige and its national representation, NAS and the group of university presidents were in a better position to bear the inevitable sore feelings of losers. As the March hearings made clear, however, JCAE members wanted the AEC, rather than URA, to handle this politically volatile issue. They felt that the AEC was more likely to make a politically responsive decision, and the JCAE had more leverage if the AEC retained control. Seaborg, Seitz and other leaders were faced with the challenge of addressing the conflicting concerns of all those involved.[75] For the next few months, a working committee directed by J. C. "Jake" Warner, president of the Carnegie Institute of Technology, explored procedures for forming the new management organization; he served as its first president. Articles of incorporation for the Universities Research Association, Inc., were filed on June 21, 1965, and a fall board of trustees meeting was planned.[76]

In the meantime, Seitz and Seaborg removed site selection from the list of duties for the new management organization. An alternate plan was drafted. Before the March hearings ended, Seaborg wrote Seitz proposing that the academy "study the site selection problem in considerable detail using facilities, information and personnel that the AEC and its contractors will make available." He identified general criteria, including (1) "suitable geology," (2) "availability of power and water," (3) "sufficient acreage," (4) "proximity to a major jet airport" and a "cultural center that includes both a large university and a well-developed research and development base," (5) "ability to mobilize the necessary staff," and (6) "regional cost variations." He noted that he expected specific recommendations for the best locations.[77]

On April 6, 1965, Seitz presented Seaborg with an outline of the plan for determining the site selection. A committee of "distinguished scientists" appointed by the NAS would agree on site criteria, identify whether specialized panels were needed, study site-related material, make site visits, and present a site recommendation report. At Seaborg's direction, the AEC had prepared a more specific set of site criteria, including a more detailed description of desirable geographic and environmental features. By mid-April, the AEC had formed a task group, with representatives from the Divisions of Research and Construction, to gather preliminary site information and screen all incoming proposals for minimum site requirements before they were referred to the NAS. By this time the NAS had also recruited several members of the site selection committee, including its chairman, IBM's Emanuel Piore. On April 20, AEC and NAS representatives signed a contract to formalize the arrangement, which specified March 31, 1966, as the deadline for the site evaluation report. An AEC press release on April 28, 1965, announced the AEC-NAS agreement and requested that site proposals be submitted to the AEC by June 15, 1965.[78]

Many physicists saw the 200 GeV site competition of 1965 and 1966 as a bizarre spectacle of opportunistic boosterism. Traditionally, siting for new accelerators had been handled by the physics community itself. Now it was clear that the matter could attract considerable public and political attention. This change was only one of many to be faced, for accelerator plans no longer proceeded within the comfortable parameters of tradition. One response was a series of songs written by Argonne physicist and musician Arthur Roberts and heard in physics circles throughout the country during 1965. Roberts, who had a knack for musical satire, played his own piano accompaniment as he sang "The 200 BeV Accelerator" to the tune of a song from a musical comedy written by Irving Berlin, "Yip, Yip, Yaphank":[79]

The AEC's new competition
Has brought forth a wondrous refrain
A new song has captured the heart of the nation
From California to Maine.
Hear Kansas City, Detroit and Decatur
Proclaiming their grace and their civilized charm
The two hundred BeV accelerator
Gives culture a shot in the arm.
Oh—Particle physics you've stolen our heart
Come all and bid for this prize.
With three hundred million to spend for a start
See how it lights up the stars in our eyes.
For the beauties of Euclid our hearts may not yearn
That hard gemlike flame in our eyes may not burn
But for that kind of money we're willing to learn
So research forever we cry.
Chicago's trees make her parks a sensation,
In Seattle the sunshine's the thing,
Texas reads only the "Times" and "The Nation,"
Mississippi loves M. Luther King.
Ballet's all the rage in Fort Dodge and Topeka,
In Hanford desire for Chagall is acute.

Los Angeles' John Birch Society's weaker,
And string quartets flourish in Butte.
Oh—Particle physics you've stolen our heart;
Now you will uplift our souls.
The two hundred BeVer will change us forever,
Transforming our tastes and advancing our goals—
We will turn from Bonanza and Gunsmoke away
Our library budgets will triple today
For three hundred million whatever you say.
So research forever we cry.
We will reach a consensus on what should be done,
With every group happy except perhaps one—
The men who can build it and make the thing run.
So research forever—
We'll part from you never
Till death do us sever—
You wives just gonna love West Texas—
We cry.

The Berkeley Design

In June 1965, as site proposals flooded the AEC, Berkeley's design group unveiled its preliminary design for the new 200 GeV machine (fig. 3.1). The Berkeley design team brought its reports together in a massive report filling two two-inch thick volume. This weighty work, worthy of the Lawrence Radiation Laboratory's fine reputation, described a machine whose protons would be accelerated in four stages. After being produced in an ion source, they would be accelerated to 750 KeV in a Cockcroft-Walton preinjector. Next they would be energized to 200 MeV in an Alvarez-type linear accelerator. They would then cascade into a rapid-cycling injector synchrotron and be accelerated to 8 GeV. At that energy they would be brought into the main synchrotron in seven successive beam pulses over one-third of a second and stored in orbits around its circumference. The magnetic field of this final "main ring" would be held constant during injection, so that the 8 GeV protons would circulate at constant radius until the entire circumference had been filled. Then all the protons would be accelerated to 200 GeV and emerge for experimental use at an intensity of 3×10^{13} protons per pulse. As the report explained, this intensity was "an order of magnitude higher than those . . . available at lower energy."[80]

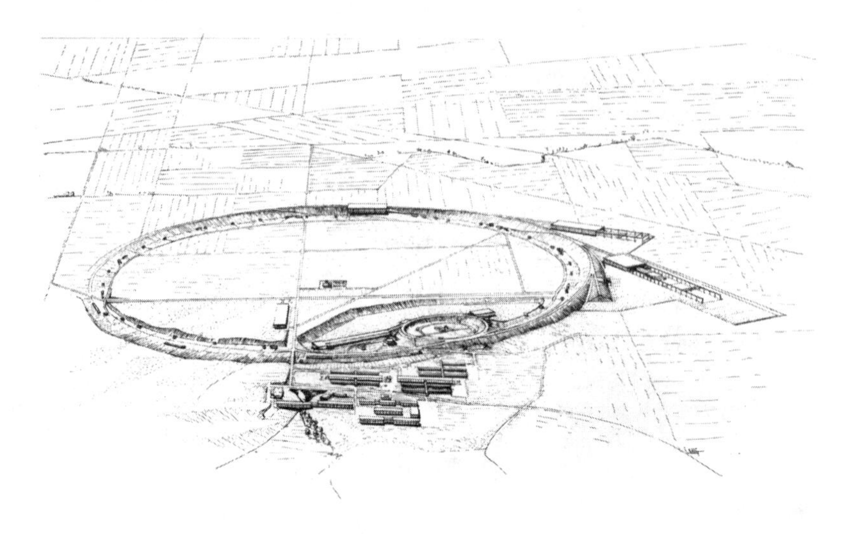

3.1 Berkeley's 200 BeV accelerator design, 1965. (Courtesy of Lawrence Berkeley Laboratory.)

Besides its booster ring and set of Collins straight sections, the new accelerator facility would provide extracted beams for fixed-target experiments, unlike previous accelerators, which only used internal targets. Obtaining efficient extraction was a new challenge, however, because of radioactivity at a high-intensity proton facility.[81] The system also offered improvements in detection, radiation safety, and the operation of the machine. The design provided for multiple experimental areas in which the 200 GeV protons would be used in twelve or more simultaneous experiments—two external beam areas and an internal-target facility. In addition to research using two bubble chambers, which would "operate nearly continuously," the Berkeley designers envisioned about "40 major experiments with other detectors" each year.[82] The report included schemes for transporting, separating, and detecting secondary beams. A section on future expansion presented ideas for increasing intensity by an order of magnitude, accelerating antiprotons, and later the addition of storage rings. Favoring the California site that Berkeley chose, the report also discussed expanded site criteria fashioned along guidelines distributed by the AEC.[83]

Detailed cost estimates and a construction schedule were given. The annual budget in the first year was estimated at $58.65 million. The construction cost estimate, which came to $348 million over seven years, was far more than the $280 million estimate given in the AEC's "Policy for National Action in the Field of High Energy Physics" of 1965, because

$19.65 million had been added to cover "possible escalation in the total project cost" over time, and $40.5 million had been added to provide "research equipment sufficient to provide for exploitation" of the accelerator, including a bubble chamber and a computer. The ample experimental accommodations, liberal cost estimating, and conservative features of the design all increased the Berkeley design's estimated cost.[84]

Berkeley took pains to avoid underestimating the cost of the accelerator. Lofgren remembered that as the design report was being compiled in early 1965, Paul McDaniel of the AEC stressed the importance of avoiding overruns because such problems had recently been encountered with reactor projects. Thus, the contingency allowance of 15% for conventional construction and 25% for accelerator construction was noticeably larger than the 10% contingency used in the 1961 preliminary Brookhaven design, and Berkeley researchers included an additional allowance for cost escalation over time.[85] Believing that the AEC was, as in the past, more interested in reliability than in cost, the Berkeley designers, like the builders of the atomic bomb, were inclined to spend a bit more money to increase the likelihood of reliable performance.[86] Berkeley designer Denis Keefe explained: "We produced the design we thought the AEC wanted to see."[87] Expecting to proceed with construction at a California site chosen by the end of the year, Berkeley formally requested that the next phase of their project be introduced into the AEC budget process.[88]

On one issue, that of national access, Berkeley researchers appeared willing to cooperate with the NAS, AEC, and URA. The design report mentioned that the facility would "be a national facility... open to all on the basis of the scientific merit of the experiments proposed." The designers believed that "approximately 70% of the experimental program" would be conducted "by visitors and 30% by the resident staff." Other than this, the design reflected a vision congruent with Berkeley's glorious past: extensive experimental areas and large, costly, C-shaped magnets set on substantial girders. As the report summarized, the design was "basically conservative, remaining reasonably close to the current state of the art, to insure reliable performance, rapid construction, and predictable costs."[89]

FOUR

Midwest Passage, 1965–1967

Scientific thinking and invention will flourish best where people are allowed to communicate as much as possible, unhampered. ENRICO FERMI[1]

Bitter controversy soon threatened the tenuous start Berkeley had made on the 200 GeV accelerator. While the AEC and URA struggled to fund the expensive machine, Washington pushed to economize. Continuing opposition from both East Coast and Midwest physicists and politicians to siting the project on the West Coast gradually loosened Berkeley's grip on the project.

Criticisms of the Berkeley Design

One of the first physicists to criticize the high cost of the Berkeley design was Robert R. Wilson. He had been invited to an instrumentation conference in Frascati, Italy, in September 1965. Initially he planned to fly to Paris soon after arriving in Rome and study drawing at La Grande Chaumière, "part of my artistic life." But then, "overcome by remorse" for having accepted the Alitalia ticket, he decided to attend the Italian meeting first.[2] Stopping in on the session at which the Berkeley group unveiled its design, Wilson recalled finding the talk "terribly dull and badly presented."[3] He found the design even worse. It fell short of the heights of creativity he expected of the important accelerator. Like the overdesigned and expensive first atomic bomb, whose reliability was mandated by its military application, Berkeley's 200 GeV machine was, in Wilson's opinion, both overdesigned and far too costly.[4]

After the session, the Italian nuclear physicist Eduardo Amaldi, the conference organizer, invited both Wilson and McMillan to lunch. As Wilson remembered, it was "a long Italian lunch, and the conversation lagged." Finally, "desperately trying to make conversation," he told McMillan that he had heard the papers that morning on the proton synchrotron. "Well, what did you think of them?" asked McMillan. After a long silence and a repeat of the question, Wilson replied, "Well Ed, I didn't like the way they were presented." When McMillan grew angry, Wilson realized that McMillan had been "under attack from a lot of other places, and I think he thought I was attacking." But rather than calming McMillan, Wilson, perhaps prompted by earlier sibling rivalry between the two in the 1930s when both had been among Lawrence's "boys," couldn't resist adding as the two arose from the table, "and I didn't particularly like what they said, either."[5]

Wilson flew to Paris after the meeting. But during his sketching classes, he could not stop thinking about the Berkeley design. "I began to doodle on the sketch pad during the *repos* time. And I began to put down a few figures and drawings. And I got completely engrossed. . . . I spent all the time in the session, instead of drawing from the model, working out theories of machines." Wilson hid his pile of accelerator doodles under a sketch of the nude. As he later explained to the 1991 Fermilab board of overseers, "this was the opposite of the usual schoolboy approach."[6]

Wilson spent several more days in Paris after the drawing classes. He continued his designing. "I went around, sitting in a cafe. . . . I think I spent the whole time going through one machine after another in a fury, making all kinds of designs. . . . It was very funny, because it was such romantic surroundings, and here I was doing this cloddish thing of designing, what even I would have considered was not a very respectable thing to do. . . . But I was consumed with a passion for making a number of designs of machines."[7] What Wilson was trying to express in his sketches was a more satisfactory design for a several hundred GeV synchrotron that followed "the philosophy I'd been using at Cornell on electron machines and transposing it to a proton accelerator just to see what I could do."[8]

Wilson's "fury" of designing incorporated ideals that he would later describe impressionistically as Gothic or Renaissance. He would identify with these ideals in his subsequent accelerator designing. For example, he asked, how "did Gothic cathedrals and accelerators both manage to express the aspirations and spirituality of their age"? This seemed to him to result from a certain unity of art, technology, and spirit. He found himself considering cathedrals and accelerators in the same terms, describing "a technological aesthetic" which included "a spirituality of the orbits," "an electric thrust," and a "magnetic counterthrust." "Both the accelerator and cathedral," he later wrote, "work in an ever upward surge of focus and function until the ultimate expression is achieved," which in the case of the accelerator is "the energy of a shining beam of particles."[9] Wilson's days of designing in Paris produced results, and, "at the end of it, I thought I would send those results to Ed."[10]

In an incendiary letter to McMillan in late September 1965, Wilson explained why he considered the Berkeley design "much too conservative" and "lacking in imagination." He acknowledged that the details were "most professionally worked out," but he was bothered that the design had been developed "without enough regard for economic factors." Wilson said that "as someone who helped disassemble radios for parts in the old days at the Radiation Laboratory," he was "offended" by its cost estimate, which seemed "ridiculously high." He feared that the high estimate would kill the 200 GeV project and endanger the future 600–1000 GeV machine.[11]

Wilson included with his letter a discussion paper, "Some Proton Synchrotrons, 100–1000 GeV," outlining his own designs for three different synchrotrons and a Linac. Expressing his frugal philosophy of accelerator designing, he urged that all components be kept "simple and understandable," as well as low in cost. He argued for the use of H-shaped magnets, which were smaller and therefore less expensive than the C-shaped magnets Berkeley proposed. He elaborated: "There is no reason to have fancy cranes, etc., nor should the tunnel be too comfortable on the few occasions when one might have access to it." He argued for Spartan guidelines as well in the design of experimental facilities and research equipment, insisting that making only minimal investment in them was actually an advantage for experimenters, because offering "expensive facilities" before the machine turned on "may tend to paralyze better developments later on." Wilson's alternate 200 GeV design, including research equipment and shielding, totaled less than $100 million, and the accelerator could, he estimated, be brought to completion in three years. The contrast with Berkeley's design, estimated at $348 million with completion over seven years, was sharp.[12]

A month later, Professor Samuel Devons of Columbia University presented another cost-saving alternative to Berkeley's design. Having been exposed as an undergraduate and as a research fellow at Cambridge University to the "string and sealing wax" tradition of the great experimentalist Ernest Rutherford, Devons was as appalled as Wilson by the high cost of the Berkeley design. Devons had held the prestigious Patrick Blackett chair in Manchester in 1953, but in 1959 he left England to accept the position of professor of physics at Columbia University. In his October 1965 position paper, Devons suggested a way to cut the cost of the new several hundred GeV accelerator by using the Brookhaven AGS as a booster for the new machine. John Adams of CERN had used a similar scheme in building CERN's SPS. An advantage of Devons's idea was that not only would the physics community have its multihundred GeV machine for only $100 million, but the AGS could continue to operate as a stand-alone accelerator. Devons suggested how to save additional money by eliminating some other projects that he considered less important, among them increasing the intensity of the AGS and ZGS. With all these savings, he argued that "prompt and vigorous development of a design for an accelerator" exceeding 500 GeV could take place within current budget limitations.[13]

Wilson's and Devons's criticism were, as expected, not appreciated by the Berkeley physicists. It was disturbing enough to be challenged about their management plan and presumed site. But given their fine reputation as accelerator builders, it was infuriating to have their *design* challenged by outside physicists. Wilson met with a chilly reception in late 1965, when he visited Berkeley to discuss his design ideas.[14]

Meanwhile, increased budgetary pressures from Washington DC affected Berkeley's plans. With the country burdened by the costs of the Vietnam War and President Lyndon B. Johnson's Great Society program, the government had begun to emphasize austerity. In August 1965, a BOB representative insisted that the AEC not make a funding commitment to the 200 GeV project during the next fiscal year (July 1966 to June 1967) because of the "rough" budget sessions currently in negotiation. In October, when Joseph Califano, special assistant to the president from 1965 to 1969, discussed the issue of the 200 GeV accelerator with Johnson, the president agreed that funding should be deferred, except for a few million dollars to support the AEC's effort to choose the site.[15] By December, White House advisers warned Johnson that unless spending declined drastically, a hefty tax increase, which the president was loathe to impose, would be necessary to avoid inflation. Also that December, after the BOB instructed the AEC to pare down the expensive accelerator project, the commission reluctantly asked Berkeley to prepare a cost estimate for a scaled-back accelerator, one less powerful, with a lower initial intensity and fewer experimental facilities.[16] The Berkeley design group responded with two ideas for a reduced-scope accelerator offering a savings of about 25%, "to avoid changes that would jeopardize the reliable operation of the accelerator." The suggestions involved building intermediate systems that could be replaced later, restoring the original design parameters.[17]

Criticism of Berkeley's design made the pursuit of government funding even more difficult, especially since "the President had not yet approved the project," as Seaborg noted on December 12. Devons meanwhile sparked the flammable situation by writing directly to the JCAE, thus drawing congressional attention to the dispute within the physics community. His letter prompted JCAE chairman Holifield to demand that the AEC formally review the issue.[18] It was not that the AEC lacked experience with negative views from physicists as a laboratory took shape: Panofky, SLAC's first director, later complained that physicists were "unenthusiastic about SLAC's creation." In the case of Berkeley's design, however, the AEC also faced BOB pressure to cut costs and the JCAE's irritation with critics. The URA was more than welcome; the AEC commissioners were already accustomed to working with a regional university consortium, since for decades Brookhaven had been operated by the Associated Universities, Inc. And the URA was well prepared to rally support for the new machine with its elected board of trustees, which included prominent scientists and other distinguished academics from all parts of the country. But as Seaborg emphasized at the URA trustees' meeting on December 12, they needed "unanimity among scientists and university administrators" if they hoped "to get the machine."[19]

Seitz and Seaborg were both upset by what they saw as attempts by Wilson and Devons to divert the course of the 200 GeV project. Seaborg also balked at the reductions the AEC had asked Berkeley to consider. In Seaborg's "personal view," attempts "to change to another energy and/or intensity of beam would be a mistake and might jeopardize getting anything because the balance was so precarious." Seitz remarked caustically in a December 14 letter to Ramsey, who was now chairman of URA's Scientific Subcommittee, that Wilson, also a member of the subcommittee, "may soon appreciate that the framework within which the big machine will proceed will probably be quite different from that surrounding the Cornell machine." In a December 20 letter to Ramsey, Seitz concluded that "it would be foolhardy to introduce any radical new proposals at this time." Despite these personal views, the AEC and URA thoroughly assessed the criticisms that were made of the 200 BeV project.[20]

During January 1966, two meetings were held to consider Wilson's and Devons's criticisms, one in New York, the other in Washington DC. AUI, Brookhaven's management organization, hosted the New York meeting, held at the Biltmore Hotel on January 15 and attended by the URA Scientific Subcommittee, AEC representatives, and about 150 high-energy physicists.[21] Wilson later characterized the New York meeting as a "sort of advertised rumble" to provide "a real thrashing out of ideas." Wilson and Devons argued that costs needed to be cut and promoted their views on how to do so. The core issue was not the relative technical merits of competing designs, but whether it was desirable to scale down the price by reducing capability or taking more risk. Predictably, members of the Berkeley Design team argued that a national facility of this scale needed a reliable design. The Scientific Subcommittee report to the URA noted that there "seemed to be no widespread strong support" for either the Wilson or Devons proposals.[22]

The AEC hosted the other, smaller, meeting in Washington on January 24, at which McMillan, Lofgren, Devons, and Wilson all gave talks. The participants could not agree on Devons's plan to cut costs using the AGS as an injector, and Wilson remembers the response to his own ideas as "generally negative." But while the participants could not settle on the optimal design, they agreed that the next accelerator should have an energy of 200 GeV. Henry Smyth, then chairman of URA's board of trustees, expressed his "personal conclusion" that none of the modifications, including that of reduced scope, seemed to "justify a re-opening of the whole question of the 200 BeV accelerator." He thought "it would be a shame to lose the momentum acquired."[23]

Shortly after the meeting in New York, both URA and the AEC took an official stand. Smyth wrote Seaborg on January 26, 1966, to explain that the URA board of trustees was in "general agreement" with his concluding statement at the meeting. The AEC, in a February review, characterized the Wilson and Devons proposals as shortcuts reflecting "a distinctly minority opinion" among physicists that neglected "the overall needs of the national program." Although the AEC would have preferred to proceed with the original Berkeley design, as indicated by Seaborg's comments in December, given that BOB pressure made it necessary to retain reduced scope as an option, the commission judged that the Berkeley reduced-scope schemes maintained a "balance between safeguards and risks" and were thus "still appropriate for a national accelerator facility."[24] With these endorsements Seaborg felt justified in sending his recommendation of Berkeley's reduced-scope design to Holifield on

February 16. Debate over the design subsided. Berkeley had yet again emerged victorious.[25]

Selecting a Site

Ten days before the AEC's June 15, 1965, deadline for site proposals, the NAS Site Evaluation Committee had met in Washington to plan the difficult site selection process. The committee decided that the AEC should provide a list of the sites that had failed initial screening with an explanation of the basis of their exclusion. The committee also planned to develop its own guidelines after receiving AEC criteria and site information.[26] Within a few days, an AEC task group led by Walter E. Hughes had devised two sets of criteria. The first was the "go, no go" set, required of any proposal. This set included minimum acreage of 3,000 acres, specified power capability, access to residential communities with a commute of one hour or less, and adequate access to highway, airport, and freight transportation. The second set called for "trade-offs between the technical and other factors" to obtain "the best over-all efficiencies and economies." According to this second set, the sites were rated according to factors such as locally dependent construction costs, the intangibles associated with the environment, and land configuration, or proximity to high-powered graduate programs in the physical sciences. Hughes admitted that the task group "expected that possibly more than 90 percent of the proposals submitted" would meet the minimum requirements of the first set of criteria. The task group rated proposals according to a 1,000 point system, in which, as Hughes explained, "300 points are applied to the criteria under the main heading for land, 250 points are applied to the criteria under the main heading of utilities, and 450 points are applied to the criteria under the main heading for environment."[27]

Soon site selection became even more complicated. The AEC announced on July 9 that the commission had received 126 proposals recommending some two hundred sites in forty-six states. Some of the proposals came from existing research labs or other AEC facilities. A group of Midwestern physicists submitted a proposal for the site of the former MURA headquarters in Madison, Wisconsin. Despite the initial opposition of Maurice Goldhaber, Brookhaven's director from 1961 to 1973, Brookhaven scientists submitted a 200 GeV site proposal, even though they had previously said they would defer and make a bid for a future 1000 GeV machine. As historian Robert Crease noted, Brookhaven physicists felt they needed to hedge their bets, because they sensed that "the rules of the game in their field were about to be entirely rewritten."[28] Proposals were also submitted from the AEC reactor sites at Hanford, Washington, and Oak Ridge, Tennessee. A large number of proposals were submitted by citizen groups with no affiliation with either the AEC or the high-energy physics community, simply because they felt the accelerator would bring prestige and economic benefit to their community.[29] For the first time, considerations outside the realm of physics were critical in the planning of a major accelerator.

The unexpected volume of proposals meant more work for the AEC and NAS site selection groups and heightened the political importance of the site contest. And as more politicians were drawn into the contest and the site selection attracted more attention, the AEC became more vulnerable to criticism. By the end of July, when the AEC had screened all proposals according to the established criteria, Seaborg became worried "about White House reaction regarding the... losers," for as he explained to science adviser Hornig, the list of thirty-four sites to be transmitted to the NAS did not include some "politically powerful parts of the country." Seaborg and Hornig tried cautiously to gauge White House reaction. Writing to President Johnson, Seaborg included the list of thirty-four sites and five alternates and emphasized that he would keep him "informed regarding the progress being made in this site selection."[30]

Seaborg received a prompt reply. On September 1, Marvin Watson, a close aide to Johnson, called to tell Seaborg and Tape that a group of White House "political animals" had unanimously agreed that the announcement of the sites should be delayed to avoid jeopardizing "passage of the President's legislative program." A few days later, on a flight to Austin, Johnson encountered an enraged fellow Texan, Congressman J. J. Pickle. Although three Texas sites appeared on the list (Houston, Dallas–Fort Worth, and San Antonio), the Austin site near Johnson's ranch had been excluded. In Pickle's presence, Johnson instructed his aide Horace Busby to pursue the matter with Seaborg. As Busby later explained to Seaborg, Johnson understood that the Austin site had been dropped because of its inability to provide adequate power, and this touched a particularly "sensitive spot" for the president "because it has been his life's work to promote and foster the lower Colorado River power down in that area." Busby therefore recommended "in the strongest terms" that Seaborg contact Johnson. "Quite frankly," he remarked, the AEC just could not "successfully argue to the President that the city of Austin could be eliminated at this stage," although this "did not necessarily mean that Austin has to get the project ultimately."[31]

Seaborg immediately issued a press release explaining there would be a delay in the announcement of the NAS list and he notified Seitz that more sites would probably be added. On September 10, two days after hearing from Watson, Seaborg and Tape met with Califano to present a new plan of action. As Califano's summary to Johnson explained, they agreed that the NAS should "examine about 85 sites, rather than the 35 it had been examining." The new list included seven Texas sites, among them Austin. Noting that the commissioners were "currently subject to considerable political heat in attempts to find out what communities are being studied and what communities have been eliminated," Califano suggested that they be permitted to proceed with the list of eighty-five sites "since they can justify the elimination of every community beyond the 85 on the list as patently inadequate as a site for the accelerator." As Califano also pointed out, the arrangement with the NAS offered another advantage. "After releasing the list," he explained, "they can refuse to answer any additional queries until the report from the National Academy of Sciences is received." On September 13, Califano told Seaborg that Johnson had approved their plan. The Chairman then sent the list of eighty-five sites to Seitz and issued a public announcement of the list on September 15.[32]

In the words of a White House information officer, the announcement attracted "considerable attention" nationwide. As Seaborg lamented in his diary, the public learned immediately of the original short list. A member of the NAS site committee, Herbert Longenecker of Tulane University, who apparently had not been informed of the change in plans, announced that the NAS committee was considering thirty to thirty-five sites. Seaborg's office was soon besieged with complaints and requests for information. JCAE member Craig Hosmer made an angry public statement. Annoyed that the site selection contest had become so large and heated, he charged that the AEC had "bungled" siting arrangements, which were now "mired in a mammoth pork barrel."[33] A skeptical tone was also heard outside Congress. "It is perfectly clear," UCLA assistant chancellor Carl York remarked after a fall URA meeting, that "the final decision on the site will be made by Lyndon Baines Johnson."[34]

Despite such pointed criticisms, Seaborg had carved out a relatively strong position, both for the AEC and the 200 GeV project. Without support from Johnson, the project could not continue; thus, the original list of sites had to be enlarged. By changing the list, Seaborg left himself open to the accusation, which was true, that politicians had been allowed to influence the site selection process. However, the AEC gained an optimally defensible position. As Califano had explained to Johnson, the longer list eliminated only sites that were clearly inadequate. Thus, it was an easier list for the AEC to defend. In addition, as he later admitted, Seaborg calculated that a larger number of site competitors would help funding prospects. With more congressmen lobbying for specific sites, congressional support for funding was bound to increase. Moreover, an intense site competition diverted attention from the issue of whether the accelerator should be funded to the issue of where it should be built. By keeping in close contact with the White House, encouraging good relations with Congress, and letting the NAS take responsibility for the sticky task of evaluating the eighty-five sites, the AEC could operate from a position of considerable strength in its efforts to arrange siting and funding for the 200 GeV project.[35]

After transmitting the list of eighty-five sites to the NAS, Seaborg and Hornig planned how to proceed with site selection. They asked Califano for further "guidance." Checking first with Johnson, Califano agreed that the NAS should choose the best three to ten sites from the list of eighty-five. Then Seaborg's task would be to "develop a comprehensive system to evaluate and analyze in depth the remaining 3–10 site locations." He stressed "the need for the AEC to make the final choice in the best interests of the country objectively, rationally and supported by clearly understandable reasons." Seaborg responded by noting that the AEC could handle a careful review, which would take three to six months, within the current budget. In the next several months, Seaborg kept the White House informed of NAS progress, including the number and locations of completed site visits.[36]

Although the AEC benefited politically when the list of sites was lengthened to eighty-five, this move greatly increased the work load of the NAS site selection committee. Seaborg assured Seitz that the commission was willing to extend the deadline and increase the budget to compensate for the greater workload and would be "happy to assist the Committee in any way." This offer proved welcome, especially since Seaborg soon pressed for visits to every site in the wake of accusations of unfairness and increased congressional interest that arose after the announcement of the list of eighty-five sites. It was subsequently agreed that the bulk of site visits would be handled by AEC-organized site visitation teams, which would include at least one high-energy physicist per team. To obtain extra advice, the committee also formed two panels of specialists—a Foundation Requirements Panel to evaluate the geological suitability of sites and a Panel of Accelerator Scientists to give advice on other physical attributes affecting construction, such as power, climate, and water.[37]

When the NAS Site Evaluation Committee began examining site data in late 1965, its members realized that their recommendations would hinge on nontechnical factors. Although the Foundation Requirements Panel urged that the accelerator be built on a site with minimal risk of geological instability, this requirement did not discount a large number of sites. As for cost, the committee members quickly decided that it was more important to choose a site where a productive laboratory could grow than a site where a laboratory could be built and operated cheaply.[38]

The site committee quickly narrowed the field to eight choice sites. They included sites proposed by the two traditional rivals—two California sites proposed by Berkeley (Sierra Nevada and Camp Parks) and the Brookhaven site in New York. The others were two sites in the Chicago area (South Barrington and Weston), and sites in Colorado (Denver), Michigan (Ann Arbor), and Wisconsin (Madison). As Piore explained, because of the expense of in-depth site evaluation, which he estimated to be as much as $1 million, the committee was concentrating on these choice sites. Some AEC commissioners were disappointed that so few AEC facilities were included among the choice candidates and that so many proposals were being dismissed so quickly.[39] The committee members were concerned about the political reactions. Goldwasser explained in an October letter to Piore. "It is foolish to think that our only job is to come up with the best possible site. Politics are involved in this decision and in its acceptance by the Commission, by Congress, by the President, by the community of scientists and by the public." He added, "Unless it is quite universally accepted and supported our exercise will have been a fruitless one."[40]

As planned, in March 1966, the NAS site committee made its recommendations for final sites. None of the sites was "ideal," but the committee found several that "in general," satisfied all important requirements. Explaining the basis for these inclusions, the report noted that the committee first determined whether "a given site had suitable physical properties." Then it "assigned paramount importance to the considerations that affect the recruiting of personnel for the national accelerator laboratory, and the participation of the nation's high-energy physicists." All the sites on the initial list of eight choice sites made the new list, except for the Camp Parks site in California, which was discounted because the committee determined that it could be affected by earthquakes. The AEC then took the lead.[41]

When the AEC received the NAS recommendation, the procedures that the staff members followed to announce the NAS list of final sites began by notifying first the White House, then Congress, and then, on March 30, 1966, the public in a press release.[42] After the announcement, Seaborg's office was barraged with angry calls from congressmen and business leaders pressuring him to reconsider a favored site. Seaborg steadfastly insisted that the selection of the final sites had been made on the basis of the site criteria and would not be reconsidered. With this settled, the expedition entered the final passage toward site selection.[43]

Seaborg and Hornig checked with the White House for further advice, but the record shows no strong public reaction from Johnson at this juncture. However, the BOB, the project's biggest foe since 1965, took action. Realizing that the first large funding request would follow the AEC's selection of the site, planned in three to six months, and given that the federal budget was tight, BOB director Charles Schultze wrote to Califano on April 22, 1966, that it was "clearly necessary that the Bureau of the Budget review the basis for the AEC's choice," since operating and construction costs would vary from site to site. Schultze also wrote to Seaborg asking the AEC to provide a study showing the cost differences in construction and operation arising from a reduction in intensity. Such considerations were "especially timely," he noted "in view of the President's concern about construction outlays."[44]

Despite rising budget pressures, Seaborg continued to take a firm stand against attempts to pare down the expense of the 200 GeV project. Although he noted in his reply to Schultze that the relative operating and construction costs would be considered in line with the sentiments of the NAS Site Evaluation Committee, he stressed that a "critical determinant of the ultimate total cost of the project . . . is the quality and capability of the scientists and engineers that can be attracted to the laboratory." Allowing for growth, future, and long life, he also noted the exciting neutrino experiments that would be facilitated by higher-intensity beams and explained that it would be much more expensive to extend experimental facilities after construction than to build them as part of the original project. "Evaluation of the differences in benefits against the differences in costs," he resolved, led the AEC to the conclusion that the original scope of the June 1965 Berkeley design study was "the most appropriate for the national program."[45]

By mid-June 1966, Seaborg had ironed out a plan for participation by both the White House and the BOB in the site selection. They agreed that Seaborg would inform Schultze "just before the Commissioners have made their final decision, but when the possible identity of the site is becoming known." Seaborg would then "get in touch with . . . Califano to get a feeling for the timing of the announcement and any possible White

House reaction to the choice." In this way, Seaborg could keep both the BOB and the White House informed and give each a chance to argue against the AEC choice before a public announcement was made, preventing potentially disastrous objections. At the same time, the AEC was free from direct intervention while the decision was being made. This arrangement pleased Seaborg and the other commissioners because it gave them the power to make the decision themselves and allowed them to deny, in good conscience, the accusation that politicians were dictating the choice of the final site.[46]

The AEC quickly arranged to send staff members to visit and evaluate the final sites, bringing along at least one commissioner. The AEC staff devised three separate methods for tabulating construction costs, and both the AEC staff and the commissioners considered ways to refine the decision criteria.[47] Local citizens' groups and politicians lobbying for each site showered the AEC with telegrams, letters, and petitions. In the aftermath of the defeated MURA proposal, Midwestern politicians pushed hard for equitable geographic distribution of federal research funds. Illinois governor Otto Kerner heartily endorsed efforts to make Illinois the new home for the accelerator, which he called the "greatest scientific prize in this century." Illinois was forced on April 5 to withdraw its South Barrington site, despite having the strongest congressional support at the time, according to an AEC tally. *Science* magazine reported that residents from the affluent Chicago suburb feared that the influx of physicists would "disturb the moral fiber of the community."[48]

In the midst of the AEC's deliberations, local politicians eagerly promoted each of the six remaining sites—Sierra Nevada, California; Denver, Colorado; Weston, Illinois; Ann Arbor, Michigan; Brookhaven, New York; and Madison, Wisconsin. The Long Island Association of Commerce and Industry announced on April 15 that their research "verified that all the competing sites have, as of this date . . . launched a major political program." The Long Island group planned meetings with local politicians and with education, labor, and business leaders to lobby support for the New York project. In Illinois, as explained in a May press release, after a thirteen-day campaign, a citizens' group proudly presented Illinois governor Otto Kerner "with 6,727 signatures of land owners and residents in support" of the Weston, Illinois, site.[49]

A petition with 114 signatures was sent to the AEC documenting the complaints of farmers living close to Weston against building an accelerator in their area. One irate farmer explained in a letter to Johnson that local farmers considered it "dastardly" to place such a facility on "some of the richest farming soil in the world." Residents of the village

4.1 Members of team inspecting the Weston site. *Left to right,* chairman of the United States Atomic Energy Commission Glenn Seaborg, with Illinois governor Otto Kerner, mayor of Weston Arthur C. Theriault, and Illinois senator Paul Douglas, April 1966. (Courtesy of FNAL Visual Media Services.)

of Weston, however, campaigned actively for the project, believing that it would bring prosperity to their financially troubled and only partially completed subdivision in unincorporated western Du Page County. When Howard Etchison, a Northern Illinois Gas Company official, contacted Weston mayor Arthur Theriault about proposing a local site for the accelerator in 1965, Theriault embraced the plan. Some of the residents, like William Wohl, wrote to the AEC in April 1966, claiming that the villagers simply wanted to sell their land as "a way out of their financial difficulties." The villagers welcomed the team surveying Weston for the new "atom-smasher" on April 8, 1966 (fig. 4.1), and lobbied for the site, which they hoped would bring "some industry to Weston."[50] Community economic development must have seemed like a new addition to the AEC agenda; at just this time the AEC was working with officials and businesses in communities around Hanford, Washington, the main AEC plutonium production laboratory, to foster an improved economy that was suffering, in that case, from the recent closure of obsolete reactors.

All the states offering proposals were expected to donate free land, but the 6,800 acres of the Weston site, sufficient for the laboratory as planned and future expansion, was especially agreeable, with suitable geological composition, utilities, and natural resources. The area offered assurances of compliance with projected costs, outstanding educational and research institutions, excellent transportation facilities, strong community and local industry support for providing human resources and housing, and progressive attitudes toward employment.[51] How did the Weston site compare with sites in California, Colorado, Michigan, New York, and Wisconsin? Its political support was, arguably, the best. An AEC tally of congressional inquiries indicated that the Illinois site had particularly strong congressional support, with twenty-seven interested congressmen. New York followed with twenty, and Michigan came in third, with nineteen congressmen. Those from Illinois, Wisconsin, and Michigan multiplied the strength and unity of Midwestern support.[52] As the first site proposed, Denver had the benefit of more than a year of campaigning by the Proton Accelerator Committee of the University of Colorado, which was aided by the Colorado Division of Commerce and Development and endorsed by Governor John Love. All members of the California delegation to the House of Representatives, with the exception of JCAE members who reserved judgment, signed a letter to Seaborg insisting that "on the merits" the California site was "the obvious and only possible choice." The Madison site was warmly endorsed by Wisconsin governor Warren Knowles, by the Wisconsin State Chamber of Commerce, and by the Madison Federation of Labor. And Michigan governor George Romney, writing about the Ann Arbor site, declared in a letter to Seaborg that his state could "match or excel others in any requirement that may be considered."[53]

The highly publicized site competition increased the funding prospects for the 200 GeV project by drawing political support, but this advantage did not come without cost. Many physicists found the public spectacle distasteful, and as many physicists had a sense that the future of their own laboratory hinged on winning the new machine, the site contest increased divisiveness within the physics community because physicists tended to rally behind their regional proposals. Emotions ran particularly high in the Midwest, where more physicists, with lingering thoughts of MURA's FFAG, joined the crusade for an accelerator to pursue the frontiers of basic physics in their region.

The AEC commissioners could not ignore the pressures within the physics community any more than they could ignore external ones, because funding prospects could be compromised without united support for the expensive project. Acting as the liaison between the commission and the physicists, Tape remembered that as the site competition continued, the commission found itself at odds with some sectors. In particular, he was disturbed when the designers at Berkeley and BNL acted as though the accelerator could not be built anywhere but at their laboratories because of their unmatched expertise. Wilson admitted that at the time he felt that the traditional prerogative of the accelerator builder should be honored and the California site chosen, and if that were impossible, then Brookhaven should be chosen.[54] But for the commissioners it would have been difficult to defend the choice of either the Brookhaven or California site unless an extremely convincing case could be made without reference to the expertise of local design groups. As Tape explained, how could they mount an expensive, time-consuming site selection process "and then turn around and say the machine goes to Berkeley, or Brookhaven because of proven competence," a fact well known at the onset?[55]

The commission had hoped to make a final site decision as early as July 1966, but in late June Seaborg received a letter from Clarence Mitchell, director of the Washington Bureau of the National Association for the Advancement of Colored People (NAACP), complaining that Illinois had a history of housing discrimination and had not passed legislation to enforce open occupancy laws. Any hope for an imminent decision vanished as the AEC mounted an extensive campaign to investigate the civil-rights compliance of finalist sites. By the end of July, all proposers were asked to provide assurances from local governments, labor unions, real estate associations, and citizen groups that minorities would not face discrimination in the communities surrounding the proposed sites. At the same time, the AEC asked for judgments on the six final communities from the Equal Employment Opportunity Commission, the Community Relations Service in the Department of Justice, the President's Committee on Equal Opportunity in Housing, the Commission on Civil Rights, the Civil Service Commission, and the Office of Federal Contracts Compliance in the Department of Labor. Detailed information about the status of civil rights was summarized for the commissioners in August 1966. In September governors of the six finalist states received copies of a summary of civil-rights objectives written by William Taylor, director of the Commission on Civil Rights. But as Seaborg explained to Schultze in July, due to "some problems regarding civil rights," the commission might need "a number of months" to reach a final decision.[56]

URA officials had been working to ease site tensions and facilitate future arrangements. After the first meeting between URA and Berkeley on January 15, 1966, Norman Ramsey met for hours with McMillan to hear his concerns. Amid the heated debate over sites, URA was careful to remain nonpartisan, constantly reminding physicists of the necessity for unified support for the machine. Ramsey remembered that the board worried that "losing groups would make a big uproar" to change the selection and "get the project killed." To avoid this catastrophe, Ramsey, who was elected URA president in October 1966, began a fall tour to meet with user groups from various laboratories. He stressed that the final AEC choice would be nonnegotiable and that the machine would be lost unless the physics community supported whichever site was chosen. In November, lists of possible leaders of the new laboratory were considered. Ramsey also prepared URA's official request to assume management responsibilities when the site was announced.[57]

After eight months of deliberation the commissioners were ready to make their final decision. The AEC staff had summarized site information into a working paper presented to the commissioners just before the final decision was announced. The "200 BeV Summary" provided an outline of the information available for judging sites according to the official criteria.[58] Organized by the categories specified a year earlier in the November 1965 description of site criteria, the 200 BeV summary showed many similarities among the six sites. All appeared suitable, but some ranked higher than others in specific categories. For instance, when land suitability was judged, all the sites had sufficient acreage, the ability to be turned over to the AEC at no cost to the U.S. government, and adequate subsurface and surface soils to support the proposed accelerator. All sites, except for those in Denver and California, required the acquisition of some farmland. In addition, all sites were within an hour's commute to communities of reasonable size. And all sites had adequate power and water supplies for the accelerator.[59]

Because any major complaint could kill the project, the AEC was under pressure to produce a site that could be defended to both physicists and politicians. The BOB would be likely to see Denver as the most attractive site because the accelerator could be built there at least expense. Denver's site was an inactive bombing range and the only site that could be acquired without disrupting farm or grazing land. Ann Arbor was the most expensive site to build on and Brookhaven was the second most expensive. But the NAS Site Evaluation Committee had judged, and the AEC concurred, that construction costs were not as important as factors that would influence staff recruitment and use of the facility. The most straightforward measure of that was the ease of accessibility to outside users. In this crucial category, the Ann Arbor and Weston sites topped the list with an estimated average travel time of 3.5 hours for all groups and one-day accessibility for 77% of users in major cities. The California site was the least accessible; with an average travel time of almost 7 hours. The Madison site had elevation differentials that exceeded the 100-foot limit set by the criteria, and the Weston and Brookhaven sites had limited expandability. In their potential to recruit staff members the California and Brookhaven sites scored highest because of the large staffs already assembled there. But, as already stated, the commissioners would have had a hard time defending the choice of either, given the expense of the site competition. Weston was generally considered the best site overall, even though it ranked poorly in civil rights.[60]

There was a notable difference among sites in the category of civil rights. The August 31 summary evaluated all six sites as having "fair employment practices laws with enforcement provisions," and the Equal Employment Opportunity Commission considered "the employment attitude at all sites 'generally progressive.'" The AEC staff felt that discrimination was unlikely at community facilities, such as hospitals and stores, in the communities surrounding all sites. But the report admitted that Illinois had "no fair housing laws," and generally "adverse comments" had been received about Ann Arbor and Weston. The NAACP expressed its opinion that in the Midwest, "Negroes suffer discrimination." Civil-rights leaders doubted "that assurance of nondiscrimination would be honored" in Du Page County, where most of the Weston site was located.[61] Taking all the different factors into account, the commissioners narrowed down the list of sites on November 29 to Weston and Madison.

On December 7, 1966, the commission selected the Weston site, after confirming that BOB and the White House knew the decision was imminent and that they concurred with it. The commissioners were concerned about Illinois's civil-rights problem but decided that Weston's advantages, especially its easy accessibility, more than offset this disadvantage. Several of the AEC commissioners acknowledged in interviews that because of previous tensions they were pleased to find that a Midwestern site was superior, though they all insisted that they would not have chosen it unless they sincerely believed it possessed superior qualities.[62] On December 16 Seaborg wrote a letter to Johnson and issued a press release announcing the site. Phone calls were made to inform JCAE members, Governor Kerner and other Illinois officials, and McMillan.[63] Twenty years later, McMillan still displayed considerable bitterness in recalling the phone call from Seaborg, another one of his rivals among Lawrence's boys. The call came late at night, about a week before Christmas. "That was my Christmas present, Seaborg calling, 'By the way, Ed, it goes to Illinois.'"[64]

Despite denials by the AEC commissioners, many believed that political issues drove the decision to site the accelerator in Illinois.[65] For instance, a few months after the decision was announced, *Newsweek* repeated the rumor that "Senate GOP Leader Everett Dirksen . . . might be brought around" to support a fair-housing bill pushed by Johnson "by the prospect of a $375 million nuclear accelerator to be built back home in Illinois." When the bill came to a vote the next year, Dirksen indeed dropped his opposition to the bill, which Congress then passed in April, 1968.[66] The oral testimony of all the AEC commissioners who were alive in 1984 supported the view that the commission made its decision objectively, but the conviction that Johnson made the site decision as part of a political deal in exchange for making civil rights part of his Great Society was conveyed in all our interviews with physicists about the site selection process, except for the interview with Gerald Tape, the sole physicist on the commission.

Frederick Seitz distinctly recalled getting a telephone call that even Seaborg did not know about: "It came directly from the White House to me." The caller told Seitz that President Johnson "was going to make the final decision." Seitz also remembered that when Crawford Greenwalt of DuPont, who had worked with Fermi during the war, "looked over the list," he said, "I bet it will be Illinois." As the president was trying to convince Illinois senator Dirksen to sign onto his civil rights legislation, it would have been logical, noted Seitz, for Johnson "to use this competition as a wedge."[67]

Written evidence that Johnson made such a deal with Dirksen has not surfaced, but the record does show that Johnson was informed of the status of decision-making efforts from the onset and was twice given the opportunity to intervene: just before the sites were submitted to the NAS in the summer of 1965 and just before the final site was announced in December 1966. Johnson clearly intervened in 1965, but we have no hard evidence that he did so in 1966.[68] As Tape suggested, it would have been easy to construe the AEC's efforts to keep the White House informed as proof that the president made the final decision.[69] Moreover, considering the disappointment felt by the losers and the natural tendency to attribute important decisions to powerful leaders, it is not surprising that the commission would be accused of yielding to political influence.[70] In any case, community-wide acceptance of this interpretation served to unite physicists at a time when unity helped to bolster stability and secure funding.

The decision to site the 200 GeV machine at Weston was almost immediately criticized. On December 19, the NAACP's Mitchell wrote to Seaborg that the decision was a "considerable disappointment" because of Illinois's lack of fair-housing legislation. He asked for "specific assurances" to "prevent racial discrimination in housing." A December 20 *New York Times* article pondered whether Weston was "picked because it was most suitable to the purpose or primarily to give a politically important area its 'fair share' of the spoils?" The newspaper asked for a detailed explanation of the basis for the choice, as did New York senator Jacob Javits, who sent the JCAE a list of twenty-nine questions with a demand that the AEC justify the dismissal of Brookhaven in light of the cost savings to be gained by using the AGS as an injector.

McMillan felt that in responding to those who were jealous of Berkeley's past success, the AEC had betrayed Berkeley. He remembered feeling that the AEC was determined not to give them anything, a feeling intensified by his tendency to see Seaborg as a rival inclined to spite him. Tape later reflected, "Old patterns had to change, and it was a painful time for all of us."[71] Ramsey, who scheduled a meeting in Berkeley just after the site announcement, reflected: "Although I had many personal friends at Berkeley, I have never spoken to such an unfriendly audience." He recalled that "among other things, I was told that if URA were a responsible organization it would refuse to manage the laboratory in Illinois and instead should investigate the AEC to determine how it could possibly have made such an incredibly bad decision."[72] Although Berkeley did not make a public complaint, McMillan refused Seaborg's request to publicly endorse the site. Berkeley physicist David Judd explained that to Berkeley ears this request sounded like "Give gladly your last drop of blood as you die!"[73]

Amid the clamor following the site announcement, members of URA and the AEC met on December 19, 1966, to discuss arrangements for the new laboratory. The commissioners were concerned about the upcoming congressional hearings, which would determine JCAE support for the first large funding allocation. Because of tensions within the physics community and the expected "thorough questioning" at the hearings, Seaborg spoke apprehensively of "pockets of resistance to the choice of the Weston site." URA representatives pledged to help prevent opposition within "the educational and scientific community." When questioned about reduced scope, Seaborg announced that the AEC had reluctantly decided to request about $240 million (eventually $250 million), sensing that "it was necessary that the future not be mortgaged too heavily." Despite the valiant arguments that Seaborg had made some months earlier against reduced scope, his attitude had changed as budgetary pressures increased throughout 1966. By the end of the year he concluded that a cost limitation was necessary to assure executive-branch support. The assessment was sensible, for at that very moment the Johnson administration was preparing an emergency request for $12.3 billion from Congress, including $4.467 billion in new appropriations, to meet the unexpected escalating costs of the Vietnam War. Ramsey then submitted URA's proposal. And the AEC awarded URA a temporary contract on January 5, 1967.[74]

The AEC defended its choice of the Weston site in a January 18, 1967, report titled "Basis for the Selection of the Chicago (Weston) Site for Location of the 200 BeV Accelerator Laboratory." Emphasizing the accessibility of Weston, which would allow work to begin quickly, the report avoided emphasizing that the site was not the least expensive. The commission downplayed civil-rights problems in Illinois, acknowledging "differing views" about discrimination in the communities surrounding Weston. On a positive note, the commission noted that it had assurances that great efforts would be made to "prevent or offset discrimination."[75] Answering Javits's questions, the commission noted that any cost savings that would be gained by use of the Brookhaven site, including the use of the AGS as an injector, were "offset by the higher construction costs prevailing in the BNL area," as compared with other sites, and that "use of the AGS as an injector would lead to a significantly different and decreased national high-energy physics program than would prevail with two separate machines."[76]

The decision to site the laboratory in the Midwest had a mediating effect. Some of the long-standing conflicts among East, West, and Midwest seemed to balance out after the contest, and certain democratic ideals, such as equitable access and civil rights, materialized at the new facility. But the hopes of Weston villagers to be part of the new laboratory (see fig. 4.2) were not realized.

Finding a Director for the Laboratory

The next step was to appoint a director for the new Midwest project. URA first contacted Panofsky, then the director of SLAC, but he expressed no interest in leading the new laboratory. URA then asked Lofgren to be director of URA design studies, hoping, as Tape remarked, that he "would influence the attitudes of those at Berkeley who had performed the original design study and would enable the AEC to capitalize on what they

4.2 "Welcome to Weston, Future Atomic Research Capitol": 1966 real estate billboard. (Courtesy of Fermilab Visual Media Services.)

had done."[77] In a January 12, 1967, letter to Ramsey, Lofgren turned down the offer, explaining that he was dissatisfied with the description of his duties. He felt that "the directorial position should not be limited to responsibility for design and construction of the accelerator facility." In addition, he was unsure that at the Weston site he could assemble the necessary staff "and develop an organization having the enthusiasm and spirit needed to make the project a distinguished success."[78]

URA was now in a tight spot. The federal budget was extremely taut and criticism was mounting over the site decision. In just a month, the JCAE would hear testimony at the annual AEC budget hearing to decide whether to authorize the first allocation of construction funding. The project could ill afford to have another candidate refuse the directorship. At just this moment "a substantive vexing issue" was, as Berkeley physicist L. Jackson Laslett explained, whether URA could effectively lead the project in light of budget reductions, "or whether the project would lead to substantial cost overruns before satisfactory completion, with consequent long-term damage" to the national program in high-energy physics.[79] On visiting Berkeley, Ramsey heard URA being "vigorously criticized for even going along" with the decision to build the accelerator at Weston.[80] URA needed to quickly appoint a strong physicist with enough energy and stature to lead the new laboratory. This leader also had to be able to build the accelerator within the given budget constraints.

Just then an interesting candidate emerged, one who would have seemed an odd choice two years earlier. Wilson's criticisms of the Berkeley design and the economizing alternatives he offered had appeared threatening to the project in September 1965. At that time no one thought of him as a possible leader of the new laboratory. Not only did his criticism appear undermining, but he had never built a large proton synchrotron. Also he was then fully committed to building the new Cornell electron synchrotron. By early 1967, however, two developments had made Wilson a possible candidate. He had completed the Cornell project on budget and ahead of schedule. Second, his economizing schemes now fit the changed budgetary context. After some preliminary deliberation, URA president Ramsey asked Wilson whether he would be willing to accept the position of laboratory director, making it clear that, unlike the offer to Lofgren, which had been limited to directing URA design studies, Wilson's offer would be to take overall responsibility for the laboratory.[81]

Before making the offer official, Ramsey gauged the reaction of both the AEC and the physics community to Wilson. The AEC commissioners were at first not enthusiastic, perhaps recalling Wilson's spirited criticism of Berkeley's design in 1965. They questioned whether the physics community would accept Wilson's appointment. He "was perhaps the worst" choice for a director capable of "securing the whole-hearted help and cooperation of the LRL team," because of his public attacks on the "judgment and competence" of the Berkeley staff, Judd explained to Seaborg in a February letter. But, as Ramsey soon concluded, most physicists outside the Berkeley circle favored the appointment of Wilson.[82]

On February 8, the AEC commissioners met with California representatives, including JCAE members Holifield and Hosmer. According to Seaborg's record of the meeting, the commissioners took pains to explain Wilson's qualifications as a "very imaginative scientist," and they stressed his attention to "cost and management details" at Cornell. Although Seaborg mentioned the meeting between Wilson and the Berkeley designers, optimistically noting that misunderstandings might soon be resolved, Hosmer pointed out that Berkeley designers would probably not join the project. At the end of the meeting, both Holifield and Hosmer stressed the importance of finding a good team to build the new laboratory.[83]

While considering his offer from URA, Wilson visited Berkeley on February 10, 1967, to discuss the criticisms that Berkeley physicists had of the Weston site. Wilson found them "completely and ostentatiously negative and unfriendly," both to him and to the site. His cost-cutting, risk-taking approach to accelerator building appeared diametrically opposed to Berkeley's careful and conservative style. They viewed the prospect of Wilson's appointment as the final, fatal blow to the Berkeley tradition. Members of the Berkeley design team were also annoyed that Wilson, who had once been one of their own, seemed unwilling to listen to their ideas. Despite this reception, Wilson pluckily insisted that he saw no reason why he couldn't build the machine at Weston, and he explained some of his ideas to his unfriendly listeners. But when he later asked who in the audience would be willing to accept a job at the new facility, only one or two hands tentatively rose from the sea of angry faces.[84]

The AEC now had another problem: several Berkeley physicists were on their way to Washington to testify at the authorization hearings. Seizing the opportunity, Seaborg organized a series of meetings to resolve the multiple tensions surrounding Wilson's appointment. At a decisive meeting on February 14, Lofgren came armed with a report prepared by the Berkeley staff which estimated that the accelerator would cost $53 million more to build at the Weston site than at the California site. An AEC representative countered that the commission expected only a $5 million cost difference. Ramsey interceded to stress that cooperation was necessary to win the machine. It was finally agreed that the project's first task would be to produce both an updated design and thorough cost estimates based on detailed data from the site. Wilson entered the February 14 meeting and agreed to a list of tasks ensuring the continued involvement of the Berkeley design group in the new design effort. By the end of the meeting a workable, if not amicable, arrangement had been negotiated so that the redesigning of the accelerator could proceed.[85]

From this point on, Berkeley accepted its loss of the project and terms were negotiated for a transition from California to Illinois that allowed continued, although limited, Berkeley participation in the project.

Later that day, Seaborg, Tape, and Ramsey took Wilson to meet Holifield, Hosmer, and their fellow JCAE member, Melvin Price, from Illinois. Wilson had not yet accepted URA's offer because he was waiting for the AEC to agree to certain conditions.[86] During the meeting, the JCAE members quizzed Wilson on his management philosophy. Wilson remembered feeling nervous, but Tape felt that by the end of the meeting Wilson had "made a hit with them."[87]

The Authorization and Appropriation Hearings

The accelerator was the top issue when the AEC faced the JCAE for the annual authorization hearings in 1967. Before deciding whether to support funding to continue the project, the JCAE demanded that the AEC justify the role of URA, the choice of Weston, and the decision to reduce the accelerator's scope. Three days were spent discussing the accelerator and directing a subcommittee to deliberate for two more days.[88] Midwestern congressmen defended the choice of Weston when California and New York representatives grilled the commission about it. Remarking that Weston was only "a cornfield" while Brookhaven was "a great working, effective laboratory," Senator Javits of New York questioned the fairness of the site selection. Senator Pastore from Rhode Island, by this time chairman of the JCAE, vehemently defended the AEC, noting it was his "unequivocal . . . irrefutable understanding that the White House had nothing at all to do with the selection of the site."[89]

The AEC was asked to defend the choice of Weston on the basis of other considerations as well, the most important being cost differentials between the sites and the question of civil rights. One session was spent on a point-by-point analysis of the site selection process, with the AEC constantly emphasizing that decisions had been made on the basis of the criteria. When Lofgren and McMillan appeared before the subcommittee, although neither expressed enthusiasm about the project, both stressed Berkeley's willingness to cooperate in the new effort and limited their negative comments to vague warnings about the need to proceed with caution. Their restraint testifies to the success of behind-the-scenes meetings with the AEC and URA.[90]

In contrast to the restraint of these discussions over site selection, emotions ran high in the testimonies about civil rights. For example,

Mitchell of the NAACP decried the choice of Weston, claiming it was an "example of how human dignity and fairplay get short shrift from some top government agencies." Edward Ruthledge from the National Committee against Discrimination in Housing demanded that the JCAE "withhold final approval of the Weston site pending specific positive demonstration that Illinois and Chicago will in fact plan, enact, and carry through an across-the-board program to open wide the doors of equal opportunity for minority families." AEC officials shuddered. If the JCAE endorsed Ruthledge's plan, the 200 GeV project would undoubtedly be deadlocked for some time, since the Illinois legislature had consistently balked at open-housing bills, despite Governor Kerner's efforts to push them through.[91]

Considerable attention at the hearings was given to the proposed reduction in scope and to the role of URA. Chairman Pastore grilled Seaborg on the reduced scope, asking what had prompted this change in plans. Although Seaborg tried to sidestep the issue, he eventually admitted that the impetus had been budgetary pressure from the BOB, rather than the consideration that a reduced-scope machine would serve as well as the original design. JCAE members were clearly annoyed that the BOB had intervened. As Hosmer later remarked, "I am not at all certain we are warranted in authorizing a nickel to proceed on this bitterly disappointing scaled down and shaken poor man's BeV accelerator."[92] Into this debate stepped JCAE member William Bates of Massachusetts, who pointed out that CERN had announced plans to build a 300 GeV accelerator. He wondered why the United States was "settling on a second-rate . . . machine when the European countries [were] planning a first-class facility." Why not think in terms of increased capability?[93] Bates's question gave the AEC the perfect opportunity to push the idea of building a machine that could be expanded later to reach higher energies.

Lofgren remembers that he had the idea of designing an expandable machine back in March 1964 during a twelve-day hospital visit for back problems. No one seriously considered implementing the idea until late 1965, when the AEC requested a reduced-scope accelerator. Expandability gave the AEC a way to bridge the gap between the JCAE's expectations and the BOB's restrictions; the JCAE would have the accelerator it originally promoted and the AEC would still work within the $250 million budget set by BOB. Because of the JCAE's support for expandability, the AEC was able to incorporate it into plans for the project, despite the fact that such a midstream change was usually forbidden. Although Berkeley designers had been first to consider ideas for expandability and were about to publish a well-honed expandability scheme, the idea became associated with Wilson because the commissioners stressed his willingness to apply innovative ideas within a budget. This association in turn improved funding prospects.[94]

The authorization process continued over the next several months. In an April 1967 report, the JCAE subcommittee applauded the expandable machine, denounced the reduced scope and reluctantly accepted the continued role of URA, which still had only a temporary contract.[95] As for the new laboratory's policy toward outside users, the joint committee stated that outside users already had reasonable access to other AEC accelerators, but they agreed with the desire to ensure access, "more as a matter of right than sufferance." In June, when the full committee made its funding recommendation to Congress, the majority report suggested that $7 million of the requested $10 million be authorized to begin construction of the new project. Chairman Pastore submitted a dissenting opinion in which he suggested that the project be deferred. He questioned whether the 200 GeV was really the AEC's top-choice machine, presumably because of the heated debate within the physics community. He insisted that at a time when the budget was swollen with high priority items there was "no urgent, compelling public need demanding an immediate beginning on the construction of the 200 BeV accelerator." Pastore's most vigorous complaint, however, was that of "the six finalists among the considered sites, the AEC chose a location in the only state of the six which does not have open housing legislation." Pointing out that open-housing legislation had been defeated in the Illinois Senate just days after money was appropriated for the 200 GeV project, Pastore angrily proclaimed: "In the name of advancing science and technology, we should not be guilty of retreating from our boasted principles of equity, equality, humanity."[96]

The appropriation bill was vigorously debated after it reached the Senate in mid-July 1967. In the midst of this debate, a *New York Times* editorial angrily articulated the mood of many Americans. "The nation is engaged in a bloody war in Vietnam; the streets of its cities are swept by riots born of anger over racial and economic inequities," cried the editorial in a year when race riots erupted in Newark and Detroit and antiwar protesters marched on the Pentagon. "It is a distortion of national priorities to commit many millions now to this interesting but unnecessary scientific luxury." Despite such complaints and Pastore's objections, Congress passed the funding bill, and on July 26, 1967, Johnson signed it. The first large funding allotment for the 200 GeV was secured.[97] The physics community would have its accelerator.

By granting $7 million of design funding for the next fiscal year, the federal government made its commitment to build the new laboratory. In turn, Congress's expectation of an expandable machine for $250 million demanded frugality in building the accelerator as well as its surrounding laboratory. It was the first time since World War II that the value of frugality was shaping a technical decision in a major accelerator laboratory. Although many challenges remained for the physicists at the frontier, it was a time of rejoicing.

By then Wilson had agreed to direct the new laboratory, writing to Seaborg on February 28, and then to Ramsey on March 1, 1967, that he was "deeply honored to accept," what he viewed as "an adventurous undertaking that appeals to me deeply. I expect to give my best to help us make it a laboratory of which we can all be proud."[98] Wilson's association with the concept of an expandable machine increased the AEC's desire to work with URA. Aware that he was in a strong bargaining position, Wilson negotiated a salary that included the maximum compensation available from the AEC plus a yearly contribution from the URA. He also obtained a number of concessions: that approval authority would be given to a local AEC office to avoid red tape and that the AEC would pledge to support construction authorization in 1969.[99] And he insisted that URA name the 200 GeV project "The National Accelerator Laboratory," to highlight its commitment to a democratic outside user policy.[100]

Wilson would thus be able to create the new laboratory his way. He initially toyed with Governor Kerner's idea of surrounding the laboratory with a kind of "science city" where scientific creativity would flourish.[101] The vision adapted itself well to a laboratory set in a suburb of Chicago, the city that Mark Twain had described in 1883 as "a city where they are always rubbing the lamp, and fetching up the genii, and contriving and achieving new impossibilities."[102] Ultimately Wilson discarded the science city idea as impractical, but NAL would prove to be one of the "new impossibilities." Like Chicago, NAL became a crucible for innovation while serving as a flagship for exploring the scientific frontier.

PART TWO

A New Frontier on the Illinois Prairie

FIVE

Wilson's Vision

We have the opportunity to build a truly magnificent laboratory . . . with beautiful architecture set in a pleasing environment . . . a significant laboratory . . . where we will have the opportunity to push the limits of our knowledge about particles to a point undreamed of 30 years ago. ROBERT R. WILSON[1]

During Wilson's first year of building NAL, he would sometimes fit an early morning ritual into his busy schedule. On the way to his office he would ask Mack Hankerson, the mail clerk who also served as Wilson's driver, to stop near a lookout point on the site's eastern perimeter. Ascending a stepladder erected on a slab of concrete, Wilson would then "look at the land."[2] As a sculptor accustomed to visualizing his finished product long before it materialized, he had no trouble imagining the transformation of the site's farmland into a scientific workplace. Many aspects of NAL's design were suggested by others, but its special character reflected Wilson's values. His aesthetics called for clean, bold, frugal, and functional components, harmoniously combined. The practical design reflected Wilson's view of himself.

Wilson's self-image was composed of three distinct yet related personae: the pioneer who pushed frontiers, the craftsman or engineer who made things work, and the Renaissance man.[3] Wilson's pioneer was an adventuresome frontiersman committed to exploring the unknown and confronting its limits. "Research plays a present-day role analogous to the role that opening of the west played at an earlier stage in our country," Wilson wrote during his first year as the director of NAL.[4] His approach to exploration was shaped by individualism, self-reliance, and reverence for observation of the world, traits shared by Frederick Jackson Turner's romantic pioneer (see chapter 1). This image had also appealed to other physicists whom Wilson admired, among them Lawrence and Oppenheimer. Unlike the brutal conquerors discussed by historians of the American West, Wilson's frontiersman was of noble spirit. His explorations were guided by democratic ideals. Wilson stated as policy in 1968: "In the course of giving a very large acceleration to our particles, let us hope that we can contribute at least a small acceleration to society."[5]

That the man who built Fermilab was born in a Wyoming town *named* Frontier and that his laboratory was in a suburb of the city where Turner had presented his frontier thesis are coincidences. But that Wilson dismissed as his institutional models both the military camp (adopted by Brookhaven) and the college campus (embraced by SLAC) was no coincidence. He wanted his laboratory to be an open space that evoked a peaceful frontier, a place where physicists worked creatively to understand and also improve the world.

Wilson believed that his identification with his second persona, the hands-on craftsman, grew from his boyhood experience of visiting the blacksmith to repair broken tools in an effort to avoid the long and lonesome ride into town to buy replacement parts. With the blacksmith he "learned how to use my hands and make things. I think that was a very useful part of my training."[6] Wilson told an interviewer that in working with the blacksmith he developed "the confidence that with your own hands you could build large contraptions and make them work." Like the blacksmith, Wilson came to trust hands-on experience over theoretical knowledge. American pragmatists such as William James and John Dewey had built their philosophies on such grounds. Their empiricism supported many characteristically American technical developments, including Edison's laboratory in Menlo Park, Lawrence's first particle accelerators, and Oppenheimer's atomic bomb effort at Los Alamos.[7]

Wilson's third and perhaps most powerful identification was with an ideal that historians have linked with the "rise of a new observational study of natural phenomena," a philosophy acknowledging a union of art, spirit, and nature which is sometimes referred to as "Renaissance naturalism."[8] Wilson's view of himself as a Renaissance man had matured by the time he was studying art and sketching accelerators in Paris in the fall of 1965 (see chapter 4). By then his vision of a frugally constructed particle accelerator had come into focus—a work of art, as well as a tool of science that could play a cultural role similar to that of the Gothic cathedrals of France.

Biographical Sketch

Robert Rathbun Wilson was born to Edith Rathbun and Platt Wilson on March 4, 1914.[9] He found strong models for his adventurous spirit and feisty independence in the western heritage of pioneer ranchers on his mother's side. Spending much time on his family's Wyoming cattle ranch, Wilson initially aspired to become a cowboy and in later years often described himself as one (fig. 5.1). He was, according to Goldwasser, "a master horseman," who could "deftly lasso any of his three sons when they were young and when needed."[10] The experience of spending long periods of time alone riding great distances with a packhorse, reading the sky for weather, and generally pitted against nature, nurtured Wilson's resourcefulness and helped him develop not only a closeness with nature, but also the belief that he was "unique in the world."[11]

Young Wilson read insatiably in the local library, finding valuable role models in works of literature. The hero of Sinclair Lewis's novel *Arrowsmith* exemplified the noble scientist seeking practical solutions in the service of humanity. Such fictional models proved more formative than any of the schools Wilson attended, a different one each year after his parents separated while he was still in grade school. Going back and forth between his parents helped Wilson develop a wide variety of educational, practical, and social skills. Edith Rathbun's family of ranchers, who moved to Wyoming after the California gold rush, maintained a high regard for learning.[12] Experimenting in his mother's attic, Wilson taught himself such practical skills as blowing glass, which he drew on in building a mercury vacuum pump and a Crooke's tube.

Platt Wilson, the son of an Iowa preacher, worked first as a civil engineer in both the coal-mining and automobile industries and then entered politics and became chairman of the Democratic Party in Wyoming. He was later elected a state senator. Although Robert remained shy and withdrawn through high school and college, rarely speaking up in class, he readily learned from Platt the value as well as the skill of rhetoric and storytelling. Robert's maternal grandmother, Nellie Rathbun, in whose care the boy was often placed, helped to instill a strong appreciation of civil rights in young Robert Wilson. She was the granddaughter of the abolitionist John G. Fee, who founded Berea College, a racially integrated school, in pre–Civil War Kentucky.[13]

Following in this tradition, Wilson pursued a college education. In 1932, he gained admission to the University of California at Berkeley, despite his undistinguished high school record. Like many physicists who began their studies during the Great Depression, he enrolled in an

5.1 Robert R. Wilson, young cowboy. (Courtesy of FNAL Visual Media Services.)

electrical engineering program, which offered the promise of future employment. Soon he discovered his passion for physics, and proceeded to teach it to himself as a freshman. One attraction was the nuclear physics research pursued at the Radiation Laboratory that Ernest Orlando Lawrence directed. One day during Wilson's freshman year, a physicist found Wilson standing and peering into the Rad Lab through a window. He invited the boy inside. When Wilson saw the cyclotron, he decided on the spot that working with such machines was what he would do in his life.[14]

The attraction of physics, especially as presented by Lawrence, proved irresistible to Wilson. By the time he entered his senior year in college, he was conducting original research on the lag time of the spark discharge. Staying on for graduate study, Wilson became one of Lawrence's "boys," the group of young scientists that included Livingston, McMillan, Philip Abelson, Luis Alvarez, and Milton White. Wilson took up a range of physics problems, including the theory of the cyclotron and the development of the waxless cyclotron vacuum seal that became known as the "Wilson seal." He later reflected that he came away from his training under Lawrence with many valuable lessons, such as, "If you want something to come true, you can make it come true just by pushing like hell," or "There are many ways of getting to a result," or "If you tried, it was a question of being optimistic or pessimistic." Wilson's doctoral thesis experiment on proton-proton scattering failed, because he could not make the cyclotron work well enough, but he could not delay moving to Princeton to begin his appointment there as an instructor in physics. In 1940 Wilson wrote his thesis up on the theory of the cyclotron, despite the cyclotron's failure. Trying until the last possible moment to make the cyclotron work, he arrived late for his wedding on August 20, 1940.[15] Jane Inez Scheyer and Bob Wilson were to enjoy a long and happy marriage of fifty-nine years, raising three sons, Daniel, Jonathan, and Rand.

At Princeton Wilson continued to pursue proton-proton scattering. He also used the Princeton cyclotron to contribute to the work of Enrico Fermi and Herbert Anderson that resulted in their demonstration of the first nuclear chain reaction at the University of Chicago. In subsequent research, Wilson showed that a resonant mode of fission occurs. He also invented the electromagnetic isotope separation method known as the "Isotron method." When the Isotron project was canceled early in 1943, Wilson's group and its apparatus were shipped to the secret laboratory in Los Alamos, New Mexico, which had been organized in 1942 to build the atomic bomb.

5.2 Enrico Fermi (*front row, second from left*) with Los Alamos, NM, colleagues including Robert R. Wilson standing behind, to Fermi's left, 1943. (Courtesy of Emilio Segre Visual Archives, American Institute of Physics.)

As a pacifist, Wilson at first felt that he could not be involved in the growing hostilities that led to World War II. But as Nazism took hold, he "grew more and more uneasy." He described his process of dealing with this malaise as a challenge of conscience: "If Hitler indeed conquered the world, could I bear to stand by and watch it happen, could I bear to think what life in such a world might be like?"[16] Reversing his earlier decision not to work for the war and its "merchants of death," he accepted the invitation to work on radar at the MIT "Rad Lab," which Lawrence was helping to organize. Wilson soon decided that he would be more useful to the American effort if he worked on problems of nuclear physics. He went west, to Los Alamos (fig. 5.2).

Wilson initially led the Los Alamos Cyclotron Group, whose responsibilities included studying nuclear properties relevant to the bomb project, for example, measuring the time delay of neutrons emitted when a neutron causes a uranium atom to fission, determining how many neutrons emerge on average per fission of an atom of uranium or plutonium, and whether neutrons are emitted from uranium instantaneously or with some delay. When the laboratory decided that the machine best suited for its research was the Harvard cyclotron, the military sent an acquisition team, which included Wilson, to Harvard to purchase the cyclotron. Their pretext was that the accelerator was needed for medical work in Saint Louis. The Harvard physicists, suspecting the team's motive, told the team it could have the cyclotron without payment if the cyclotron would be used in the fission project. Bur the army group obeyed its orders and paid Harvard a large sum for the cyclotron.

Wilson was back in the saddle in the austere environment of northern New Mexico. He flourished in the tense military context and felt comfortable employing the practical methodologies that governed the research of this wartime laboratory (e.g., successive approximations, trial and error, multiple lines of inquiry, numerical methods, or building scale models) to compensate for the incomplete theories and spotty experimental knowledge that confronted them. The approach appealed to Wilson's persona of the hands-on craftsman. Later, in the summer of 1944, Wilson succeeded Caltech's Bacher as head of the Experimental Physics Division in Los Alamos. Following in his father Platt's footsteps, Wilson dabbled in Los Alamos politics, serving on the town council and later as the council's leader.[17]

In July 1945 Wilson experienced the Trinity Test of the first atomic bomb in the Alamogordo desert of New Mexico as a withdrawal "into reality," a "re-awakening from being completely technically-oriented." He subsequently criticized the decision to drop the atomic bomb on Japan, became a founding member of the Federation of American Scientists, and completely gave up weapons development, resolving henceforth to work on nuclear energy only as "a positive factor for humanity."[18] When he left Los Alamos, shortly after the war ended, he joined the faculty at Harvard as an associate professor and designed Harvard's 150 MeV cyclotron. In 1946 he also explored the use of protons emerging from this cyclotron for cancer therapy,[19]

In February 1947 Wilson moved to Cornell as a full professor, again following Bacher, this time as the director of Cornell's Laboratory of Nuclear Studies. At Cornell, Wilson made one of the first applications of the Monte Carlo method, a procedure offering approximate solutions to problems by performing statistical sampling experiments on a computer; the method grew out of calculations made in developing the atomic bomb. Wilson developed a way to produce very high temperatures in plasmas by producing an imploding shock wave in an ionized gas. Conducting elastic electron-nucleon scattering experiments, he separated the nucleon's electromagnetic form factors. Wilson melded the empirical approach he experienced at Berkeley and Los Alamos with his aesthetics of frugality to yield his philosophy "that something that works right away is over-designed and consequently will have taken too long to build and will have cost too much."[20] Applying this philosophy over his twenty years at Cornell, he built four new electron synchrotrons, each more powerful than the previous one. They culminated in Cornell's 10 GeV synchrotron.[21] The 1.5 GeV synchrotron in this series (later upgraded to 2.2 GeV) was the first accelerator based on the strong-focusing principle.

Although Wilson was happy at Cornell, by the early 1960s he yearned to express his emerging aesthetics of Renaissance naturalism. He found the right project while attending the Frascati meeting at which he learned about Berkeley's 200 GeV accelerator design (chapter 4). The burst of creativity that Wilson experienced in Paris in 1965 formed the basis for his subsequent design of NAL.

Bringing Wilson's Frontier Vision to Life

Wilson found that he needed help getting organized to build and direct the new laboratory in Illinois. Still at Cornell during the spring of 1967, he was reluctant to ask members of his staff to work on the new project. Like the pioneers settling on the American frontier, Wilson drew on his friends. The well-known theoretical physicist Hans Bethe and his wife, Rose, had been close friends of the Wilsons since Los Alamos. (Later that year, Bethe would win the Nobel Prize for his theory of energy production in stars and for the theory of nuclear reactions.)

Rose's organizational skills had been of use to Oppenheimer when he had created the Los Alamos laboratory. Wilson now asked Rose whether she could help him organize NAL. She remembered "a period between vision and fact" when Wilson needed "to work out what he wanted to do."[22] He particularly needed help with writing letters and answering the telephone. Rose was happy to offer such assistance for a short time. From Ithaca she served as "hostess," making travel arrangements for the scientists who would be attending Wilson's design conference in Oak Brook, Illinois, in June 1967. She already knew many of the people involved.[23] Another Cornell friend who helped Wilson in this period of organization was the engineer Robert Matyas. He was "very efficient as a batman," Rose Bethe recalled. Matyas in turn remembered that when Wilson asked Rose to create the letterhead for the laboratory's stationary, he asked Rose to emphasize the fact that URA managed NAL. Rose pointed out that it would not be suitable to place those two acronyms side by side.[24]

Some months later, when it was time to move to Illinois, Wilson again followed a precedent set by Oppenheimer. He hired Priscilla Duffield as his principal administrative assistant. Priscilla's husband, Robert Duffield, had recently been appointed the director of Argonne National Laboratory. Even earlier Priscilla had worked as Lawrence's administrative assistant in Berkeley. She joined NAL on November 1, 1967, when the new laboratory's offices were based in Oak Brook. Years later, Duffield compared the experience of working for Wilson in Illinois with working for Oppenheimer at Los Alamos. "There was the same close feeling of a group of people in a strange land, . . . a group of people who were isolated from the rest of the world and trying to do something special." Duffield said that moving to the Weston site evoked memories of the esprit de corps at Los Alamos, and "that did really make it a frontier." She also recalled the feeling of independence and adventure in those early days on the Weston site. "We had a flagpole and a big fancy colorful tent was put alongside the house" (fig. 5.3). As she explained, "It was a place to sit and have a meeting," but it also "was a symbol." And at both labs, "everyone did sort of everything—there was a crisis every day." Duffield emphasized: "Bob managed to give people the feeling of tremendous urgency, of getting the thing done and getting it done fast and getting it done cheap."[25] Later, secretary Barb (Rozic) Kristen recalled that many new employees considered it hard to know just what to do in those early days, but Duffield, unlike everyone else, knew exactly what was needed and how to get things done. Many other gifted coordinators would follow in Duffield's shoes, including Cynthia Sazama, Jean Plese, Judy Ward, Jackie Coleman, and Helen Peterson.[26]

The Weston site became a blank canvas on which Wilson expressed his frontier ideals. From the nineteenth century on, farmers had tamed this windswept prairie, with its wetlands and black soil, into rich farmland. Earlier, Native Americans had appreciated the area's abundant wildlife and natural resources, particularly around the forest called the Big Woods, in the northwest quarter of the site.[27] Wilson saw the terrain's intrinsic qualities as symbols of the American heartland. He manipulated many aspects of the site, including the barns and simple houses, to create a vision of the frontier. He considered that an apt backdrop for the work at NAL—an open space eventually complete with a herd of bison. He stated that he "loved playing boy designer . . . moving houses around, improving the appearance of the place."[28] He tried to preserve the laboratory's integrity by planning its facilities to fit their natural surroundings. He encouraged preservation of natural habitats and open space. His ecologically sensitive goals, presented at a time when America's nascent environmental movement was taking root, helped to assuage local fears that the new government installation would negatively affect their quiet lives.

5.3 Flag raising in NAL Village, September 1968. (Courtesy of FNAL Visual Media Services.)

As for the accelerator, the reason for the laboratory's existence, Wilson wanted this to be more than a tool for pushing frontiers, a creation of "great beauty" which would "add to the satisfaction of our lives."[29] The spare but powerful machine would be hidden from view in an underground tunnel, its surface covered by an earthen berm well blanketed with native plants. The berm would alter the topography just enough to make the ring visible from high altitude.[30] He planned accessible experimental areas with scarcely visible underground beamlines.[31] He wanted affordable yet elegant architecture in both physics and society. The laboratory would be governed by democratic values, and offer civil rights for all. It was to be "primarily spiritual,"[32] contributing to society "not only in a technical but also in an aesthetic, social, and philosophical sense."[33] Wilson later wrote: "I envisaged the Laboratory as a utopian place where physicists coming from all parts of the country—and from all countries—would be doing their creative thing in an ambiance of well-functioning and yet beautiful instruments, structures, and surroundings that would reflect the magnificence of their discoveries and theories. All this to be done in a scientific climate of mutual respect and responsibility; it would be a place where, according to the Chinese ideal, 'all would be happy to do what they had to do, and would have to do what they were happy to do.'"[34]

As his deputy director Wilson chose Goldwasser, a physicist whose administrative and communicative skills he admired and whose liberal social views matched his own. Goldwasser, based at the University of Illinois, had come to the forefront of the movement for equitable outside user access to particle accelerators during the construction of Argonne's Zero Gradient Synchrotron (ZGS). He had in fact organized one of the first users' groups in high-energy physics. Known for being a patient, systematic, and reliable communicator, Goldwasser had ample administrative experience, having served in many advisory positions, including positions on the URA board of trustees, the Ramsey Panel, and the 200 GeV project's Site Evaluation Committee.[35]

Goldwasser visited Wilson in Ithaca early in 1967, a week after Wilson telephoned him asking him "to consider joining him in some undefined capacity." The two spoke at length, devoting "an appreciable fraction of our time (to) discussing the opportunity we would have to take affirmative action to bring motivated young Blacks into solid employment at the growing laboratory," Goldwasser recalled.[36] They soon formed a relationship they both characterized as "complementary."[37] (See fig. 5.4.) While Wilson preferred an informal management style, preferably without organization charts, Goldwasser favored "lots of people and lots of administration." He tried to attend every meeting that Wilson held. They worked so intimately together that "if Wilson had to leave," Goldwasser recalled, "I would take over half completed administrative tasks without a word being exchanged between us."[38] Wilson reflected, "it was often hard to say afterwards who did what."[39]

Drawing on his expertise as an outside user advocate and as an active experimentalist, Goldwasser would play a leading role in planning NAL's research program, selecting the members of the powerful Physics Advisory Committee (PAC), which would judge the experiment proposals requesting use of the accelerator. As NAL's deputy director, he worked to establish good communication with users at NAL, organizing annual summer studies to solicit user advice, especially about the development of experimental facilities (chapter 7).[40]

With Wilson, Goldwasser planned ways the laboratory could inspire solutions to a broad range of social, economic, and cultural problems. Like Oppenheimer's Los Alamos, NAL engaged as many women as possible in building its community programs. Many were the wives of laboratory physicists. As the first lady of the laboratory, teacher-writer-poet Jane Wilson hosted social activities at the Wilson's home in Chicago's

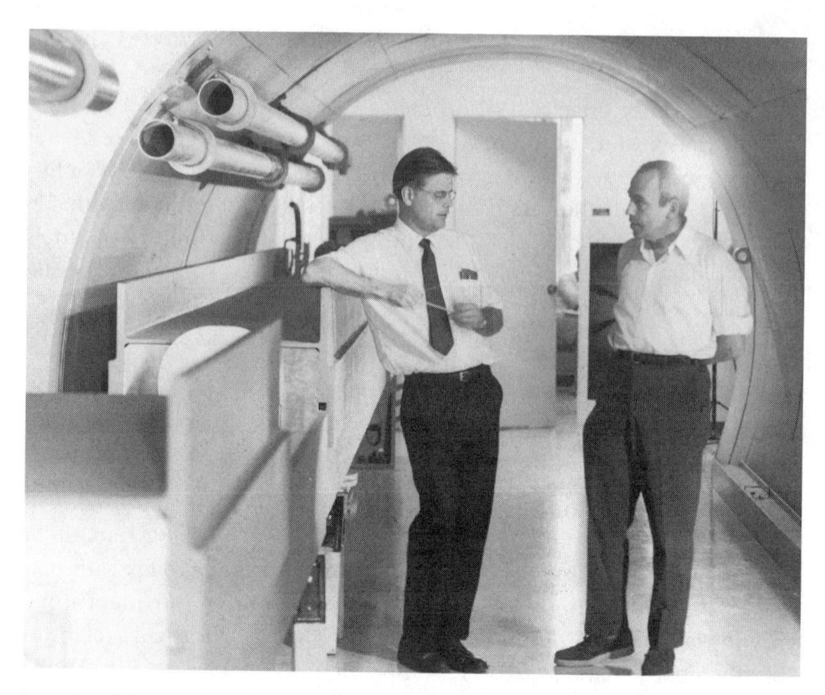

5.4 Robert R. Wilson and Edwin L. Goldwasser in ring mock-up at the first NAL offices in the Oak Brook Executive Office Plaza, 1967. (Courtesy of FNAL Visual Media Services.)

Hyde Park, and later at their on-site director's home, site 29. She also enlivened and stimulated the intellectual environment, and coordinated cultural events at NAL with Liza Goldwasser, who was known for organizing memorable picnics for staff and visitors. Musician Janice Roberts, married to Argonne physicist and musician Arthur Roberts, became involved in 1970 in making housing arrangements for the first NAL-based summer study. With their considerable knowledge of the abundance of offerings from the Chicago arts world, Janice and Art Roberts developed the NAL arts and auditorium committees. Janice also created the Guest Office, which assisted the families of new recruits to the laboratory. These and other dedicated women were instrumental in establishing NALWO (the National Accelerator Laboratory's Women's Organization), a women's organization created to support community life at the new laboratory.

When Wilson and Goldwasser learned in June 1967 that civil-rights activists were planning to protest the government's decision to spend federal funds in a state lacking fair-housing legislation with a "tent-in" and a march to NAL's entrance from the Cenacle, a religious retreat house in neighboring Warrenville, they embarked on a campaign to address the complaints.[41] Wilson sent Martin Luther King, who was expected to attend the demonstration, a telegram: "We scientists now designing the 200-BEV accelerator to be located at Weston strongly support the struggle for open housing in Illinois. Science has always progressed only through the free contribution of people of all races and creeds."[42] The protest march took place peacefully the next day.[43]

Wilson had for some time been writing to Illinois governor Kerner to explain the laboratory's position on civil rights and to discuss his problems recruiting physicists because Illinois had no open-housing legislation. During NAL's authorization hearings in 1967, NAACP representative Clarence Mitchell had opposed constructing the laboratory at the Weston site because of open-housing concerns. Kerner and the Illinois Department of Business and Economic Development understood both the economic importance of bringing the 200 GeV project to Illinois and the relationship of open housing to racial unrest. In August 1967, President Johnson appointed Kerner to chair a commission, the first study of its kind, to examine racial issues in America. The Kerner Commission's report in March 1968 described a worsening of racial tensions and concluded "that racism and economic inequality," not outside agitation, "spurred the riots the previous summer."[44]

To specifically address local racial discrimination, Goldwasser and Wilson designed NAL's human rights policy "to seek the achievement of its scientific goals within a framework of equal employment opportunity and of a deep dedication to the fundamental tenets of human rights and dignity."[45] The document emphasized that

> the formation of the Laboratory shall be a positive force . . . toward open housing . . . [and] make a real contribution toward providing employment opportunities for minority groups . . . . Special opportunity must be provided to the educationally deprived . . . to exploit their inherent potential to contribute to and to benefit from the development of our laboratory . . . .
>
> Prejudice has no place in the pursuit of knowledge . . . . It is essential that the Laboratory provide an environment in which both its staff and its visitors can live and work with pride and dignity.
>
> In any conflict between technical expediency and human rights we shall stand firmly on the side of human rights. This stand is taken because of, rather than in spite of, a dedication to science.[46]

These civil-rights efforts brought Mitchell to reverse his earlier position against NAL. In mid-1968, when the laboratory's funding allocation was cut, Mitchell wrote a strong letter in support of NAL. Within a year open housing was Illinois law.[47]

NAL's commitment to civil rights was further implemented in early 1969 by the laboratory's pilot Training and Technology Project (TAT), led by Kennard R. Williams, NAL's equal employment and community relations officer. TAT was designed for minority men to learn skills qualifying them for employment at the laboratory. As part of an interagency agreement between the AEC and the U.S. Department of Labor, young men from the inner city of Chicago were sent to Oak Ridge National Laboratory for four to six months of technical training as machinists, draftsmen, welders, or electronic or mechanical technicians at NAL.[48]

Another of Wilson's continuing social concerns was applying high-energy physics research to medical problems. In 1946 he suggested using accelerated particles in the treatment of cancer.[49] During the mid-1970s he worked with oncologists Lionel Cohen and Frank Hendrickson to develop a neutron therapy facility at NAL with funding from the National Cancer Institute. By September 1976 consenting patients were being treated at Fermilab's Neutron Therapy Facility.

The negotiation talents and administrative experience of Norman Ramsey offered another useful complement to Wilson's passionate and sometimes argumentative style.[50] Ramsey's experience organizing large laboratories reached back to the 1940s, when he had helped to establish Brookhaven National Laboratory. Wilson and Goldwasser enlisted Ramsey to help head off complaints that might be brought to the URA, for example, about Wilson's resistance to writing annual reports.[51] "We agreed that the URA Annual Report would also serve as the laboratory's annual report," Ramsey explained.[52] As URA president for three terms between 1966 and 1981, Ramsey was consulted not only for scientific, political, and technical concerns but also for personnel and salary matters at NAL, noted Goldwasser. Despite Ramsey's busy schedule, which included a professorship at Harvard, he managed to visit NAL once a week to confer with Wilson and Goldwasser. "During these sessions Bob and I made it a point to inform him about everything that was going on—successes, failures, personnel problems," Goldwasser recalled. "Our aim was never to let him be surprised by anything that had happened at the laboratory." As Wilson recalled in 1981, "Ned, Norman, and I . . . ran the laboratory."[53]

Ramsey's intervention was crucial in 1968 when Wilson submitted a letter of resignation as an expression of his frustration over initial funding delays. Instead of accepting the letter, Ramsey worked with AEC officials and Wilson to fashion a workable budget, a task that became easier when the allocation was raised.[54] Ramsey often acted as a buffer for Wilson when the temperamental director interacted with government officials, university representatives, and leaders of the physics community.[55] In late 1971, when a campaign from dissatisfied users and some URA representatives tried to force Wilson's resignation after a series of accelerator difficulties, Ramsey brought the problem under control by March 1972 with the help of Cornell accelerator physicist Boyce McDaniel, who came to Fermilab to help solve the problems.[56] "Mac" would serve as Wilson's troubleshooter at several crucial moments discussed in later chapters.

Wilson faced an ocean of skepticism in 1967 when he began to seek staff to join his new accelerator-building project. While its technical feasibility was no longer an issue, many questioned whether the proposed expandable 200 GeV accelerator could be constructed for only $250 million. The Weston site seemed an unsuitable and unappealing cornfield. Wilson struck an urgent and unusually personal note in his March 20, 1967, letter inviting potential staff members to his April and May 1967 planning meetings at Argonne National Laboratory. "We are going to have a tough time," he stated. "We will need the best creative effort of the physics community to do this. Won't you please come and help with this start?"[57] In his talk he exclaimed, "I feel lonely and exposed and have felt that way for about a month." After describing his plans for administering the laboratory, he asked prospective employees to make a commitment: "Most of all, I want some one to come and say, 'I'm going to work on this job, and I am going to live in this area, and this is the laboratory with which I'm going to identify, and I am going to be committed to its success.'"[58]

In his staff, Wilson said he tried "to find people that were just the opposite of me" so that he could incorporate elements foreign to his own experience and temperament. At the 1967 Argonne users' meeting, Wilson referred to his own accelerator building skills as "somewhat amateurish" and said he hoped to hire "the most professional kind of accelerator builders." He wanted staff who "had broad experience in large national laboratories." Most of all he wanted "good physicists" who had had "direct experience working in national laboratories where proton physics is done."[59] He appealed to accelerator experts: "I hope that experts will give us the benefit of their advice under any conditions," adding "the deeper the involvement, the more likely I am to follow the advice."[60] The fact that Wilson openly valued commitment and involvement over experience explains, perhaps, why not many recognized accelerator builders stayed with the project after the Main Ring's design phase. Wilson fared best with those who were "young and innocent" enough to want to gather at the feet "of the emperor of the sixty-eight hundred acres," in the words of J. Richie "Rich" Orr, who left a postdoctoral appointment at the University of Illinois to join NAL in 1970.[61]

As advertised, Wilson established a nonhierarchical organization that proved so fluid that under his tenure there was never an organization chart stable enough to send to the AEC. In this freewheeling atmosphere, opportunities abounded for those willing, because of their youth or temperament, to accept Wilson's terms. Orr explained: "Wilson delegated fully. If you told him what you were doing, he didn't steer you, so long as you worked within the constraints of the project."[62] Two other young recruits, Ernest Malamud, a physicist who left UCLA in 1967, and Jack McCarthy, who left the University of Illinois in 1969 after earning a BS in electrical engineering, agreed that Wilson granted a great deal of freedom and decision-making power to the committed, especially those who were physicists. As McCarthy and others noted, "Wilson felt physicists should run everything—the business office, personnel, and working groups for accelerator components." Yet even for nonphysicists, like McCarthy, the rewards more than outweighed the limitations. "I was simply too young to mind," McCarthy explained. "What mattered to me was that I was given far greater responsibility than I would have had at another job."[63] Since jobs were difficult for young physicists and engineers to find in the slumping U.S. economy of the late 1960s and early 1970s, especially with the disbanding of MURA and the imminent closures of the Princeton-Pennsylvania Accelerator and the Cambridge Electron Accelerator, Wilson had little trouble finding eager staff members; from mid-1967 to early 1974, the number of employees rose steadily to 1,152.[64]

To keep his staff involved, Wilson drew on the great storytelling tradition of his youth. As Wilson's driver Hankerson remembered, Wilson would convene a meeting of the entire staff roughly every six months in the early years of the project. At the first meeting, about two dozen employees gathered in an open area in front of the Curia, Wilson's office in the Fermilab Village (fig. 5.5). A campfire, as in Wyoming, would have been impractical. Instead, Hankerson recalled, "people took benches and chairs. Some even sat on the hoods of their cars" as they listened to Wilson's inspiring rhetoric.[65] At a staff meeting in October 1969, when Wilson's staff was struggling to meet his ambitious building schedule, he described building the accelerator as "the easiest part" of the job, "because that's just building an instrument.... We've got to do a lot better than that." He explained that "We have a higher obligation ... to make a Laboratory ... in which discoveries in physics—significant discoveries—

5.5 Robert R. Wilson addressing NAL employees in front of the Curia, 1968. (Courtesy of FNAL Visual Media Services.)

can be made. A Laboratory where we can add to our understanding of the world around us." Wilson portrayed the job of building the laboratory as an uplifting mission that "affects all of us much more personally . . . you've all come here to stay . . . to see this job through."[66]

Wilson instructed his employees to be concerned about "the physical beauty of the place." They would soon be "walking up and down the roads, through the trees thinking, and talking to one another." He was convinced that it was important to form the "proper kind of environment" for stimulating creativity and productivity.[67] The environment would be infused with team spirit; he invited all his employees to bring problems and suggestions to his attention, noting that they would "have a reasonable kind of administration" only if all employees took "responsibility for the administration of the Lab." He added, in a humble tone: "When you see problems arise, don't assume that there is a beneficent, omniscient director working [on them]. He is probably right in the middle of some terrible goof and doesn't really know where the real problems are."[68]

Again and again Wilson would draw on frontier rhetoric in an effort to inspire his staff. Speaking of the primitive working conditions in 1969, he described the "pioneering effort" of the Linac group, the first to move to the site from Oak Brook, and he encouraged those building and using the accelerator to share the adventure of primitive working conditions and risk taking. "Right now, things are inconvenient. You have to be pioneers to come out and go through a period of inconvenience." Referring to the uncertainty of funding, he noted: "Your coming here is a kind of adventuresome thing." Later, when the accelerator was ready but when experimenters were facing hardships (described in chapter 7), he told experimenters they were "ready to embark on [an] adventure." Nature's mysteries, he proclaimed, were "just as much of a challenge to the experimenter" at this juncture "as they were when our pioneer forebears started at the beginning of the century."[69]

In Wilson's first weeks as director he explained some of his aspirations to those attending his April and May 1967 planning sessions at Argonne National Laboratory.[70] Before addressing any of the administrative or technical issues, he said he first thought more abstractly "about the laboratory, about what it might become, even about the conditions for it becoming a good laboratory."[71] He took the same approach working on a sculpture or a physics project. "Although I keep changing my mind about the details, I constantly keep a vision of the whole thing in my mind." That vision was "to build a truly magnificent laboratory . . . a significant laboratory . . . where we will have the opportunity to push the limits of our knowledge about particles to a point undreamed of 30 years ago."[72]

Theoretical physicists Geoffrey Chew of the University of California at Berkeley and T. D. Lee of Columbia University explained in more detail how these limits would be extended in the physics research at NAL when they addressed the first meeting of what would become the NAL Users Organization, held in April 1967 at Argonne National Laboratory. They spoke in broad terms of the possibility of reaching an understanding of the strong and weak interactions in experiments conducted at higher energies. Berkeley physicists Glen Lambertson and Denis Keefe spoke there about the promise and anticipated problems of advances possible with "extendible energy" synchrotrons. Robert Ely, also of Berkeley, identified the difficulties of bubble chamber experiments in very high energy experiments.[73]

In January 1968, as the deadline for NAL's design report approached, Wilson and Goldwasser worked alongside secretaries and support staff in the editing, copying, and collating of pages for final assembly. The production of the report, as well as much of the writing, was managed by the assistant laboratory director for technical affairs, Francis T. Cole, one of the few Berkeley physicists who joined NAL. Assisting Cole was technical editor Rene Tracy. Cole's job description included acting as liaison with the architect-engineers, and he also had the responsibility for all the laboratory's planning, scheduling, and technical reports. These broad duties occasionally brought Cole into conflict with Wilson. The design report responded to the larger objectives of giving the laboratory its government funding, as expressed in the hearings held before the JCAE from 1965 to 1967. In these, as the report stated, "witnesses have developed the idea that this instrument will enable physicists to probe more deeply into the structure of matter than ever before. It can be expected with confidence that the limits of our knowledge of the physical world will be greatly enlarged."[74]

The report explained to NAL's government patrons: "That nature is more complex than first expected is a challenge rather than a disappointment. In building higher energy accelerators to study these complexities, all kinds of exciting and fundamental discoveries have been made. Not only have various kinds of new elementary particles been observed, but also new physical laws have been discovered while old laws have been observed to be violated." And Wilson saw to it that the humanistic aspects of research were stressed in the document as well; for example, "Pure science, the search for understanding, is as important for its effect on the minds of men as it is for its eventual contributions to his standard of living. Man's effort to achieve a better comprehension of the world in which he lives will continue to have a profound effect not only on his philosophy, not only his well-being, but also on his whole social organization."[75]

Then the report listed a few of the "many questions" of a scientific nature that would be addressed by using the particle accelerator at NAL:

Which, if any, of the particles that have so far been discovered, is, in fact, elementary, and is there any validity in the concept of "elementary" particles?

What new particles can be made at energies that have not yet been reached? Is there some set of building blocks that is still more fundamental than the neutron and the proton?

Is there a law that correctly predicts the existence and nature of all the particles, and if so, what is that law?

Will the characteristics of some of the very short-lived particles appear to be different when they are produced at such high velocities that they no longer spend their entire lives within the strong influence of the particle from which they are produced?

Do new symmetries appear or old ones disappear for high momentum-transfer events?

What is the connection, if any, of electromagnetism and strong interactions?

Do the laws of electromagnetic radiation, which are now known to hold over an enormous range of lengths and frequencies, continue to hold in the wavelength domain characteristic of the subnuclear particles?

What is the connection between the weak interaction that is associated with the massless neutrino and the strong one that acts between neutron and proton?

Is there some new particle underlying the action of the "weak" forces, just as, in the case of the nuclear force, there are mesons, and, in the case of the electromagnetic force, there are photons? If there is not, why not?

In more technical terms: Is local field theory valid? A failure in locality may imply a failure in our concept of space. What are the fields relevant to a correct local field theory? What are the form factors of the particles? What exactly is the explanation of the electromagnetic mass difference? Do "weak" interactions become stronger at sufficiently small distances? Is the Pomeranchuk theorem true? Do the total cross sections become constant at high energy? Will new symmetries appear, or old ones disappear, at higher energy?

And just after this catalog of research questions in the NAL design report came the optimistic philosophic statement "Nature in the past has always surprised us. It is probable that, as we take the step up to an energy of 200 BeV, more surprises await us."[76]

Access to the Weston site was limited during 1967–1968, the first year of NAL, because the acquisition efforts by the state of Illinois proceeded more slowly, and with greater difficulty, than anticipated. Plots of the 6,800-acre site were acquired piecemeal. While awaiting access, the physicists worked in rented temporary offices on the tenth floor of the Oak Brook Executive Plaza, about halfway between Weston and Chicago. Three experienced managers carried out Wilson's and Goldwasser's plans: Donald Getz, the assistant director for administration, smoothed the transition and forged alliances for the new laboratory, Charles Marofske handled personnel issues, and business manager Donald Poillon instituted AEC procedures and arranged financial matters. A two-lane state highway, Butterfield Road (Route 56), would bring the pioneers west from their temporary staging ground to their Weston settlement in the fall of 1968.

Obtaining the farmland was problematic. Illinois had established "quick take" provisions, allowing the property of the Weston residents and farmers to be handed over to the state for transfer to the AEC, even if the price were negotiated later. The landowners were expected to vacate, but several challenged the process in 1968 by complaining to state representatives that their land was being appraised below its true market value. When some filed suit over the issue, quick take was upheld in court, but the protracted negotiations were difficult and consumed precious time. Not until late 1968 had enough of the land been obtained to allow staff to begin to settle on the site. A marker commemorating the contributions of the residents of the village of Weston was placed near

5.6 Robert R. Wilson with Glenn Seaborg at NAL Linac groundbreaking ceremony, December 1968. (Courtesy of FNAL Visual Media Services.)

the directorate's offices.[77] That year the laboratory's public information office began a project to preserve the history of the site's farm families, including the "Pioneer Cemetery."[78] Despite his frustration with delayed construction Wilson showed his concern for the land's heritage by having a collection of farm machinery preserved and displayed as an outdoor exhibit near NAL's eastern entrance.[79]

The official groundbreaking of the laboratory occurred at site 21, the place Wilson chose for the Linac, the first component of the accelerator

5.7 Illinois governors Samuel H. Shapiro (*left*), Richard B. Ogilvie (*second from left*), and Otto Kerner (*far right*), with Robert R. Wilson (*second from right*) and AEC chairman Glenn Seaborg (*center*) at the NAL site conveyance ceremony at Chicago's Palmer House hotel, April 1969. (Courtesy of FNAL Visual Media Services.)

to be built.[80] On the cold and snowy morning of December 1, 1968, buses brought 1,200 guests, including physicists from many countries and federal, state, and local officials, among them AEC chairman Seaborg. Guests were transported from the new NAL offices in the recently transferred houses of the village of Weston, on the eastern side of the laboratory in Du Page County, to heated tents pitched on the recently vacated Schimelpfenig farm just across the Kane County line. A jubilant Seaborg spoke of the promise of the new accelerator as he broke the ground with a golden shovel (fig. 5.6). The Linac was where "the protons . . . [would] begin their trip through the accelerator," Seaborg announced. This new accelerator "will help man advance significantly into new frontiers of knowledge."[81] He went on to remark, "Symbolically, we could say that the spade begins our deepest penetration yet into the mysteries of the physical forces that comprise our universe."[82]

Not until March 31, 1969, was the full acquisition of the entire site completed, with the official transfer of the title to the AEC. During a ceremony to celebrate the event, held on April 10, 1969, at Chicago's Palmer House(fig. 5.7), Governor Richard B. Ogilvie transferred the Illinois site to Commissioner Seaborg and read the words on a ceremonial plaque: "This gift of the good land of Illinois to the high energy physics community is symbolic of Illinois's sincere interest in the revelation and communication of new and vital knowledge for all mankind. The National Accelerator Laboratory will be a significant cathedral for research and learning and an international house for scholars. It promises a firm foundation on which the future of Illinois, and of mankind, can be anchored in an academic as well as economic sense."[83]

Wilson saved money by using many of the modest Weston homes and farmhouses for offices, laboratories, and housing for visitors (fig. 5.8). Inexpensive prefabricated laboratories were constructed adjacent to the Weston houses.[84] Some of the sturdier farmhouses were moved between 1971 and 1974 from their original locations to "the Village" for visitor housing. "The Curia," the offices used by Wilson and his staff, named after the ancient Roman house of the Senate, was an assembly of one-story Weston houses connected by interlinking hallways. According to Priscilla Duffield, his idea of joining houses together stemmed from a similar custom at Los Alamos to build "lean-to's." There "you were not allowed to build a new building but you could always make a lean-to. So the lean-to's got bigger and bigger and bigger."[85]

Wilson contracted DUSAF, an architectural-engineering consortium of the firms Daniel, Mann, Johnson, and Mendenhall (DMJM) and Urbahn, Seelye, and Fuller, to build the laboratory's facilities (fig. 5.9). Earlier, Berkeley had hired DUSAF to make cost estimates for its 200 GeV project. Wilson gave much of the credit for DUSAF's success to Colonel William Alexander, who supported Wilson's plans.[86] In a review held in Oak Brook, when Berkeley engineers challenged Wilson's design, Alexander boldly

5.8 Aerial view of NAL offices in the former village of Weston, 1969. (Courtesy of FNAL Visual Media Services.)

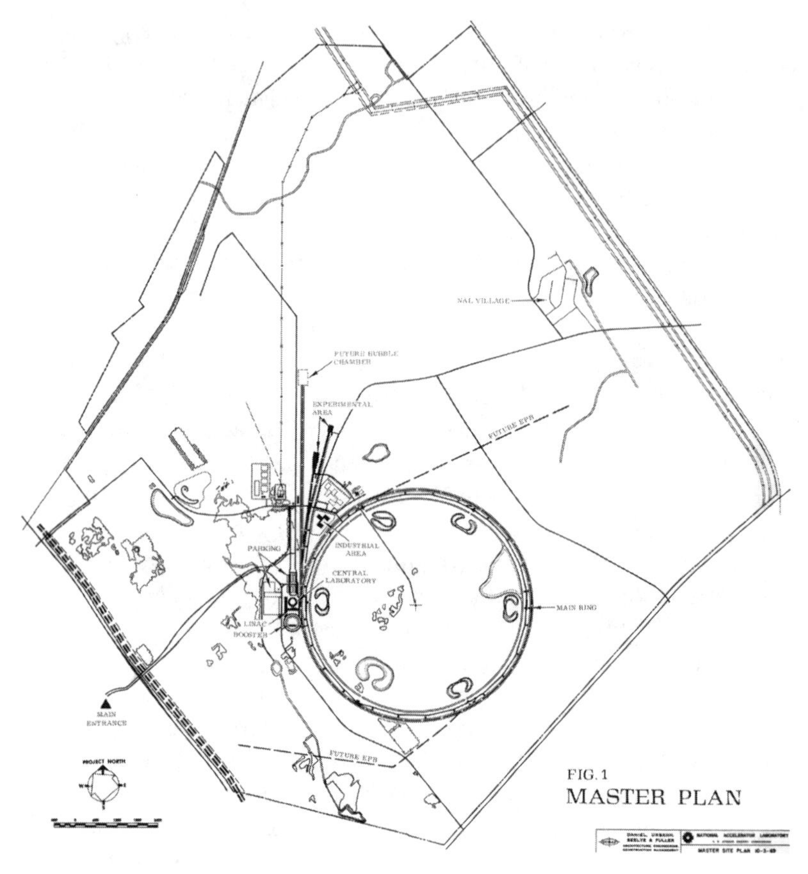

5.9 Master plan of facilities for NAL's 200 GeV accelerator project, drawn up by DUSAF, 1968. (Courtesy of FNAL Visual Media Services.)

defended Wilson's project and DUSAF's estimates. Wilson's relationship to DUSAF was unconventional; he worked alongside its architects in designing many aspects of the laboratory. Wilson later recalled the dismay of the DUSAF engineers when his artistic input took the form of temper tantrums that included tearing up unacceptable design plans, throwing them on the floor, and stomping on them. Wilson insisted, "They came to love my displays of temper because they showed that I cared."[87]

Wilson wanted the laboratory's cathedral-like central office building, later renamed Robert Rathbun Wilson Hall, to unify the users and staff, while at the same time it served also as "a magnet telling people: something important is happening here."[88] Just as the Main Ring was being completed in 1972, Wilson announced a "Workfest" to find the best design and layout for this building. The suggested designs included reflecting ponds and a footprint area. But after this bow to democratic process, Wilson proceeded to ignore the suggestions of his staff, and he worked on the design himself during the winter of 1969–1970 along with Alan H. Rider, who had been brought in by E. Parke Rohrer, DUSAF's project manager.[89] Rider had been the architect for the John F. Kennedy grave site and was attracted to the idea of building a "symbolic focus" for the world's most powerful research laboratory. The High-Rise was designed to offer a panoramic view of the open Illinois landscape.[90] Wilson even considered methods for slowing the elevators to allow time and opportunity for staff to interact and share ideas. Physicist Ernie Malamud later observed that "in fact, many such chance encounters happened over the years while waiting for the elevators, sometimes for a long time, to arrive."[91]

Initially the idea of a high-rise sparked complaints from many physicists, and even from the URA board of trustees, who pointed out that it would be cheaper to build a cluster of smaller buildings. Wilson disagreed that building a high-rise would cost more and proceeded with his plans. "A building doesn't have to look cheap and ugly to be inexpensive," he insisted.[92] He suggested an atrium on the ground floor, as in the Ford Foundation Building in New York City. To determine the best height, he hired a helicopter and "plotted the aesthetic factor as a function of height," settling on approximately 250 feet so that several floors on top could share the best view.[93] In 1975 the Society of American Registered Architects acknowledged the building with an award of excellence as an architecturally significant landmark (fig. 5.10). While much of Wilson's earlier emphasis had been on recycling and short-term use in the changing experimental program, the new central laboratory was a step toward creating a lasting laboratory.

Wilson's care in designing his laboratory extended to the laboratory's postal address, whose box number, 500, signaled the energy in GeV that his accelerator was designed to reach. According to laboratory lore, he went to neighboring towns to ask for box number 500. When he was told at the post office in Warrenville, on the eastern side of the site, that they didn't have box numbers that high, he went west to Batavia, where he was immediately accommodated. It is for this reason that Fermilab is based in Batavia.[94]

In a few cases, Wilson's desire to create a beautiful laboratory overrode his urge to economize. DUSAF representative George Laposky wrote that a road was rerouted to avoid a pair of 250-year-old twin oaks. Elsewhere on the site, the pavements were planned to bypass clumps of trees. "The roads will be less efficient, but more harmonious," Wilson proclaimed.

5.10 The central laboratory building, Robert Rathbun Wilson Hall with obelisk sculpture *Acqua alle Funi* (see chapter 8), 1980. (Courtesy of FNAL Visual Media Services.)

Sometimes he found ways to achieve beauty while solving a practical problem. When the excavation for the pump storage reservoir for the water supply demanded moving tons of earth, Wilson ordered that the dirt be piled into a large hill, named Mt. Taiji (after NAL experimenter Taiji Yamanouchi), erected at the end of the Meson Area. It served as protective shielding for an experimental beamline while adding contour to the flat Illinois landscape.[95]

One of the more active steps Wilson took to restore the site's prairie and woodlands was to institute in 1969 NAL's annual Arbor Day communal planting of hundreds of trees and shrubs. Later, in 1974, the Prairie Restoration Project in the center of the Main Ring accelerator drew on employees and local volunteers working with the nearby Morton Arboretum and with Northeastern Illinois University professor of biology Robert Betz to reintroduce seeds to reestablish the original prairie vegetation (fig. 5.11). When Wilson asked Betz how long the prairie restoration plan would take to complete, Betz replied that it could take one hundred years. Wilson replied, in a Jeffersonian planter style: "Well, we'd better start now!"[96] The restored prairie would grow into one of the largest in the United States.

Wilson's concern for aesthetics also led to his decision, unprecedented for a national research laboratory, to hire an "aesthetic watchdog." Wilson knew the artistic work of Angela Gonzales from her time at Cornell where she served as a draftsman. He came to trust her artistic eye and talent for graphic design. She joined NAL in 1967 as assistant to the

5.11 Beginning the Prairie Restoration Project: Robert and Jane Wilson (*left and center*) with Northeastern Illinois University professor of biology Robert Betz (*right*), 1974. (Courtesy of FNAL Visual Media Services.)

5.12 Evolution of the Fermilab logo, 1968. (Courtesy of Angela Gonzales.)

director for art and design. Working with Wilson, Gonzales oversaw NAL's public appearance, including the design of the site and the laboratory's publications, posters, and artwork. He gave her special assignments that helped fit the design of the laboratory to his aesthetic vision. The two collaborated in developing the laboratory's logo (fig. 5.12), which was based on Wilson's drawings of magnets. Wilson did not typically tell Gonzales what to do, but, as she recalled, when she showed him her ideas and designs, "he approved everything."[97]

As an artist, Gonzales could identify with Wilson's strong feelings about the site's appearance—for example, his wish not to allow trailers (referred to by the physicists as Portacamps) to "mushroom all over the site." But at the same time, he did not wish to appear dictatorial. So to "hide his dictatorial efforts regarding the design and development of the site," she explained, "Wilson asked me to devise the laboratory's color scheme." She had had a similar responsibility at Cornell, where she created a color scheme to improve a battleship grey Cornell physics building. Gonzales remembered the brightly colored blocks of her childhood as "a way of separating structure from nature. The best thing is to contrast."[98] For NAL she chose four elementary colors: orange, yellow, blue, and red, avoiding green because of its abundance in nature. Designing the NAL Village became for Gonzales like "a game of blocks," in which the former Weston houses, painted in bright colors, were picked up, moved, and connected like children's building blocks to give NAL's early offices a distinct look.

For a member of the staff who needed a poster or book cover, Angela was "one of the major hurdles" to overcome, reflected Bruce Chrisman. "She wielded a great deal of power in her sphere of influence."[99] The cover she designed for the laboratory's original 1968 design report created a furor in the political arena, for it featured an adaptation of Leonardo da Vinci's sketch of man, a nude with all his parts, inscribed within the circle representing the Main Ring accelerator (fig. 5.13). Despite the furor, "Bob remained adamant" about its use, Goldwasser recalled. To avoid immediate rejection, Goldwasser proposed sending the first set of copies to Congress with plain white covers and an explanation that the cover "wasn't quite done yet." Subsequent copies had the revealing original cover. "Bob uncharacteristically agreed to that compromise," reported Goldwasser.[100] Duffield recalled that while the cover indeed "caused a crisis" and many vocal objections," the design appeared "almost immediately . . . on practically everything . . . it became almost a cliché."[101]

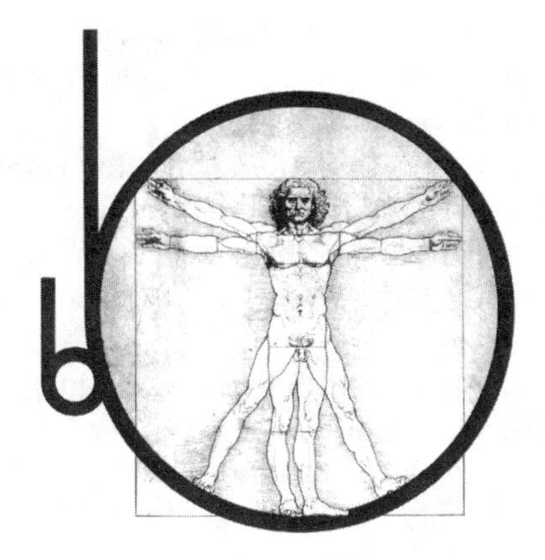

5.13 Cover of the NAL design report inspired by Leonardo da Vinci's drawing, 1968. (Courtesy of Angela Gonzales and FNAL Visual Media Services.)

Over the course of his term as director, Wilson came to recognize many of the conflicts and tensions inherent in his vision for NAL—the intrinsic conflict between the large collaborative venture of high-energy physics and his idea of physicists as lone, self-reliant frontiersmen pursuing their individual initiatives. In promoting his vision of a frontier laboratory, Wilson used such terms as "adventure" and "heroic" so often that to some, as physicist Jim MacLachlan remembered, the task of building NAL "sometimes . . . felt like a search for the holy grail." Malamud looked back on the time of building the Main Ring, "as an incredibly exciting adventure."[102] Physicist Drasko Jovanovic, who first encountered Wilson at the April 1967 users' meeting at Argonne, recalled the moment when Wilson stood up and told the crowd that on inspecting the Weston land, "like any farmer, I took a piece of soil and I chewed it a little." Jovanovic's first reaction was "There was this guy talking about eating dirt who made decisions rapidly." It appeared a "childish game." Jovanovic soon changed his mind and joined the NAL staff. Later, when the staff faced its crisis of the Main Ring magnets (chapter 6), Jovanovic thought, "There was something heroic" about Wilson's attitude. And "the adventure of it added to the fun."[103]

Wilson may never have recognized all the conflicts, ironies, and paradoxes within his vision for Fermilab, for instance, that there is something contradictory about conducting small, frugal, "nook and cranny" experiments at the highest energies. The irony that Wilson's accelerator would become a vehicle for the rise of megascience struck Wilson painfully by the time he stepped down as director in 1978, but the realization that there was a mismatch between building a lasting laboratory and erecting mainly temporary facilities on its site may have escaped him. In an essay published in 1970 in the journal *Daedalus*, Wilson wrote that "as a young man" he had "accepted the cliché" that individual research was "creative, poetic, and enduring," while team research was "superficial, uncreative and dull." Because of this conviction he had sought the life of "the lone scientist in pursuit of truth," a search that led him "deep into the nucleus of the atom."[104] And in this work, although he was in the center of the first generation of large-scale teamwork in physics, he continued to imagine the work of an experimental physicist as a holistic experience that arose from designing and building one's own equipment and carrying out an experiment from start to finish. Experimental collaborations, he strongly felt, should never be so large that each participant could not understand all parts of the apparatus.

As the experiments at Wilson's laboratory grew larger and larger, and as the energies of the protons rose, due in large part to his efforts, Wilson could not escape the reality that "it is almost as hard to reach the nucleus by oneself as it is to get to the moon by oneself." Large numbers of collaborators were needed to complete a successful experiment in high-energy physics, yet as "a witness—and even something of a participant—in the growth industry, physics," he admitted feeling "ambivalence" and a certain "prejudice" against large collaborations, especially against the increasing bureaucracy that arose in the large collaborations.[105] "I couldn't begin to sympathize with the large experiments that were going on," he told an interviewer, "because I couldn't imagine myself wanting to do one of these experiments." By the time he ended his term as director, a thoroughly conflicted Wilson saw his role as merely that of a facilitator for others to do large experiments. This kind of "voyeurism," he said, left him cold. His passion lay in building and designing experiments and experimental apparatus. "Unless I was doing that I was a goddamned administrator! It couldn't be worse," he lamented in 1992.[106]

Wilson was considered a strong director, whose leadership style was characterized by clear priorities and sometimes abruptly made decisions. At the time he accepted the position, URA president Norman Ramsey had assured him he would be "the key person in the entire undertaking."[107] The idea that a high-energy physics laboratory would be controlled by an authoritative director was in line with Lawrence's tradition as well as with Lederman's notion of the "truly national laboratory." In the 1967 hearings of the JCAE, Ramsey explained that URA leaders intended to "work through" the director, rather than "spending our time in a Washington office second-guessing him."[108] Wilson could be a harsh taskmaster. Yet in calculating his response to encourage rapid progress, he coaxed his machine and his troops into operation. In the end Wilson's staff marveled at his abilities and celebrated by his side. They recognized that his firm decisions were often questions in disguise, or as Don Young, who was designing the Linac, called them, "tentative decisions"; they were "decisions that he wanted further input on.... He expected to receive arguments when decisions were so labeled."[109] Wilson believed "that a bad decision is better than no decision, for even a bad decision is a basis of action and eventually it can be corrected."[110] As he put it, "I learned from Lawrence to define what you want and then, damn it, make it come out that way... you don't think about what other people regard to be impossible... you work as hard as you can... you don't ever say no, ever."[111]

But Wilson's style contrasted with Lawrence's in an important aspect. Lawrence characteristically built the largest, most powerful accelerator he could afford; cost and waste were not important issues for him. The historians John Heilbron and Robert Seidel have pointed out that the rectangular magnet design for Lawrence's 60-inch cyclotron "was wasteful of material." Lawrence justified its size simply by noting, "We can get the money for [it]." For Wilson, on the other hand, a spare, clean design that cost less, and wringing out the last drop from what was on hand were not only aesthetically pleasing but virtuous. Accelerator designer Tom Collins explained: "It was Wilson's style to whittle." If two designs looked equally promising, "he'd take the smaller one, every time."[112]

Such contradictions between Wilson's vision and reality filtered into his relationships with the rank and file of the laboratory. Sometimes he insisted that staff members conduct their individual research, pursue their own ideas, and make their own decisions. For example, in his April 1967 talk to users at Argonne, he promised that those who made a commitment to build the machine would "have the right to make the decisions, for they will be the ones who will have to implement those decisions."[113] He elaborated on this picture of an egalitarian mix of individual researchers in his 1970 *Daedalus* article. Describing his stint as a division leader in wartime Los Alamos, he noted: "I could not and did not ever even try to give a direct order to a colleague. . . . I found that . . . if I listened carefully" to employees, "my own ideas as well as their ideas and objectives could be clarified and improved." He tried "to emulate" his idols, Fermi, Lawrence, and Oppenheimer, "who, by the clarity of their thinking, automatically provided leadership." But a few sentences later in the same article, Wilson contradicted his endorsement of consensual decision making: "Most important of all" for a director, he explained, "is a quality which allows one to make an arbitrary decision rapidly."[114]

These perhaps irresolvable conflicts within Wilson's leadership style were hardly noticed in the early days of NAL, but over time even he could not help but realize that his aesthetics, his rhetoric, and even his goals sometimes conflicted with the realities of the growing program at Fermilab. As a boy, in listening to his male relatives in Wyoming spin tales around the campfire, Wilson had learned that "it was how you told the story that mattered."[115] The main point was to communicate a shared tradition. Only in his later years did Wilson experience how flexible tradition had to be in developing a large laboratory. The story had constantly to be revised to fit changing realities.

SIX

Constructing the Ring, 1968–1972

Something that works right away is over-designed and consequently will have taken too long to build and will have cost too much. ROBERT R. WILSON[1]

In the period of Fermilab's creation, American government officials considered frugality a virtue. President Johnson urged his staff to reduce its electricity bills. Many in the United States viewed Wilson's philosophy of taking risks to save money as an appropriate, noble, or even patriotic response. The disappointed Berkeley designers mistrusted Wilson's lack of experience in building large proton synchrotrons, but Wilson in fact knew how to build frontier accelerators with limited resources. He had applied his own version of the "less is more" philosophy in building electron accelerators at Cornell during the 1950s.

The ways Wilson went about saving money at NAL had much in common with the approaches he had experienced at Ernest Lawrence's laboratory in the1930s and at Robert Oppenheimer's laboratory in Los Alamos during the 1940s. In the periods when Wilson was at these three laboratories, all three labs were new. Each arose from the ambitions, dreams, and sense of urgency of its strong and charismatic leader, who was also its founder. All three laboratories operated in the shadow of external crisis—too little money in

An early version of this chapter was published as part of Westfall and Hoddeson 1996.

the case of the early Radiation Laboratory and NAL; or war in the backgrounds of early Los Alamos and early NAL—which set limits that defined, consolidated, and helped motivate the effort of scientists as they pressed for rapid success. Many of the researchers were young, talented, and relatively inexperienced scientists who had not yet made their marks or committed themselves to any particular approach, style, or organization. And all three laboratories shared a sense of adventure and optimism about conquering the unknown and the impossible.

Unlike most of his Berkeley and Brookhaven colleagues, Wilson did not believe that accelerators should be built by engineers. He felt that accelerator designers should be practicing physicists who "could design things that would be much prettier, sparser, and cheaper than what the professionals would design." In fact, he believed that the orderly, specialized efforts at both Berkeley and CERN had been overtaken by "engineering types."[2] As the CERN historians Dominique Pestre and John Krige explained, at CERN engineers were often "the 'real bosses' of the laboratory" because of their expertise with large-scale projects. But their professional concern with engineering and innovation often caused them to remain "detached from the urgency of research and the needs growing from it." Wilson believed that as builders, physicists were far more likely than engineers to focus on research goals than on technical innovations.[3]

One of the problems that historians of European physics have grappled with is explaining why in the 1950s and 1960s CERN's high-energy physics program lagged behind that of American physics laboratories. The standard answer attributes this lag to the fact that in the 1950s Europe had only recently emerged from a devastating war. In their history of CERN, Pestre and Krige suggest that a factor of key importance was that the aim of CERN's engineers to achieve technological perfection caused the physicists to suffer by having to wait for the "perfect" model. They needed "an 'imperfect' piece of equipment ready at the right moment." American physicists had the advantage, at least initially, because the gap between physicists and engineers closed in the United States between the 1930s and 1960s, in what Pestre and Krige call a "profound symbiosis previously unknown in basic science, a fusion of 'pure' science, technology, and engineering." The result was "a new kind of researcher" exemplified by people like Fermi, Lawrence, Oppenheimer, and Wilson, "who can be described at once as a physicist, i.e., in touch with the evolution of the discipline and its key theoretical and experimental issues," and as a "conceiver of apparatus, and engineer, i.e., knowledgeable and innovative in the most advanced techniques . . . and entrepreneur, i.e., capable of raising large sums of money, of getting people with different expertise together, of mobilizing several kinds of human, financial, and technical resources."[4] To this critical mix of physics, engineering, entrepreneurial, and organizational talent, Wilson added his unique aesthetic sensibility.

Designing the Accelerator System

Wilson began the job of designing his accelerators in the spirit of an artist planning a sizable canvas. In an effort to find the largest possible diameter for the Main Ring, he sketched the whole complex, with its Cockcroft-Walton preaccelerator, the Linac, the Booster, and the Main Ring, drawing a number of circles on a map of the site. "Most professionals divide the work into parts and then sit down and do a professional job on each part," he explained. That approach made as much sense to him as creating the sculpture of a human head by first parsing out the design of the eyes, the forehead, and the nose, and later assembling these parts. Only after conceiving the whole picture did Wilson feel that he could develop each to enhance every other part. He tried to keep his design fluid, so that innovative ideas could easily be incorporated.[5]

At one point, Wilson was so taken with the beauty of the Cockcroft-Walton preaccelerator that he decreed it should be on display as a showpiece in the atrium of the central laboratory, the High-Rise. Donald Young, who had worked for about five years at MURA designing its 200 MeV Linac, and Phil Livdahl, the two physicists struggling to link the components into a working accelerator (fig. 6.1), protested because placing the preaccelerator in the atrium would mean building a Linac with a right-angle bend in it. Such a "bend between the accelerating cavities in the Linac was unthinkable," according to Young, "considering the state-of-the-art in beam transport design and the lack of suitable beam dynamic computer programs," not to mention the lack of "computers to execute them at NAL at that time." When Young and Livdahl threatened to resign, Wilson conceded. This episode became the source of much humor in the physics community, yielding cartoons of the Leonardo drawing on the cover of the NAL design report showing the Vitruvian man's phallus bent at a right angle.[6]

Wilson considered the acceleration of protons to 200 GeV to have been "Berkeley's job." He claimed that "from the beginning I was thinking about 1000 GeV."[7] As the goal of building at minimal cost was crucial to Wilson's design work, innovation was important. More than a thousand magnets would be needed in the Main Ring. To reduce magnet

6.1 Donald Young and Philip V. Livdahl conferring on NAL plans, 1968. (Courtesy of FNAL Visual Media Services.)

costs he decided to incorporate expandability into their design, using magnets capable of being ramped up to higher fields at a later time. The concept was unlike Berkeley's expandability scheme, in which a substantial fraction of magnets would have been left out and added later for the accelerator to reach 400 GeV.[8]

Wilson held his design workshop during the summer of 1967, in the Oak Brook offices. He invited prominent accelerator designers from all over the country. They came for varying lengths of time. Typically some twenty-five participants were present. Robert Serber, a prominent theorist, agreed to lecture on theoretical topics, repeating a service he had performed in 1943 in the opening weeks of work at Los Alamos and earlier at Berkeley. As at Los Alamos, Serber's series of lectures were called "Serber Says."[9] And the design work proceeded with the hurried pace that had characterized most aspects of work at the two earlier laboratories.[10]

Wilson and those attending the workshop enjoyed the outlook over the countryside from their tenth-floor offices in the Oak Brook Executive Plaza. These minimally furnished quarters, open and without walls, reflected the open-ended quality of the discussions. Priscilla Duffield, Wilson's assistant, described the Oak Brook offices as "open in a very barren way," with offices for Wilson, Goldwasser, and Getz, a secretarial office "not actually separated off from a large open space," a conference area, and a "huge big table in just empty space and Angela alongside it." Around the open space were "little cubby holes that had physicists in them." While Duffield did not consider the Oak Brook office complex "an attractive place," it was "nice . . . to be able to get up and have a view and look at the sky. . . . [It] kind of relieved the curse of the middle west."[11]

With an eye toward completing the design of the accelerator by early 1968, as scheduled, Wilson froze major features of the Main Ring in late July, only a month into the workshop. Preliminary calculations suggested that the radius would need to be about 1,000 meters. Surprising those who expected some leeway in developing the design, Wilson abruptly announced that the radius would be exactly 1,000 meters. When he explained that a round number would be easier for everyone to remember, all discussion of the issue ended.[12]

For the Main Ring magnets, Brookhaven physicist Gordon Danby brought to Oak Brook a "separated-function" design described as early as 1952 by Toshio Kitagaki in Japan, as well as by Milton White at Princeton. In this scheme cheaper magnets were used for bending the beam while more expensive ones focused the beam. The Berkeley physicists had rejected Danby's suggestion of separated-function magnets, but as the design allowed 15%–20% higher magnetic fields in the aperture at less cost than using conventional "combined-function" magnets, Wilson decided to use separated-function magnets for the Main Ring "almost without discussion," as Arie Van Steenbergen reported.[13]

In casting for ways to achieve the highest field while reducing the size and cost of the Main Ring's bending magnets, Wilson hit on the idea of modifying a conventional "window frame" construction by fitting an added coil into the unoccupied space on either side of the poles. By containing the beam very precisely, the extraction of the beam could be limited to a few discrete points, a novel feature for a synchrotron. Two kinds of bending magnets (B1 and B2), of different sizes, were made to cut down the overall dimensions needed for matching the width of the beam and to save power.[14] As the radiation level is high in this part of the magnet, the risk of potential radiation damage to the coil's insulation is substantial in this design. And because epoxy, the only organic material in the magnets, is particularly sensitive to radiation damage, Wilson insisted that the magnets be assembled using as little epoxy as possible. This risky move would contribute to much trouble in 1971, at least according to some interpretations offered later in this chapter.[15]

In an unexpected, yet characteristic, move during the fall of 1967, Wilson increased the energy goal of the NAL accelerator, originally 200–300 GeV, even more. When Alper Garren and Lee Teng designed the so-called Main Ring lattice, the arrangement of components, they calculated that the acceleration could reach 400 GeV using the newly designed bending magnets run at 18 kilogauss in a ring with a radius of 1,000 meters.[16] Wilson argued that if they could find a way to stretch the performance of the bending magnets to 22.5 kilogauss, the machine would be able to reach 500 GeV! That became the new goal.[17]

To achieve this goal at the minimal cost, Wilson aimed to reduce the overall dimensions of the bending magnets and find simplified methods for insertion and fabrication of the coils. Instead of using the C-shaped magnets specified in the Berkeley design, his staff chose the smaller, but higher-field, H-shaped magnet design. Sizing the magnet aperture to match the beam's properties allowed using a smaller vertical aperture for half the bending magnets. Wilson planned to fabricate both types of bending magnets (B1 and B2), as well as the quadrupole magnets, using die-stamped laminations that were aligned and mounted in self-supporting girders.[18] While previous proton synchrotrons had used coils that were mounted mechanically in the core, Wilson hoped to save manufacturing costs by using glued-in coils.

Taken together, these various design changes brought considerable savings. Cost comparisons made later that year showed that although NAL's Main Ring design called for almost twice as many magnets as in Berkeley's design, the Berkeley magnets required almost 10 tons more steel per magnet and cost a total of almost $6 million more.[19] But the use of pared-down magnets was inherently risky, because larger magnets produced higher fields more reliably. Further savings came from the fact that NAL's smaller magnets allowed building a smaller tunnel enclosure—only 10 feet wide rather than the 14–16 feet that experts considered the minimum acceptable width.

Ramsey later explained that Wilson took risks on about twenty aspects of the accelerator's design, saving about $5 million per risk. "We knew something would fail," he noted, "but we figured it would be much less expensive to fix the failure than to play it safe with all 20 items."[20] Wilson's critics disagreed with his cost-cutting and risk-taking philosophy. Considering the importance of the project, they insisted that Wilson had taken unacceptable risks. In July 1969, Wilson justified his cost-cutting philosophy in his Sanctimonious Memo 137: the "money and effort that would go into an overly conservative design might better be used elsewhere. Failure should be designed into a successful machine. . . . A major component that works reliably right off the bat is, in one sense, a failure—it is over-designed."[21]

The many other limitations that Wilson imposed, either in the interests of aesthetics or cost, included dispensing with the use of unsightly and expensive cranes, costly air conditioning in the machine's smaller and less expensive tunnel, and not anchoring the tunnel to bedrock with pylons, perhaps the riskiest of the cost-cutting gambles. Instead Wilson asked Collins to design a tunnel that "floated" on the glacial till just beneath the topsoil of northern Illinois. When accelerator experts told Wilson the scheme would never work because the beam would disappear into the walls of the tunnel, he answered, according to Goldwasser, that if the tunnel was not stable "he would simply install motorized jacks on every magnet and actuate the motors with signals from detectors that would be 'servoed' to the proton beam." As the glacial till proved stable, this system was not needed. Wilson himself never believed there would be any problem, noting the stability of the massive city of Chicago.[22]

Wilson also cut costs by accelerating the building schedule, thus reducing the cost of personnel. The Berkeley group had estimated seven years for the construction of their machine. Wilson planned to complete his by June 1972, in just three and a half years. That ambitious goal, Wilson felt, encouraged creativity.

One reason that Wilson found broad acceptance for his early design, despite some criticism within the physics community, was that its low cost fit the funding context of its day. Federal funding for physics, which had risen steadily since the end of World War II, was no longer growing in 1968. That year, the funding for high-energy physics was 40% lower than the AEC had projected in 1965. But while the AEC commissioners realized that the funding environment would not improve in the next year, Congress still expected an expandable accelerator. Both the commissioners and JCAE were thus inclined to support Wilson, since both had sponsored the initial funding of NAL and thus had a political stake in the project. Wilson strengthened the backing of both groups by presenting a design costing $248 million for a 200 GeV accelerator that could be expanded to as much as 500 GeV.[23] To avoid red tape and expedite construction, Wilson insisted that the AEC grant approval authority to a local AEC office led by Kennedy C. "K. C.," or "Casey," Brooks. With the streamlining of federal administrative procedures, with URA, AEC, and JCAE approval, and with the necessary staff, Wilson was now poised to build the world's most powerful accelerator.[24]

An Academic Start: Modeling the Main Ring, 1968–1969

The preparation stages of building an accelerator are similar to those of any large construction project. After the preliminary designing, the focus is on building models to test the basic characteristics of parts of the larger structure, often in miniature.[25] On the basis of these models, refinements are made in the design. Then full-size prototypes are constructed. After further development, the construction specifications are defined and building can begin.[26] In the Main Ring group the model-building phase dominated attention until the middle of 1969. In the holistic approach that Wilson favored, modeling many components almost full scale allowed checking the overall dimensions and performance of the accelerator's many elements. Understanding how the parts fit together helped the staff anticipate fabrication problems.[27]

The pace and tone of the decision making were set in this early period by Frank Shoemaker, a veteran accelerator builder then on leave from Princeton. The accelerator theory group, led by Lee Teng throughout the construction years, contributed to the conceptual design of the Main Ring. A technical services group, under Henry "Hank" Hinterberger, provided engineering support as well as machine shop and drafting services.[28] Like the other group leaders at NAL, Shoemaker was faced with a lack of facilities. Because the Oak Brook offices could not accommodate model building or computing, and because the Weston site was not available until later in 1968, the new laboratory had to rely heavily on other institutions to support construction efforts. The first models of a vacuum chamber and the 2-foot H-magnet were constructed at the Physical Sciences Laboratory in MURA's former headquarters at the University of Wisconsin. Most of the early magnet models were built and tested at Argonne. Computer modeling for beam dynamics was done using data telephone links to Argonne and New York University.[29]

Wilson supported Malamud's idea of furnishing a 200-foot-long model of a section of the Main Ring tunnel known as the Protomain with real-size components. Professional model builder Jose Poces installed the Protomain, complete with full-scale mock-ups of magnets, vacuum system, and water and power connections, in the NAL Village. The Protomain contained a complete cell of eight bending magnets, two quadrupoles, and vacuum systems. The Protomain's tunnel served as a prototype for the Main Ring tunnel.[30] By spring 1969, the Protomain had been completed and the group was ready for the full-scale prototype stage.[31]

To be able to work on the magnets and vacuum system before construction funds for a building had been allocated, the Main Ring group purchased and, in June 1968, installed an inflatable building, dubbed the "air building."[32] The problems that developed with this building demonstrate the enormous difficulties faced by those working in the frontier-like conditions at the new laboratory. In August a portion of the vacuum chamber model had to be extended through a hole in the air building because the structure was not long enough to house the entire unit. On one windy day, the building collapsed, fortunately causing no damage because a unistrut framework had been built inside it. Later, during the subsequent Illinois winter, the group discovered that the building stayed only 30°F warmer than the ambient temperature, instead of 50°F warmer, as advertised by the manufacturer. Freezing temperatures inside the building were prevented by installing gas heaters. Although Shoemaker hoped to replace the air building with a more substantial metal structure by the next summer, the group used the air building until early 1970 because of the continued shortage of funding and space. By this time the group, which was gearing up for full-scale magnet production, had spread out to several buildings on the site.[33]

The design of the vacuum system also advanced in this period. After reassembling the model built at the University of Wisconsin in the air building, work began during the summer of 1968 on developing a fixture for welding together flanges of adjacent magnets. As it was assumed that magnets would occasionally need replacement, Walter Pelczarski developed a clever and efficient means for cutting these flanges apart, an innovation that would later prove more useful than anyone imagined at the time. By September the 50-foot vacuum model had been equipped with small ion pumps and the hardware necessary to isolate the straight sections. In March 1969, a month after the mock-up vacuum system was set up in the model tunnel, the vacuum chamber model performed as expected during pulsed-field magnetic measurements using digital recording and analyzing equipment.[34] Meanwhile, arrangements were being made for an innovative magnet power supply system designed by engineer Richard Cassel. Wilson decided to use the Commonwealth Edison Company's network in West Chicago as a power storage system, pumping electrical energy from their grid, but with poles of his own design. Referring to the system, Wilson later remarked, "the power comes in and the protons go out [to the experimental areas]."[35]

Despite the group's accomplishments, some felt that the Main Ring effort had been too relaxed in the period of 1968 to mid-1969. Malamud and Ryuji Yamada remember the Main Ring effort having an "academic" atmosphere that seemed at odds with Wilson's stress on speed and economy. "We spent endless hours philosophizing," Malamud noted. Visible progress in other groups added pressure. In April 1969, when the Linac group accelerated the first beam with its prototype preaccelerator, and with construction underway on the Linac's building and the Booster's enclosure, Wilson began worrying that the Main Ring effort would lag behind the rest of the project.[36] At just this point, Shoemaker announced that he would return to Princeton. Wilson seized the moment to make an irregular move. He appointed himself Shoemaker's successor. To ease the administrative burden of simultaneously acting as laboratory director and as a group leader, Wilson formed a Main Ring management "troika" of Malamud (who suggested the troika idea) and the two engineers, Cassel and Hinterberger. The troika lasted until summer 1971.[37]

Cascading Accelerators, 1968–1969

Always seeking to streamline, Wilson looked for, and found, many timesaving short cuts in building the accelerator's components. For example, rather than going back to the drawing board in designing the radio frequency (RF) system of the accelerator (which imparts energy to the particles and controls the particle beam), he accepted a design that Berkeley had selected after considering three alternatives: a system made of ferrite-tuned cavities. He hired a former member of the Berkeley team, Quentin Kerns, to direct the construction of NAL's RF system. Although Kerns reduced the power and the number of cavities to accommodate the NAL design, he built ferrite cavities along the lines of the Berkeley design.[38]

As Cockcroft-Walton accelerators were available from a commercial manufacturer, Wilson saved time and funds by purchasing, rather than building, such a machine. For the Linac, he drew on the work of an earlier study of Linacs by a group from Brookhaven, the Los Alamos Meson Physics Facility (LAMPF), and MURA. Don Young was asked not only to build NAL's Linac but also to work to coordinate the system of Linac and Cockcroft-Walton. As Young remembered, Wilson decided to use the Brookhaven design to take advantage of the savings in cost that would come from this move. Young liked the idea because the Brookhaven Linac "was more conservative, and I felt the Linac should be a very reliable thing, because everything depends on the Linac." Although this reasoning contrasted with Wilson's risk-taking approach, Wilson let Young build the Linac without interference because, according to Young, he "didn't want to pay any more attention to it. He had other problems." Wilson had allowed Kerns the same latitude in building the RF system. By adopting certain ready-made components and designs that had been developed by others, Wilson saved time for the project and reserved his own time for the many challenges the Main Ring would face.[39]

After setting up initial operation in a rented garage in Downers Grove, a town about fifteen miles from the Weston site, Young's Linac group occupied three abandoned tornado-damaged buildings in the village of Weston. Other residences nearby were still occupied. Young remembers needing to take "defensive action against their unleashed dogs." On November 11, 1967, the Federal Savings and Loan Insurance Corporation (FSLIC) granted permission to URA to construct a "butler-type" building for the Linac group, with the promise to restore the land if the accelerator project was not completed. They constructed a 10 MeV prototype to check the design and develop subsystems. To reduce cost, components and materials were borrowed from many other laboratories, including MURA, Argonne, Minnesota, Livermore, Brookhaven, and Los Alamos.[40]

Under the pressure of Wilson's time schedule, Young streamlined the standard construction process, skipping both the modeling and prototype phases in building the Linac. When the group set up the first major pieces of the Cockcroft-Walton and Linac—the Cockcroft-Walton's ion source, voltage system, and accelerator tower; the transport system; and the first Linac tank—the system worked! Young explained: "We called it a prototype, but we all knew it was going to be used in the final machine."[41] With this shortcut, the first Linac cavity produced protons at 10 MeV, its design energy, by June 1969. Less than six months later the equipment was dismantled and installed in the new Linac Gallery.[42]

The Linac group had to assemble more tanks before the accelerator could produce 200 MeV protons and reach the design energy of the complete machine. Nonetheless, by that time components for the final system were already ordered and the Linac group was in a position to proceed rapidly. By April 1970 five tank sections were being assembled in the Village: two had been installed in the Linac building, permitting the permanent system to produce a 10 MeV beam. The end of the summer of 1970 saw beam accelerated to 66 MeV through three out of nine Linac tanks as well as plans to have six tanks installed, so that the machine would be able to accelerate a 139 MeV beam.[43]

The Booster effort was initially led by Arie Van Steenbergen, who was followed by Paul Reardon in fall 1968, Roy Billinge in 1969, and Helen Edwards, who brought it into operation. Development was fast paced and streamlined. The work resembled the building of the Main Ring in that the influence of Wilson's whittling style was very much apparent. Billinge had come to NAL in 1968 from the Rutherford Laboratory and became leader of the Booster group in 1970 when Reardon took the helm of the Business Office. Edwards was associate leader, and she succeeded Billinge as head in December 1970, when he moved on to lead the Meson group, many of whose members he had worked with previously on the Booster. The next year he went to CERN.

Because Billinge and Edwards were intent on reducing the dimensions of its magnets, and because that offered considerable risk, the Booster went through the full set of standard construction stages. Modeling of the Booster magnets had started in summer 1968, and prototyping was well underway by the fall of 1969, when two complete modules consisting of four magnets were tested successfully.[44] By late 1970 the construction of the Booster was finished. In January 1971, the final design of the Booster magnets was tested by placing magnets in the Booster tunnel. A proton beam was injected from the Linac and coasted without increasing energy around the entire Booster. A week later, protons were accelerated to 1 GeV in the Booster, and when this beam was fed into the completed section of the Main Ring, the beam coasted through, demonstrating that the Booster would work as planned. The Booster reached its full design energy of 8 GeV in May. When beam was guided through the whole system for the first time, on June 30, 1971, Wilson declared the Booster completed.[45]

Fermilab's innovative beam extraction system was enabled by the imaginative development of an electrostatic septum by the prickly and independent Alfred Maschke, who took on the daunting technical challenge of directing the beam from the Linac to the Booster and then to the Main Ring. Maschke's idea was for a septum to extract protons using an electric field and then deflect them out of the accelerator beam with a magnet. Formerly of Brookhaven, Maschke joined NAL during the summer of 1967 and became head of the Beam Transfer Division, charged with extracting the beam from one accelerator into the next and on to the Switchyard, where it would be separated and sent to targets in the experimental areas. Together with Tom Collins's long-straight-section concept, by this time considered a brilliant innovation, Maschke's septum formed an extraction system with an efficiency of 99%, much higher than that achieved in previous accelerators.[46] According to Jovanovic, without Maschke's contributions, "Fermilab would not work."[47] Unfortunately, Maschke had countless arguments with Wilson and in 1971 resigned and returned to Brookhaven. The members of Maschke's group were placed under Wilson's managers, at that time led by Boyce McDaniel. This group had to deal with the frightening Main Ring magnet crisis that surfaced in the summer of 1971, discussed later in this chapter.

An Experiment in Theory, 1969–1974

Wilson and Goldwasser felt that there should be an interplay between theorists and experimenters in the research program at NAL. Following Serber's early series of lectures and theorist Robert Marshak's advice, they hired five theoretical physicists who specialized in high-energy phenomena. They arrived at NAL in the fall of 1969 and worked at 27 Sauk Boulevard in one of the remodeled Weston houses in the Fermilab Village. Early in 1970 they moved into the Curia complex adjacent to Wilson's office. David Gordon from Brandeis represented the group as its acting head; he had worked on an early version of string theory. The other four theorists were Louis Clavelli from Chicago and Yale, Pierre Ramond from Syracuse, Jim Swank from the University of Illinois, and Don Weingarten from Columbia. In a November 6, 1969, statement, Goldwasser informed the NAL staff that the five new "post Ph.D. theorists are available to our experimentalists in connection with the current experiments and with their formulation of plans for facilities to be provided for the 200 BeV research program."[48] Soon Gordon, Ramond, and Clavelli were directly involved in extending Gordon's string theory study. As Clavelli later wrote, "The theory group was a unique experiment at Fermilab . . . designed to answer the question of what might happen if you assembled five, hand-picked and pedigreed, young physicists with no senior staff, gave them complete freedom to explore the frontier of theoretical physics with no computer facilities or archival library, fifty miles from the nearest physics department, but with a virtually unlimited budget to bring in expert consultants."[49]

With the new accelerator more than two years from turning on, the new NAL theorists at first found it difficult to engage experimental physicists in more abstract theoretical discussions. For additional support, the laboratory started a "professor of the month program," with weekly lectures on current topics of theoretical physics. A distinguished professor would spend one day a week at the laboratory for a period of two to six weeks. The first such professor of the month, in December 1969, was Jun John "J. J." Sakurai, from the University of Chicago but then based at the University of California at Los Angeles, delivering the lecture "Vector Mesons and Electromagnetic Interactions and Hadrons." Subsequent professors of the month included Chen Ning "C. N.," or "Frank," Yang (Institute of Advanced Study, Princeton), Tsung Dao "T. D." Lee (Columbia), Gabriele Veneziano (MIT), Geoffrey Chew (University of California, Berkeley), and Stanley Mandelstam (University of California, Berkeley).

From the vague language of their appointment letters, the members of the theory group had deduced that their group would be a permanent component of NAL's program and that "we were expected to continue at Fermilab," recalled Clavelli. They planned to grow into a much larger group, similar to CERN's theory group, with a staff of permanent senior theorists to offer direction to the younger physicists.[50] But as the theory group approached the end of its first year, "something did not ring true," according to Clavelli. Sam Trieman of Princeton was invited to NAL for a two-year visit starting in the fall of 1970 to act as the group's temporary head while a new crop of postdocs was hired—Steve Ellis, Manny Paschos, and Tony Sanda. Clavelli noticed that at this point the directorate began referring to the original theory group's members as "postdocs," a title that had not been used earlier.

In November 1970, a little over a year after the original theory team arrived at NAL, four of the five members (all but Gordon) received letters announcing the termination of their postdoctoral appointments. Not only had the temporary status of their appointments not been made clear, but their letters stated, "We had hoped that considerably stronger interactions would develop between you and the experimental physicists than has been the case."[51] They were puzzled because they recalled that back in December 1969, Goldwasser, who was in charge of the theory program, had told them, "Bob Wilson felt that the primary work of the theory group was expected to be directly related to the experimental program." But Clavelli recalled being told that "if some truly outstanding work of a more formal nature could be accomplished, that was OK too," Thus, in working on their "dual model" the group's members had not worried about their job security. They had been excited about the significance of this work, which contributed to subsequent theories of superstrings and supersymmetry. The group had also investigated certain questions of experimental research on bubble chamber measurements and had helped with the first round of experimental proposal reviews. The four terminated theorists were especially hard hit when they discovered that the job market had collapsed during their year at NAL.[52] Fortunately, all found positions during one of the worst physics hiring periods of the late twentieth century. While the four members of the first theory group started new jobs, Gordon focused on modeling the scattering amplitudes being studied at the end of the era of Regge Pole theory. In 1973, he was also discharged.

The turmoil in the theory group was not apparent to Jeremiah Sullivan, who joined the group for 1971–1972 some weeks before Clavelli, Ramond, Swank, and Weingarten left in September 1971. He found an active theory group working in a "very interactive and pleasant" environment, with visitors, seminars, and "lots of activity." By then, experimentalists were thinking about the design of their detectors and "came around all the time looking for theorists to talk to." Sullivan, who had held a postdoc at SLAC and then became an assistant professor at Stanford, had moved to the University of Illinois largely because Weston "had just been chosen as the site of the new high energy physics accelerator in the United States." He recalled that Jerry Almy, then head of the physics department at Illinois, "marched down the street to the office of the Dean of engineering and managed to get four new positions, two in theory and two in experiment. . . . I'm sure the department argued that NAL would make Illinois in the Midwest an international center for high energy physics."[53] During Sullivan's time at NAL the theorists were excited about the high flux of high-energy neutrinos to be produced in the fixed-target experiments. Neutrino physics was clearly going to be an important part of NAL's higher-energy physics program, because the neutrino cross section grows as the square of the energy.

During March 1972, when Sullivan was serving as Trieman's "stand-in," the group received a phone call "saying they were expecting to get beam around the entire accelerator . . . anytime this morning." All the theorists rushed to the control room. "And when they got the first bunch around there was great celebration. We all signed the logbook." In recognition of the event's importance, Sullivan signed his name in big John Hancock letters.

Before Trieman departed, he recruited Henry Abarbanel and Martin Einhorn into the theory group. In August 1972, J. D. Jackson of the University of California at Berkeley arrived for a one-year term as temporary head of NAL's theory group. In addition to restarting the visitors program, Jackson arranged for the move of the theorists from the Village house to the third floor of the newly constructed central laboratory building, the High-Rise. A Theoretical Physics Steering Committee was appointed to advise the group. Jackson and Einhorn came up with the idea of instituting a stimulating Friday afternoon program, initially called "The Experimental Theoretical Seminar." It was later officially named "The Joint Experimental-Theoretical Seminar" and was casually referred to as "the Wine and Cheese Seminar." It was a huge success and quickly became an NAL tradition.[54] A January 1973 conference, "Quarks and Partons," featuring the iconic Richard Feynman from Caltech, was hosted by NAL's theory group.

Benjamin Lee, a physicist from the State University of New York at Stony Brook, agreed to lead the group for the 1973–1974 year and he joined Fermilab's permanent staff in 1974, when new postdocs Bob Savit and Shirley Jackson arrived. Lee brought Chris Quigg, his young associate from Stony Brook, to Fermilab in 1974 and Hank Thacker followed in 1976. William Bardeen had moved from Stanford to join the group in 1975. The group was doing well until June 1977, when Lee was killed in a car accident while driving to Aspen with his family to attend the annual Fermilab PAC meeting. His untimely and tragic death was a shock not only for his family but also for his colleagues in the international high-energy community. Quigg succeeded Lee as head of the group.

The First Flush of Success (1969–1971)

Wilson infuriated his critics by ignoring the standard procedure of freezing the design of a technology when prototyping begins. As late as mid-1969, he was still injecting fresh designs intended to improve, simplify, or lower the cost of the Main Ring magnets. He considered this effort an important part of a laboratory-wide campaign against "heavy-footed over-design." For the Main Ring magnets, this campaign lasted until December 1969, when the first production dies were made and stamping began. One result of the campaign was a more compact design for both the quadrupole and the bending magnets. The group developed a new method of fabrication for bending magnet coils, in which the inner and outer coils were constructed separately. In this way, outside manufacturers could continue fabricating outer coils, while the more exacting window frame fabrication of the inner coil could proceed under the supervision of Wilson and his staff.[55] Other elements of Main Ring design that were made simpler or more economical were the water and power distribution system inside the tunnel and the welded joints in the vacuum system.

The alignment system originally proposed for the Main Ring was a risky system of stretched wires, dubbed the "spider web." It stretched between alternate quadrupoles, with small pickup coils on the intermediate quads to allow accurate alignment of any given quadrupole with respect to its two neighbors.[56] The staff believed that had it been built, it would have attracted spiders. Gordon Bingham from Australia worked for over a year on this system. He subsequently left the laboratory after a group, which included Wilson, his "troika" (Malamud, Cassel, and Hinterberger), Don Edwards, Tom Collins, and Bingham, decided late in 1969 to abandon the complicated and expensive spider web alignment system and replace it with a much less expensive laser alignment system suggested by Chuck Schmidt of the Main Ring Section. By the end of that year, Schmidt had demonstrated a precision of ±0.005 inches on the laser system in the Protomain.[57] Unfortunately, changes in the tunnel temperature (due, for example, to colder regions, where the tunnel had been filled with frozen earth during the winter) prevented the system from working well. The group eventually gave up on its elegant laser scheme and summoned help from Argonne's Bill Testin, whose Research Division Alignment Group, working under NAL's Ed Blesser, aligned the ring properly.[58]

Wilson's goal of achieving a spare, clean design was reflected in the modifications that Hinterberger made to the magnets in May 1969, expressing, as Wilson later noted, a sense of design which was "both elegant and economical." In the 1968 design, the steel yokes of the magnets, which were made of thousands of thin laminations, were secured with a steel I-beam, but these 20-foot structural models sagged. Hinterberger realized they could eliminate the I-beam, because the laminations could themselves secure the yoke. He inserted angle girders around the four outside corners of the magnet. In this "box girder" design, the magnets weighed less and their properties improved because the girders became part of the magnetic flux return circuit.[59]

Unlike Berkeley's or CERN's specialists, NAL's engineers tended to be nonspecialized workers who could blend skills from engineering, research physics, and accelerator science. The important contributions made by engineers at the laboratory—such as Hinterberger, Cassel, Willard Hanson, the electronics and control engineer who headed the magnet factory, or Anthony Glowacki, who designed the magnet water-cooling system—show that both engineers and physicists could live up to

Wilson's ideal of creative accelerator building. The possibilities of Wilson's nonspecialized approach to accelerator building are well illustrated by the work of physicist Ryuji Yamada, then a young experimentalist, who succeeded in boosting the capability of the accelerator from 400 to 500 GeV. In late 1969, the accelerator theory group considered the prevailing bending magnet design, which yielded an 18 kilogauss field, to be optimal. But Yamada reassessed the calculations to see if a higher field could be achieved. Assuming the role of theorist, he explored alternate calculations of design. Thinking as an engineer, he decided how the new design should be fabricated. Then, becoming an accelerator physicist, he constructed a short model based on his calculations, made magnetic measurements of the model, and plotted the data. He repeated the process, creating a further revision based on the results. After many iterations he emerged with a new design in which the decrease in the magnetic field at the edge of the magnet was offset by saturation effects due to tapering the pole tips.[60]

Yamada's approach proved remarkably effective. Wilson later concluded that Yamada's improvement, which depended on the earlier decision to place the coils inside the gap, was "one of the biggest innovations" of the entire accelerator. In a December 3 staff meeting Wilson noted that the resulting "pole shape was very similar to the ideal pole shape worked out many years ago by H. A. Bethe."[61] By December, calculations predicted that the field shape with Yamada's design would be "acceptable without further correction up to 21 kilogauss." In the next few months tests proved that Wilson's challenge posed two years earlier would be met, because the magnets were in fact good up to 22.5 kilogauss. This meant that the Main Ring would perform at 500 GeV.[62]

Wilson's focus on economy continued to suit a funding environment in which the laboratory consistently received smaller allocations than requested. As the historian Spencer Weart has explained, if all figures are adjusted to reflect 1972 dollars, federal support for physics peaked in 1967 at about $350 million. It then began to fall and by 1975 had dropped to 70% of that peak, rising only slightly after that.[63] The new laboratory was destined to struggle for its funding. In 1968, NAL requested $75 million, battled against an allocation of only $7.1 million, and subsequently received $14.7 million. In 1969, the laboratory requested $102 million and eventually received $70 million. Although approval for the entire $248 million budget was authorized in July 1969, the same month as NASA's lunar landing, by early 1971 NAL was still awaiting $93 million of its construction budget.[64] By then there was little fear that funding would be terminated, but the limited annual appropriations forced Wilson to

6.2 Groundbreaking for the Main Ring, October 1969. (Courtesy of FNAL Visual Media Services.)

economize throughout the entire construction period. Cost overruns would have made the project unpopular in Washington, jeopardizing future funding.

Wilson sometimes used the budgetary limitations to justify tactics aimed at obtaining optimum efficiency from his staff. Knowing that if time were wasted, the project could not be completed within the shrinking budget, he shortened the time before deadlines and made the goals more ambitious. For example, when Wilson became Main Ring group leader in 1969, he advanced the deadline for installing Main Ring components by six months, from January 1, 1972, to July 1, 1971. This dramatic move, which moved the schedule for completion up by an entire year, heightened Wilson's efforts to portray the building of the Main Ring as a risky frontier adventure, like building the transcontinental railroad or landing on the moon.

In October 1969, when NAL broke ground for the Main Ring tunnel enclosure (fig. 6.2), the team's prototype program was spurred on by the ambitious new deadline of having eight bending magnets and two prototype quadrupoles installed and operating in the Protomain by March 20, 1970. They had to work frantically.[65] The process of large-scale mass production, essential for building more than one thousand magnets for the Main Ring, appealed to Wilson because it was a way to keep the permanent NAL staff as small as possible. He viewed keeping the staff small as both an economy measure and a means to achieve creativity: he felt people worked more efficiently in smaller groups. But Wilson's approach—to leave the design flexible as long as possible and to insist on economy—differed from the industrial procedures of mass production, where all specifications that would be costly to modify are set as soon as possible.[66]

Procuring the services and material to produce the Main Ring components rapidly and economically was no trivial task. Malamud remembers that he, Hinterberger, and Robert Sheldon traveled all over the world to award coil contracts, saturating the international market for coil manufacture. In the never-ending quest for efficiency, savings, and speed, Wilson used a trick. As he explained to his staff in 1970, they would "give one-third to one producer; one-third to another; and whoever gets through first gets the remaining one-third."[67] By January 1970, arrangements had been made for factory production and for materials procurement. At the peak of production, coils were being manufactured in nine factories: one in France, one in England, and seven in the United States. NAL's factory in West Chicago was managed by Sheldon.[68]

Wilson's daring challenge aligned with one of his standard strategies for speeding the work up by devising two different work schedules. One was for the public, the AEC, and the users. The other, an "excessively optimistic" one, was for the accelerator builders. This internal second schedule was meant to "increase the probability of at least meeting the public schedule." The users, on hearing about Wilson's updated schedule for the accelerator builders, worried that the laboratory's internal staff might be able to use the accelerator before they could and thus "skim the cream of the new discoveries." In April 1970, Wilson risked his reputation when he confidently announced to the users that the optimistic internal schedule would also become the public schedule and that the laboratory would "have an accelerated proton beam by mid-1971, a year earlier than the originally scheduled date." Not only would the laboratory offer beam earlier, but the machine would eventually reach energies close to 500 GeV, at reduced intensity.[69]

At this point, Wilson shifted his emphasis to the building effort. The tighter schedules he set added much pressure to the constant risk taking, the primitive working conditions, and the continuing efforts to find further economies. The result was an exceptionally tense environment. Wilson deployed his powerful rhetoric to remind his employees of their "higher obligation" and their shared pioneering mission.[70] The rapid progress made over the next year and a half proved the power of Wilson's strategies.

In May 1970, just as magnet assembly began on-site, Wilson announced yet another goal which the group scurried to meet in the next four months. By October 1, the group was to have one-sixth of the Main Ring ready to handle an injected beam. By August, 200 magnets and most of the water piping and magnet power system had been installed in the tunnel. The milestone was reached by the end of September 1970.[71]

Wilson had thus far encouraged individuals to have strong loyalty to their own groups. Now with the whole accelerator almost ready to function, he changed the focus to coordination among groups, with the operation of the whole ring as everyone's goal. The September monthly progress report announced the goal, dubbed "Oktoberfest," of "accelerating protons to 139 MeV in the Linac and transporting them through half the Booster and into the Main Accelerator."[72] This announcement annoyed some of the staff, because the scheme caused a number of artificial technical problems. The components were not designed for such a low energy. When the long string of Main Ring magnets was powered by a car battery, remnant fields dominated at that low field.[73]

The Oktoberfest was not a complete technical success. Although the 139 MeV beam from the Linac and Booster reached the Main Ring on October 9, the proton intensity was too low to perform beam tests. Some judged the Oktoberfest little more than a stunt, but many noted that the exercise had caused a large amount of equipment to be installed in record time. Moreover, the entire NAL staff had been forced to work as a team for the first time since the beginning of construction.[74]

The Main Ring group spent the next five months working feverishly to complete installation. By the following March, in 1971, eight hundred magnets had been installed. The water and power bus systems were complete at the tunnel level, although at this point the water system was not yet working. In April, the last magnet was installed.[75] The group enjoyed a pleasant surprise when it found that advances in the technology of thyristors, used as rectifiers in the magnet power supply, coupled with unexpected savings in the cost of transformers, allowed the laboratory to install a power supply that would in time prove capable of achieving 500 GeV for less money than had previously been estimated for 200 GeV.[76]

The trials the group suffered during this period ranged from mere inconveniences to brief, but severe, crises. For example, meeting notes from the installation of the Protomain reported "one black eye and several sore heads" resulting from the repair of water leaks that unexpectedly erupted from the power- and water-manifolding system after the magnets were in place. More distressing was the need to wade through water and mud in the Main Ring tunnel to install magnets in the fall of 1970. Installation, aided by portable dehumidifiers, continued nevertheless, and no time was lost.[77]

Another crisis born of haste came in late December 1970—six months before Wilson's accelerated target date for installing the Main Ring. Malamud found that the plaster being used to fill an empty space in the magnets was "sopping wet." Yamada's tapered pole tip had left an empty space because the original design had been square. Magnet production halted abruptly. The solution to this problem offers another example of the importance of engineering skill in the Main Ring effort. Hanson, the engineer, suggested that the void be filled with epoxy by the technique of vacuum impregnation, which had been used successfully for the Booster magnets. Wilson opposed the change, claiming that the use of epoxy was undesirable because of the increased danger of radiation damage. Hinterberger and Sheldon agreed with Wilson. At Malamud's suggestion, however, Hanson was given two weeks, which included the Christmas holidays, to show that his scheme would work. Laboring day and night, Hanson set up a vacuum impregnation system that was capable of accommodating the magnets. In the end, Wilson was convinced, and magnet construction was altered to safely include vacuum impregnation.[78] Despite these setbacks, Wilson reported that he was quite "confident" about the laboratory's prospects as his mid-1971 deadline approached.[79]

The Main Ring Magnet Crisis of 1971

With the many risks that had been taken, NAL and URA leaders expected some aspects of the accelerator to fail. But they were not prepared for the traumatic technical disaster that occurred during the summer and fall of 1971. The colossal failure would cast doubt on the value of Wilson's building style and dash his dream of completing the world's most powerful accelerator a year ahead of schedule.

The Main Ring magnets had been installed in the midst of severe winter weather. In the spring of 1971, water condensed on the magnets when the ventilating system brought in warm humid air. As a result, almost a quart of water needed to be removed from *each* of the wet magnets. That May, when the NAL staff tried to bring the Main Ring magnets into operation under these conditions, many shorted out. By summer 1971, magnets were failing at an alarming rate and the staff was not sure where the trouble was coming from.[80] Wilson called on Boyce McDaniel, his friend and former colleague from Cornell, to come to his rescue and help solve the serious magnet problems in the Main Ring. He also asked "Mac" to manage the completion of the commissioning of NAL's accelerator complex. Helen Edwards recalled the spring and summer of 1971 as a "grim time." The disadvantages of Wilson's style of accelerator building were becoming apparent to many of those involved with NAL. The problems seemed to support the earlier criticism of Wilson's building style by Berkeley and Brookhaven. Wilson and Goldwasser heard rumors of a campaign to convince URA to oust Wilson.[81]

The Main Ring's magnet crisis caused considerable trouble for outside users. Some had arranged sabbatical leaves to prepare experimental equipment based on Wilson's optimistic projections. CERN physicists were quick to conclude that NAL's magnet problems revealed the superiority of CERN's emphasis on technological perfection. Lederman remembers CERN officials gloating over Wilson's misfortunes. Other difficulties surfaced as well. In July, Malamud identified electrical and mechanical problems which were likely to interfere with the NAL accelerator's performance; they included misaligned magnets, malfunctioning ion pumps, a piece of copper in the beam pipe, and a plastic cap found in a quadrupole. Also, the Linac and Booster were running "with an efficiency of about 50%." Although this figure was considered "reasonable for such new accelerators," the Linac and Booster problems complicated the Main Ring's development because beam studies could be performed only when the first two accelerators were operating well. Problems arose due to poor power supply functioning and because construction debris cluttered the Main Ring vacuum tube. To examine this debris problem in the Meson Lab, the NAL staff tried to train a ferret, affectionately named Felicia, to drag a cleaning cloth through roughly 300 feet of vacuum tube, but this strategy proved futile (fig. 6.3). Frustrated by the length of her new tunnel, Felicia accepted an early retirement. Eventually Hans Kautzky developed a robotic mechanical spear able to pull a magnetic cord through the 2,650 feet of vacuum tube.[82]

In an attempt to rally his staff in this time of serious problems, Wilson devised a new management plan that emphasized flexibility and lack of hierarchy, the very organizational concepts he had celebrated earlier when describing his vision of the laboratory. He merged the groups responsible for accelerator theory, operations, radio frequency, and beam transfer to form an Accelerator Section, headed by himself. Next he appointed three strong managers, Orr, Richard Lundy, and Livdahl, who would ultimately make possible the commissioning of the Main Ring.

6.3 Felicia the ferret, attempting to solve technical problems in the Meson Lab, 1971. (Courtesy of FNAL Visual Media Services.)

In parallel, he assigned each component of the accelerator to one of the "commissioners." The commissioners (among them Malamud, who served as the "senior" Main Ring commissioner, Helen Edwards, Yamada, Shigeki Mori, and Dave Sutter) were, as Wilson explained at the time, "expected to identify work problems and to come to one of the Managers to get the work force . . . to do the actual work." Orr later noted that the work assignments focused the efforts of technical experts on solving technical problems, leaving organizational decisions to those with managerial skills. Wilson met daily with the commissioners and the three managers, as well as with other key leaders, so that the effort was tightly coordinated. To further expedite problem solving, he reassigned workers from other groups, such as those working in the experimental areas, to increase the pool of those working on Main Ring problems.[83] During the crisis, approximately 350 magnets out of 1,014 failed, causing six months of "lost" time, and about 10% (approximately $2 million) of the original cost of the magnets was spent to overcome the difficulties.[84]

More than three decades later, experts still do not agree on what caused the Main Ring magnets to fail, or even on the relative severity of this problem. The standard explanation for the magnet failure blames the epoxy insulation, which had been made very thin to decrease the possibility of radiation damage. Veteran accelerator builder Collins blamed the shorting on the decision to glue in the coils, stating that the considerable thermal and mechanical stress resulting from temperature cycling cracked the epoxy in the glued-in coils, allowing water to get in the cracks.[85] He felt that the B-1 magnets were too small, causing "field quality and systematic errors." Jeff Appel noted that the plaster of paris, which absorbed the water, had to be removed and the magnets rebuilt without it. Jovanovic suspected that the power supply was the major source of difficulty, because the new method of using thyristors introduced short, high-voltage spikes on the line. Other early pioneers of the Main Ring point out poor mechanical engineering. Despite the best efforts to keep good records, in many cases the relevant documentation has vanished, making it impossible to offer a complete explanation of the problems that plagued the Main Ring in 1971.[86] In any case, the problems began to abate once the three managers focused the entire laboratory on solving them. In time much of the Main Ring was filled with reconditioned or newly built magnets with a lower failure rate. Some magnets continued to fail (even decades later), but workers learned how to replace them quickly, a job facilitated by a clever pipe-cutting scheme invented by Walter Pelczarski.

Critics of Wilson's approach point to the high costs of his risky approach to building accelerators. CERN's history provides a telling contrast to NAL's. In line with CERN's greater emphasis on reliability, researchers of the European laboratory were inclined to proceed more cautiously when confronted with accelerator problems. This approach, these critics say, allowed a more methodical assessment better suited for solving complex technical problems. Yet as Ramsey later reflected, Wilson's approach succeeded in building the Main Ring in a time of budgetary crunch. And even though many of the riskier elements failed, many more succeeded. Much more money was saved on the whole project than was lost on all the failed elements.[87]

The Triumph of Frugality: Setting the Indoor Proton Speed Record, January–March 1972

The NAL staff made steady, if sometimes painstaking, progress through the rest of the winter of 1972. A short memorandum by Jovanovic reveals the difficulties and frustrations faced as well as the uplifting role Wilson played during this challenging period. Jovanovic's operations report of January 21, 1972, includes the following comments: "8:30 Trouble, troubles. Beam doesn't go all the way out." A bit later: "11:50 Note—due to the phone being installed in the toilets, you can't hear them ring in the service building. Therefore we have lost our most reliable communication system. Suggest installing extension phone bell outside of toilet."[88]

The next day Jovanovic asked a question: "How was the gloom and doom of Jan 21 transformed into success only 24 hours later?" He answered with a description: "On Jan 21 at 22:30 RRW walked into a dimly lit control room. Few people were about while the RF group was changing a booster RF station. Out he produced a small book and read to us in French." As Jovanovic remembered, the passage Wilson read came from the medieval "Song of Roland," a *chanson de geste* (epic ballad) whose original intent had been to inspire the armies of Charlemagne. Jovanovic noted that the "ancient French modified by a timbre from Wyoming echoed in the central room. It was strange . . . not understanding the verse but understanding the occasion." In a memorandum written in 1977, Jovanovic added: "Only now, five years later, did the significance of this event become obvious."[89]

On the next day, January 22, 1972, a 20 GeV beam seemed stable from pulse to pulse. The energy of the beam continued to rise. Orr remembered someone commenting that Wilson's efforts in early 1972 were aimed at "trying to set the indoor proton speed record."[90] On February 11, a 100 GeV beam was reached; 200 GeV was achieved on the afternoon of March 1, 1972, surpassing the 76 GeV record at the Serpukhov machine in the USSR and recapturing the lead for the United States (figs. 6.4, 6.5). Photographs were taken, champagne was uncorked, and over one hundred staff members assembled to witness the milestone event and sign the logbook declaring the achievement. Wilson triumphantly announced to the AEC and JCAE that the project had come in ahead of schedule and under budget.[91]

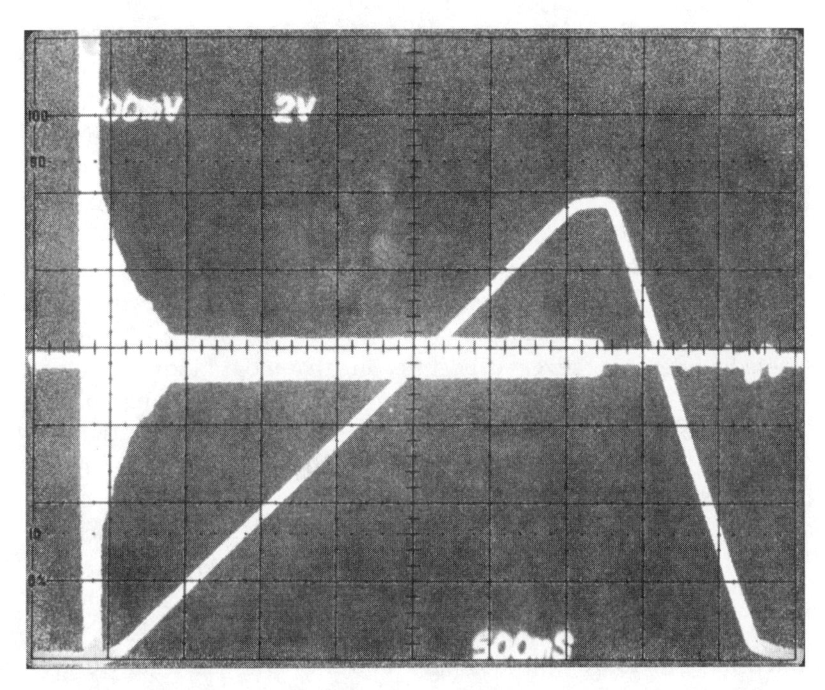

6.4 Oscilloscope display indicating achievement of design energy, 200 GeV, March 1, 1972. (Courtesy of FNAL Visual Media Services.)

6.5 Robert R. Wilson toasting his staff at the celebration of the achievement of 200 GeV. (Courtesy of FNAL Visual Media Services.)

Wilson's announcement was well received in Washington, especially since by 1972 at least one other large federally funded project had fallen on hard times. That year planners of what would become the Hubble Space Telescope had to cut their project from $700 million to around $300 million to produce "a cheaper, and thus politically more feasible, telescope." To meet this goal, the Hubble team had to employ many of the budget-cutting devices that Wilson used, for example, stimulating competition between contractors to lower costs, skipping the prototype phase, designing cost-effective components, and transferring costs from the design stage to the operation stage. Nonetheless, in June 1974 Congress denied funding for the telescope, throwing that project into jeopardy.[92]

In contrast, not only was NAL alive and well in early 1972, but the laboratory and its director had become the toast of Washington. JCAE members joked about considering Wilson for sainthood. In 1972 JCAE member Hosmer reassured Wilson that magnet troubles and delays didn't tarnish "a bit the brilliance with which this whole effort was conceived and constructed."[93] By 1974, four experimental areas (proton, neutrino, meson, and internal target) were in operation. A full-scale research program was underway by 1975, with routine operation at 400 GeV, a beam intensity of 1.84×10^{13} protons/pulse, and only 28% unscheduled downtime.

On May 11, 1974, NAL was formally dedicated as Fermi National Accelerator Laboratory, in line with a 1969 AEC decision to honor Enrico Fermi.[94] Stormy weather that day dissipated only moments before the scheduled ceremony on the steps of the newly constructed central laboratory. Government officials (including Chairman Dixy Lee Ray of the AEC, Senator Charles Percy, and Congressman Melvin Price) gathered with URA officials as well as physicists from many countries (figs. 6.6, 6.7). Laura Fermi, the widow of Enrico Fermi, also attended. Over the gusts of wind, Leon Lederman gave the dedication address, speaking on behalf of the NAL User's Organization and inspiring those in attendance with his confident words about "this marvel of engineering and organization." Explaining that every new accelerator "has yielded discoveries totally undreamed and unmentioned in the plans and in the designs," he said that this machine "cannot fail to maintain the tradition." At the end of his remarks, Lederman played a 1952 tape recording of Enrico Fermi speaking to "constructors of high energy accelerating machines" about the challenges of understanding the cosmos.[95]

Few would characterize Fermilab's original Main Ring (fig. 6.8), in which the first 500 GeV beam finally coursed in May 1976, as an elegant instrument, for its minimal magnets had relatively poor field quality

6.6 Dedication of FNAL on the steps of the central laboratory building, with special guests (*left to right*) Leon M. Lederman (*second from left*), AEC chairman Dixy Lee Ray, Representative Craig Hosmer, Norman Ramsey, Representative Melvin Price, and Laura Fermi to the left of the podium, May 11, 1974. (Courtesy of FNAL Visual Media Services.)

6.7 Robert Bacher, Senator Charles Percy, Mrs. Laura Fermi, and Edwin L. Goldwasser at the 1974 dedication festivities. Norman F. Ramsey is in the background. (Courtesy of FNAL Visual Media Services.)

6.8 Interior of the Main Ring, 1972. (Courtesy of FNAL Visual Media Services.)

and periodically needed to be replaced. Interruptions in the operation of the machine introduced uncertainty in scheduling experiments. And the accelerator cost more to run than it would have cost had the initial investment in copper been greater. Escalating electricity costs during the energy crisis of the 1970s made this disadvantage even greater. Despite these drawbacks, the Main Ring, with its Booster and Linac, did come to function as a world-class accelerator at almost twice its projected energy. The laboratory's next challenge would be to mount a successful experimental program within the scope of Wilson's frugal vision.

SEVEN

A Users' Paradise, 1968–1978

The main application of the work here is spiritual, if you will. It's because, in a philosophical sense, in the tradition of Democritus, we feel we have to understand, in simplest terms, what matter is, in order to understand who we are.
ROBERT R. WILSON[1]

Wilson believed in the need for "a kind of balance" between the laboratory's staff and its users. This balance included division of the responsibilities and investments by each in the work of the laboratory. He felt that the laboratory's staff should provide the accelerator and a "hospitable physics environment"; the universities should supply most of the resources for mounting and conducting experiments. As the users "would do most of their work at universities,"[2] Wilson felt justified offering them minimal facilities for their experiments. Back in 1965, when Wilson wrote his letter to McMillan criticizing the Berkeley design (chapter 4), he argued that a laboratory was better off providing minimal facilities and experimental equipment, because "providing expensive facilities . . . may tend to paralyze better developments later on." He favored leaving "plenty of space" and building flexibility into the facility, for he believed that "techniques and research interests change so rapidly that it is better to let the major part of these areas grow out of the actual ideas and demands of the experimenters at the time they use the machine."[3] He explained his attitude in a more light-hearted tone three decades later: "In a frenzy of saving big bucks, I had a fantasy of not putting up (or down) any laboratory building at all. Instead the idea was that once an experiment had been accepted, an outline of the necessary space would be drawn in an empty field at the end of one of the proton beams, then steel interlocking piles would be driven along the outline down to the necessary depth to protect against radiation. Then the experimental equipment would be lowered to a luxurious graveled floor, and finally a removable steel roof would be covered with the requisite thickness of earth. Once the experiment was finished, the pilings were to be pulled up, the earth filled in, and then the next experiment would be ready to receive its tailor-made enclosure."[4]

Applying this minimalist philosophy regarding experimental facilities at NAL meant that most of the experimental areas would be, as Wilson described them, "rough and ready places." Cold and damp from the Midwestern climate, or flawed in other ways, they "had the reputation of being, not Paradisios, but rather Purgatorios."[5] The consequences would breed widespread criticism from experimenters who were frustrated by having to conduct smaller or simpler experiments than they felt were necessary to compete successfully with other laboratories. Contrasting NAL's experimental areas with those at SLAC, the anthropologist Sharon Traweek noted that SLAC workplaces lacked decoration and had a "clean, functional grey metal and glass decor" which created "a strong impression of stoic denial of individualism and great preoccupation with the urgent task at hand."[6] Nevertheless, the physicists there worked in finished, modern buildings. Only after Wilson stepped down did all the Fermilab experimental areas even have floors.

In the balance of responsibilities that Wilson envisioned between users and staff, the university physicists were responsible for educating students, building apparatus, and serving on committees at their universities. The laboratory's scientific staff was expected to spend roughly half its time in service work, which included operating and improving the facilities.[7] While this division of the research at large federally supported facilities would later be taken for granted, it was unprecedented in federally supported research laboratories in the 1970s. In some research groups at Brookhaven, Berkeley, and SLAC, senior-level researchers could hold 100% research positions.

The details of Wilson's balance between users and staff were presented in NAL's design report in 1968, which emphasized that "a reasonable sharing" between staff and visiting scientists would allow the resident staff to "conduct about twenty-five percent of the laboratory's total program." As it turned out, this level was rarely achieved, given the privileged status of outside users at NAL.[8] The in-house physicists sometimes joked about the laboratory's policy, for the laboratory was so short-staffed that its "half-time" service jobs typically required a full-time effort. Employees muttered that they were allowed to spend their second forty hours per week on experiments. Experimental physicist Joseph Lach noted that NAL experimenters who took the time to work on experiments tended "to get punished in salary reviews."[9] Moreover, since the early NAL staff was grouped according to service responsibilities, the laboratory lacked an administrative framework on which a concerted in-house research effort could be built. Staff members were encouraged to collaborate with visiting users, rather than with each other. As Wilson later commented, Fermilab "only hired physicists to grease the wheels for outside users."[10] The situation that users faced at NAL was thus very different from the one Lederman had criticized in his 1963 TNL paper (see chapter 3).

Wilson saw himself primarily as an experimenter and firmly believed, according to Goldwasser, that "the people who built accelerators should be the people who wanted to use them."[11] It was, therefore, natural for him to build accelerators. But by the time he became the director of NAL, accelerators had become so complex that those who built and those who used accelerators were members of separate communities. He identified more with the builders.

While Wilson focused on leading NAL's accelerator developments, thus "providing for the future," Goldwasser took responsibility for the experimental program. "I was really guarding the on-going program," he recalled.[12] Important parts of the work were supporting the outside users by planning their organization and encouraging the universities to invest in equipment. Initially, as he soon realized, Wilson "was quite resistant" to the idea of forming the users' group because there was no machine to use for several years. But Wilson eventually agreed that a users' group "was the politic and probably the essential thing to do" in establishing the partnership he envisioned between universities and the laboratory.[13] Ramsey announced the plan for such a users' group in his letter inviting participants to Wilson's April 1967 planning meeting at Argonne National Laboratory. The next users' meeting was held on December 9, 1967.[14]

Goldwasser was central as well to the organization of annual summer studies in which a select group of users would offer advice on such facilities as experimental areas and detectors. NAL had only a small staff of physicists, and they were preoccupied with building, and later maintaining and improving, the accelerator. Because university physicists were unlikely to abandon their professorships and work for NAL during the academic year, or to use their summer break to do so in the muggy Illinois weather, Wilson, Goldwasser, and Ramsey looked into creating a summer program in a more inviting location. They arranged a joint NAL–Aspen Institute program with workshops in the Rocky Mountains. At first Wilson and Goldwasser were concerned that their plan of meeting in the picturesque town of Aspen, Colorado, a popular ski resort, might strike the AEC as extravagant. But they discovered that during the summer season housing was actually cheaper in Aspen than in suburban Illinois. Goldwasser thus had no difficulty explaining the rationale for meeting in Aspen to Seaborg. With the incentive of Aspen's beautiful location, NAL could recruit large numbers of eminent participants to NAL's summer studies, despite the fact that most were required to pay for their lodging and meals from their own research budgets.[15] The advice offered by those who attended the summer studies held in 1968 and 1969 was incorporated into the planning of NAL's experimental program and facilities.[16]

James R. Sanford, who from 1969 to 1971 served as the head of NAL's Experimental Facilities Section, assisted Goldwasser in developing the research program. The former Brookhaven physicist later became NAL's associate director for program planning and the editor of *NALREP*, the publication that succeeded NAL's early "Monthly Report of Activities."[17] Goldwasser and Sanford spearheaded plans for the laboratory's Physics Advisory Committee (PAC), later renamed the Program Advisory Committee, which was organized by December 1969, several months before the first call for proposals in March 1970. The group of external advisers who constituted the PAC reviewed the experiment proposals submitted to the laboratory and provided Wilson with general advice about the experimental program. As with the users' group, Goldwasser persuaded Wilson "that he needed that kind of committee" to support the laboratory's policy of treating all users equally. In this case too, Wilson soon agreed that such a committee would help rally the support of users.[18]

For the initial nine-member PAC, Goldwasser recruited a distinguished group that included three Nobel laureates (Owen Chamberlain, Murray Gell-Mann, and Tsung-Dao Lee) and one future one (Val Fitch). Goldwasser acted as the primary liaison between the committee and the users. He would usually offer Wilson a summary of the arguments presented by experimenters and the PAC, supplemented by his own opinions.[19] Almost ninety proposals were received within six months after Wilson sent out the first request for proposals to the NAL users on March 26, 1970.[20] The PAC met in Aspen in early August 1970 to review the proposals.

A 1975 survey showed 184 approved experiments; they would be conducted by researchers from 127 research centers in the United States, Canada, Europe, and Asia. There was "a lot of pressure to approve a large number" of minimally funded experiments, reflected early PAC member and experimental physicist Thomas Kirk.[21] By 1976, half of the 500 experiments proposed since 1970 had been approved. By 1978, 250 of the 300 approved proposals had been completed.[22] Fermilab researchers were far from predominant in these early experiments. Only two simple and short-lived experiments, a monopole search and a beam dump study, were performed by the in-house staff alone. The experimental program dominated by outside users offered a blueprint for the power relations existing in the laboratory for the next several years.

The Experimental Areas

The minimal experimental facilities at Wilson's laboratory were arranged in four basic experimental areas: the Internal Target Area, known as CZero (at the CZero straight section of the Main Ring), and three external areas fed by beamlines: the Neutrino Area, the Meson Area, and the Proton Area (fig. 7.1). The facilities included a 30-inch bubble chamber, which had been built by MURA for Argonne (one example of many such collaborations with Argonne in these early years), the laboratory's 15-foot bubble chamber with its superconducting magnet, and a small computing center.[23] In addition, the University of Chicago's 1950s-era cyclotron magnet was recycled for use in the Muon Laboratory of the Neutrino Area.

Goldwasser and Sanford had proposed the 15-foot bubble chamber to encourage university investment in the experimental program; they argued that in the long run the bubble chamber physicists would return to their home institutions to perform their data analysis and thus primarily use university rather than laboratory resources.[24] Goldwasser later admitted that neither he nor Wilson had been convinced that a large bubble chamber "was likely to be a competitive detector for exploring the physics of the next generation," but the physics community included a "rather large core of physicists who had been engaged in bubble chamber physics and who wished to embark on Fermilab physics using their familiar bubble chamber techniques." This group of "bubblers" included both physicists, such as William B. Fowler, and engineers, such as George Mulholland and Hans Kautzky, who "were politically strong because they were numerically strong. . . . We couldn't afford to ignore them,"

7.1 The Fermilab experimental areas, with Fixed Target areas (*A*, the Meson Lab; *B*, the Neutrino Area; *C*, the Proton Area) fanning out left to right in tangent to the Main Ring, 1977. (Courtesy of FNAL Visual Media Services.)

Goldwasser explained.[25] Goldwasser had recruited Fowler from Brookhaven in 1970 to design and build the 15-foot bubble chamber after the AEC turned down the laboratory's 1969 proposal for a 25-foot chamber to be built in collaboration with Brookhaven.[26] The chamber operated at the laboratory from 1973 until it was finally decommissioned in 1988, having taken three million photographs for research physicists from thirty-seven institutions (fig. 7.2). A close-down party for experiment 632 (E-632) on April 8, 1988, marked the event. By then the physicists had used the device to publish more than one hundred technical papers.[27]

The Internal Target Area was home to NAL's first experiment, E-36 (see the appendix), a U.S.-USSR collaboration that studied proton-proton scattering (fig. 7.3). With the help of an innovative real-time telex communication system (an early form of E-mail) the scientists at NAL were connected with those at the Joint Institute for Nuclear Research (JINR) in Dubna (near Moscow), Russia. The collaboration was unique in being established at the height of the Cold War.

The Meson Area, constructed from 1971 until 1973, supplied six secondary beams which supported a variety of experiments, such as quark and monopole searches, total cross-section measurements, emulsion experiments, and elastic and inelastic scattering studies. Wilson himself designed the distinctive architectural feature of the Meson Lab, its blue and orange roof made of huge, inverted half sections of culvert pipe (which, unfortunately, leaked, making the conditions in the Meson Laboratory rather miserable).

The Neutrino Area, the first external experimental region to be developed, supplied neutrino and muon beams for many experiments during the Wilson years that explored neutrino interactions, cosmic rays, emulsion scattering, scaling, charm, and neutral currents (see the appendix).[28] A striking geodesic dome covering Lab A in the Neutrino Area was designed by DUSAF model builder Jose Poces and Angela Gonzales. It was built with materials suggested by engineer Robert Sheldon. For the dome's roof, Sheldon proposed sandwiching metal cans, collected and donated by neighboring communities, between thin sheets of multicolored plastic in triangular shapes.[29] The panels were assembled during lulls while waiting for supplies. To repair subsequent leaks the panels were clad in copper in 1982.

The Proton Area consisted of three smaller beamlines (P-West, P-Center, and P-East) where experiments with protons, hyperons, photons, mesons, and electrons were conducted starting in 1975 (fig. 7.4). The so-called Proton Pagoda, the architectural highlight of the Proton Area, built in 1976, was based on a design for the central laboratory building

7.2 Scanner looking for unusual particle interactions in a bubble chamber experiment photograph, 1978. (Courtesy of FNAL Visual Media Services.)

7.3 In the Internal Target Area at CZero, collaborators on E-36, the first NAL experiment, including physicists from NAL, Rockefeller, Rochester, and JINR in the USSR, 1972. (Courtesy of FNAL Visual Media Services.)

that was rejected. This Pagoda housed the area's control room and featured a distinctive double-helix staircase.

Alternating Currents (the Neutrino Area)

A major study in the Neutrino Area during the early days of NAL was concerned with finding evidence for neutral (as opposed to charged) currents. These currents were the focus of experiment-1A (also called HPWF, as it included Harvard, Pennsylvania, Wisconsin, and Fermilab), one of the first experiments in the NAL program.[30] The experimental basis remained shaky, and given the lack of evidence, many physicists assumed that neutral currents did not exist, despite the conclusions to the contrary in many theoretical papers written between 1968 and 1971 on neutral currents or the electroweak gauge theory (e.g., by Steven Weinberg [1967], Abdus Salam [1968], Sheldon Glashow, John Iliopoulos, and Luciano Maiani [1970], and Martinus Veltman and Gerard 't

7.4 Users of the FNAL Proton Area, including Wonyong Lee (*front row, left*), John Peoples (*back row, second from left*), Leon Lederman (*front row, second from left*), Lincoln Read (*back row, with hand raised*), Jim Sanford (*second from right*), with Brad Cox (*front row, third from right*), 1975. (Courtesy of FNAL Visual Media Services.)

Hooft [1971]). In 1970 Andre Lagarrigue and his collaborators, who had been searching for neutral currents since 1963 at CERN, assembled a freon-filled heavy liquid bubble chamber called Gargamelle, which ran using the 20 GeV PS. For the next two years, CERN performed the laborious task of scanning and analyzing 100,000 Gargamelle bubble chamber pictures. After presenting their results at the Rochester Conference held at Fermilab in 1972, the collaborators continued to scrutinize their data knowing that E-1A (HPWF) was closing in on them.

The Fermilab group used a different method, in which a 20 GeV wide-band neutrino beam impinged on various detector systems (a liquid scintillator and hadron calorimeter and a muon spectrometer with large spark chambers). Having access to much higher energy particles than Gargamelle, E-1A obtained more interactions. In the course of planning the publication of E-1A's first results, which included nearly one hundred events, Carlo Rubbia, as a spokesman for E-1A, wrote to Lagarrigue about the status of Gargamelle's work on neutral currents. He asked Lagarrigue for reciprocal referencing. In poker parlance, he "called his hand." But rather than "fold," Lagarrigue called Rubbia's "bluff." On the next day, July 19, 1973, Lagarrigue sent Gargamelle's results to be published. They were announced at the Bonn Conference in August 1973, where E-1A's results, which were close to Gargamelle's, were also made known. For a more complete discussion of the neutral-currents work at NAL and CERN we refer the reader to the excellent historical treatments by Peter Galison and Donald Perkins.[31]

In the meantime, E-1A encountered pitfalls. A borrowed trigger did not function as hoped. Although they had already distributed their results as a preprint in late July 1973, the experimenters mistrusted their results enough to hold back publication while they rearranged their detectors and took a second look. As Galison notes, there were many justifiable reasons for the E-1A group to hesitate. E-1A's second run had problems with a shield of insufficient strength, which caused their data to be misread. Moreover, the physics was not yet understood. The results from E-1A's newly adjusted detector cast so much doubt on its earlier findings that by November 1973, the group prepared a letter to *Physical Review Letters* stating its tentative conclusion that there were no neutral currents. Following Wilson's advice, the group did not submit the letter, but Rubbia delivered the draft letter to Lagarrigue, causing even more anxiety for everyone involved. In the face of much criticism, Gargamelle's experimenters trusted their experimental work and their analysis and held firm; they were first to claim the discovery. By mid-December 1973, the fog over HPWF lifted and the evidence for neutral currents was clear. When E-1A's paper was finally published in the spring of 1974, it merely confirmed Gargamelle's discovery. The legend of "alternating" neutral currents would be told far and wide.

Although the E-1A neutral-currents experiment could not count as a victory for NAL, it did demonstrate the fledgling laboratory's ability to compete with the older, more established CERN. The discovery of charm at Brookhaven and SLAC in 1974 explained the absence of neutral currents, encouraging confidence in the evolving foundations of the Standard Model and making clear that electroweak theories had to be taken seriously.

Finding Beauty in the Bottom of the Bog (the Proton Area)

Four years after missing the discovery of neutral currents in the Neutrino Area, the laboratory found success on another front. Researchers in the Proton Area discovered the "bottom or beauty quark," one of the six quarks of the Standard Model. This was a major find. Every experiment has its story worth telling in detail. But to include all of Fermilab's experiment stories in a single book would form an unwieldy product. As the discovery of the bottom quark is considered to be Fermilab's crowning achievement during Wilson's time as director, we will focus on this case in our effort to project the character of research in this period. Our account of this project illustrates some of the limitations of experimental work during the laboratory's early years and highlights what was needed to make a major discovery in that context.

The experiment's spokesman, Columbia physicist Leon Lederman, personified Wilson's model researcher in many ways. A clever experimenter driven to explore the highest energies, Lederman was willing to make the best of the laboratory's primitive conditions and limited resources. His group "didn't bargain on the frogs, or the ditches, or the roof leaking," but as his team made its bold grab for discovery, its members accepted Wilson's minimal support of experiments as part of "the pioneering spirit of doing experiments at early Fermilab."[32]

Lederman was among those who put forth an experiment proposal at the first NAL summer study in Aspen in 1968. Building on research that he was then conducting at Brookhaven, he proposed a search of the particle debris left after protons struck a beam dump (a block of matter thick enough to absorb all the incident protons). He planned to look first for single leptons (electrons and muons) emitted at large angles to the incident proton beam and then for pairs of leptons. The interest in this search came from the fact that the weakly interacting particles (such as muons), created during collisions in the dump could signal the occurrence of rare processes. The group reasoned that when an energetic proton collides with a nucleon in a target, the collisions create various particles (e.g., pions and kaons) and occasionally virtual photons that decay almost immediately into pairs of leptons. Lederman hoped that by studying lepton pairs his group might find a new particle. The emission of virtual photons (energy bundles that decay immediately into particles and antiparticles) could indicate "unexplored domains" inside particles.[33]

The same basic idea had guided Lederman's recent study of muon pairs at Brookhaven's AGS. The data there had revealed some ambiguous evidence for a "ledge," known as "Leon's Shoulder," in the plotted data around 3 GeV. This ledge in an otherwise smooth distribution of muon pairs seemed to indicate a new and very massive particle. The smooth distribution appeared to confirm Richard Feynman's parton model (a version of the later quark theory). The production of lepton pairs promised to probe hadron structure and spurred a series of so-called Drell-Yan experiments, based on Drell and Yan's 1969 formulation for the process of production of virtual photons when a quark and antiquark in colliding particles are annihilated. These experiments were of wide interest because they promised to test the new theory of quantum chromodynamics (QCD) and provide basic information about hadron structure.[34]

Aiming to extend this search to the higher particle masses that higher energy allowed, Lederman proposed two experiments to study the leptons that emerged from collisions, both singly and in pairs. He submitted one proposal to CERN and the other to NAL. In the CERN experiment, to be conducted at the ISR, by measuring the high-energy continuum spectra, his team hoped to detect resonances in the mass range up to about 28 GeV, including the exotic heavy photon predicted by Lederman's Columbia colleagues T. D. Lee and Gian Carlo Wick.[35]

The proposed NAL experiment, P-70, had two parts. The first planned to study neutral pion and photon production and search for the W particle by detecting single electrons or muons of high transverse momentum. The second part would focus on pairs of leptons—first, electron pairs, because electrons can be detected directly as they emerge from the collision point and provide the best determination of the mass of any observed resonance, while muons would suffer scattering as they were filtered through meters of beryllium absorber material. In the words of experimenter Hans Jöstlein, who later joined Lederman's group, although such an experiment did not have "very good odds . . . it was clear if they did find something it would be worth the trouble." They were "playing [the] lottery for a large jackpot."[36]

P-70 was just the kind of imaginative and daring experiment that Wilson had in mind when he envisioned the new laboratory. Most of the resources needed were simple and would come from outside the laboratory. The proposed detector would be composed of four subdetector systems: magnets, scintillation counters, lead glass Cerenkov counters, and hadron calorimeters. The first part of the experiment would employ a detector with one arm of detection apparatus; the second part, focusing on lepton pairs, would need two arms. To allow for a beam survey, they began to refer to the three phases of the planned experiment: Phase I, the beam survey; Phase II, the single-arm experiment; and Phase III, the double-arm experiment. The proposal was approved and became known as experiment E-70.

As was typical at NAL, E-70 included both laboratory and visiting physicists. From NAL were Lincoln Read, John Sculli, Tom White, and Taiji Yamanouchi. Besides Lederman, those from outside the laboratory included Wonyong Lee, Jeff Appel, and David Saxon from Columbia and Mike Tannenbaum from Harvard. In line with Wilson's wish that other institutions shoulder the lion's share of the experimental apparatus expense, a December 1970 addendum to the proposal pledged that the entire apparatus (with the exception of some magnet construction) would be provided by Columbia University's Nevis Laboratory, where Lederman served as director.[37]

Lederman's group economized in setting up its experiment during 1971 and 1972, while Lederman did what was necessary to place his team first in line for beam. These strategies put him in a favorable position, although sometimes this advantage came at a cost to others. For example, in procuring lead glass for the experiment, Lederman undertook an extensive search for a producer. Determined to "shave" the final cost "by another 5–10%," Lederman spoke with four prospective sellers and engaged in competitive bidding with them for the exotic item. He admitted that the winner "must have lost its shirt."[38]

Lederman had another effective way to cut costs. Even in the physics community, where reuse of equipment was the rule, Lederman was known for being "a great believer in borrowing." In the words of Charles "Chuck" Brown, who later joined the Lederman group, "The apparatus we put together included borrowed pieces from all over the place."[39] The negotiations through mid-1971 included failed arrangements for borrowing magnets, which were among the largest and most expensive pieces of equipment for the experiment, from Brookhaven's AGS and from the defunct Cambridge Electron Accelerator. Sometimes Lederman's borrowing attempts fell through.

One of Lederman's most important strategies in coping with the limited resources at NAL was to take advantage of resources outside—for example, those available at Nevis. Besides having access to shop equipment and some extra funds, the experiment benefited from the contributions of Yin Au, the Nevis chief mechanical engineer, and William Sippach, a Nevis electronics engineer. These talented experts were assigned to help with Lederman's NAL experiments because, in Appel's opinion, "Lederman was one of the big wheels who got the best." Sippach's talents proved crucial, as Bruce Brown later explained, because the experiment involved very high fluxes of particles; the electronics had to distinguish between individual particles coming at high rates.[40]

In February 1971, while the E-70 team members were planning all three phases of the experiment, Lederman had an inkling that the NAL accelerator might not provide a 200 GeV beam as early as expected. In a February 16 letter to Wilson, Lederman wryly noted "some slight sluggishness in your otherwise well-oiled progress. . . . I can see our caravan of trucks arriving from Nevis, groaning with lead glass, drenched with the sweat of our feverish preparations, only to be prevented from unloading" in their assigned place in the Proton Area "by the masons, stone cutters and other artisans, still emplacing the last gargoyles, weeks behind schedule." The group delivered a signed copy of the formal document specifying the obligations of the experimenters and the laboratory. "All we expect of NAL," Lederman joked, "is 10^{13} 500 GeV protons, a 40-foot× 200-foot enclosure, tons of shielding, money, magnets, engineering assistance, gargoyles, and love."[41] Lederman could not really hope for a beam of such high energy and intensity, but he could have expected the magnets and other planned provisions for the experimental area. Unfortunately NAL could not provide all it had promised when its already minimal allocation of manpower and funding for experimental resources had to be tapped to help solve the devastating magnet crisis that erupted that spring of 1971 (see chapter 6). The result was that E-70's experimental capabilities had to be reduced. The researchers used smaller magnets and a new experimental setup that placed the beam dump close to the target area, avoiding the expense of beam transport.[42]

Lederman responded with good humor and without complaint to the changes made in his experimental setup.[43] But he did not completely abandon self-interest. Noting that the "opportunity of early involvement at NAL [was] too exquisite to forgo," he asked Wilson in a June 7 letter to approve plans for a makeshift beam transport system so the group could quickly do a photon survey with two lead glass blocks and a small collection of other small devices mounted to a 3-foot × 6-foot rolling table. This would mean that "a few of us will infiltrate quietly in July, complete with sleeping bags and sterno."[44] Although the experimenters insisted they were still making pre–phase I measurements, Wilson questioned Lederman's underlying motive.[45] In a reply written a few weeks later, Wilson pointed out that the requested survey appeared "to be, in fact, phase I" of their experiment "moved forward in time and space." Wilson also noted that "the demands upon the laboratory seem to be negligible and the time . . . required for the run . . . extremely short." Wilson responded positively, however, but he emphasized that he would not allow the survey to be further expanded. He responded in the friendly tone that Lederman had also used, concluding: "How could I reject such a proposal?"[46] Thus, in return for accepting Wilson's terms Lederman was among those granted first access to the beam. But even for Lederman, actual beam time lay far in the future, for the crisis of the Main Ring magnets would continue for almost a year. And the 200 GeV beam, first produced in March 1972, had lower intensity than expected and was available during only about 60% of the time scheduled through September 1972.[47]

The arrangements for the experimental area, under the direction of Sanford and Read, proceeded more slowly than planned because of the accelerator problems. According to Orr, one of the three managers during the magnet crisis (see chapter 6), it was by this time apparent that NAL had "too many things going . . . too many ideas . . . too many experiments for [the] budget." The laboratory was understaffed for doing "all the things we were supposed to do, but on the other hand, we didn't have enough money to pay all the people we had, so we were in a triple bind."[48] Lederman's group was scheduled to have beam for Phase I, the beam survey, in September 1971 and for Phase II, the single-arm spectrometer experiment, a few months later. The group actually obtained beam for the first time in late fall 1972, roughly a year later than expected, and they began taking data for Phase II in the spring of 1973.

The 1972 URA annual report characterized the research in this period as "rough and frustrating." The report complained about unrealistic scheduling, the delays in obtaining usable beams, the "primitive nature" of facilities and support, and the fact that the accelerator problems were so severe that "machine development often required a higher priority than even on-going experiments."[49] In reporting on the first beam to Lederman, Appel said that "we get only dribs and drabs!" He noted that the "spill was short—unfortunately," and he was concerned that there was "no NAL beam commitment" for further stages of the experiment. In October and November 1972, at just the time when the group needed beam to calibrate the lead glass and otherwise check and fine tune their detector, Read, one of the NAL physicists on the E-70 proposal, and Peoples, the NAL physicist in charge of the Proton Area, were battling over beam-tuning problems, complicating the group's efforts to extract and steer the beam into the Proton Area. Although in November 1972 these problems were solved, allowing a 24-hour 200 GeV run for Lederman's group, difficulties with the Booster accelerator "scrubbed the rest of the run." As Appel told Lederman, the schedule called for "finally 40 hours for the Proton [Area]," but a series of problems with the Main Ring and Booster prevented the staff from extracting beam "until 36 out of 48 hours was gone." And then, "after (about) 4 hours" consumed by extraction and beamline tuning, "a massive power failure . . . curtailed operations." The brief beam spill that did emerge swamped the counters they hoped to check. In any event, the "beam was not too stable for any length of time, even in the Main Ring." And so the run "was a complete washout."[50]

In December 1972, the accelerator operated at 400 GeV for the first time, but as the monthly progress reports stated, the varying energies and rapid schedule changes "reduced significantly the amount of beam actually available for physics experiments." Laboratory employees worked on improving beam transport to the Proton Area, where a number of experiments were being set up. Besides the Lederman group in P-Center, three other groups were setting up in P-East: E-100, a particle production measurement to search for particles led initially by James Cronin and later by Pierre Piroue; E-87, a photoproduction search for heavy leptons led by Wonyong Lee and later Thomas O'Halloran; and E-25, a photon total cross section experiment led by David Caldwell. In P-West, Jimmy Walker was working on E-284, a survey of particle production in proton collisions, and Mel Shochet was working on E-258 measuring particles produced at high transverse momentum by pions. The experimenters in the Proton Area received no beam that month, and laboratory attention shifted to providing beam in January and February 1973 for the Neutrino Area, where E-1A, the neutral currrent experiment of Rubbia, David Cline, and Al Mann, and E-21A, a study of neutrinos at very high energies led by Barry Barish, were underway. (In this period, only one experimental area could run at a time.) Finally, during March 1973 beam for E-70 began to slowly trickle in.[51]

As time went on, many of the physicists who had originally been listed on Lederman's P-70 proposal dropped out. Lederman and Appel formed the core of the Columbia contingent, which expanded over the next few years to include two additional Columbia physicists, Jeffrey Weiss and John Yoh, as well as three European visitors, Maurice Bourquin, Jean-Marc Gaillard, and Jean-Paul Repellin. Appel shuttled back and forth between New York and Illinois to direct much of the design and construction of equipment at Nevis, until he moved to NAL in February 1973. In addition, five graduate students or postdocs, Irwin Gaines, David Hom, Hans Paar, Daniel Peterson, and David Snyder, took part in the experiment. One of the original NAL physicists, Yamanouchi, remained and was joined by Bruce Brown, Chuck Brown, Walter Innes, and Jon Sauer.[52]

While waiting for beam, members of the group installed equipment for both Phase I and Phase II. They encountered many difficulties because of NAL's primitive conditions. The experimental pit housing their sensitive instruments had standing water due to improper drainage and an inadequate roof. NAL had promised a finished roof by November 1972, but that month Appel reported there was not much "besides sheet metal

(1 layer) between us and the elements!" Yamanouchi doubted that Wilson would divert resources to address temperature and humidity control in the area.[53]

In January, in letters to Read and to Peoples, Appel laid out the requirement that the temperature for Phase II needed to be kept above 40° and humidity between 25% and 50%.[54] Making light of a situation that obviously concerned him greatly, Peoples wrote to Appel that on a day when the temperature was about 15°F, with the inside temperature "certainly not warmer," he descended into the experimental pit. He observed the snow "falling through the clever fresh-air skylights in the P-Center roof. Enraptured by nature's harsh display, I lifted my head skyward to watch the swirling flakes only to slip and fall on the ice in the pit. I couldn't respond to your specifications just then." On a more serious note, Peoples admitted that although they could probably keep the humidity below 60%, temperatures would vary from minus 15° to 105°F. Sensitive equipment would need a special enclosure. While Peoples understood "the nature of your requirements" and sympathized "with their spirit," he needed a detailed justification, since elaborate provisions for temperature control were "out of step with the style of NAL." Appel wrote back to Peoples with a detailed explanation of the adverse effects of temperature on their lead glass system. By the next month, Peoples had gained approval from Wilson to begin constructing minimal but adequate enclosures. They constructed a hut within the enclosure for controlling temperature there.[55]

Despite these rough conditions, the Lederman team continued to act as a cooperative partner with NAL. In a January 1974 letter to Wilson, Lederman referred to his group's "patina as one of the more hardened, early believing, pioneers of NAL."[56] Earlier, Appel had aptly defined the drill in a memo to Lederman: "General Program for making [Proton-Center] usable is (as everything at NAL): Put it together, try it, fix it, add heaters used in the tunnels etc.—it may work—add as necessary." By falling in step with the risk-taking, cut-and-try approach of Wilson's style, Appel and other group members endured the risk of possible damage to their detector and were rewarded with service and support, albeit limited, at a time when both were in short supply.[57]

In September 1972, Lederman started to move equipment into the Proton Laboratory, having made a request to conduct an experiment, dubbed "Phase 0.8" of E-70, that would search for heavy, long-lived masses produced in the forward direction.[58] Requiring only two weekends of beam time, this experiment relied on radio frequency bunching and time-of-flight measurement. It used equipment which was already on hand and required no extra resources from the laboratory or from the group, although they did make some new counters, including a "cheap beam-line differential Cerenkov counter." As Appel later noted, the experiment "was a long shot," but "it was fun."[59]

It now appeared that the measurements planned in the original P-70 constituted more than a single experiment. The laboratory assigned Phase 0.8 a new number, E-187. Thus, E-70 became one of the first NAL experiments to be part of a "string," a series of related experiments that used much of the same equipment and continued the original pursuit with some of the same team. As Yamanouchi remembered, it was typical of Wilson to offer quick approval of an experiment proposal when dealing with physicists who had earned his esteem. If such an experimenter "walked into Wilson's office with a great idea, . . . suddenly it was approved." The anecdote highlights Wilson's determination to live up to his promise to foster creativity and avoid undue bureaucracy. It also suggests how it happened that under Wilson so many experiments had to compete for the laboratory's minimal resources.[60]

Lederman had another strategy for dealing with the unpredictables of NAL life. While remaining accountable for pursuing the experimental goals mentioned in his proposals, he also retained the prerogative to use the beam he obtained to his best advantage. When he told Appel to relay the good news that "Phase 0.8 has been approved for 1 or 2 weekends," he went on to say: "I suggest that we don't use up the first 3-day run for this but try to do some Phase I," at the same time trying "some aspect of [E-] 187" and perhaps take a measurement that would also "calibrate the Pb (lead) glass" for Phase II. When this running period proved to be a washout, the group abandoned its plans for Phase I.[61] Lederman's strategy, and Wilson's willingness to indulge it, suggests that Wilson extended the same terms to experimenters that he extended to employees: those willing to do business his way were awarded considerable freedom.

First View and the Advent of E-288

Between mid-1973 and late 1974, Lederman's group installed its equipment and began to take data both for Phase II of E-70, the single-arm experiment, and for the time-of-flight experiment, E-187. At the same time, they prepared for the double-arm experiment, Phase III of E-70, revising their plans for apparatus. New data-taking ideas arose as the group reacted to unexpected physics results and limitations imposed by the laboratory.

Lederman continued to play the role of Wilson's ideal group leader, one who encourages creativity while keeping his group focused on a common goal. Subordinates followed because they respected Lederman. While individual group members were left to their own tasks, Chuck Brown noted that decision making was done in an environment "of anarchy that somehow moved toward consensus." Unlike Wilson, who was sometimes autocratic, Lederman persuaded rather then decreed. He made the general arrangements, such as negotiating for beam time, but "he didn't take detailed charge of the situation," according to Yoh. Jöstlein, who became a collaborator on a related experiment, added that "if something was a toss up," Lederman would "be the ultimate authority." Once he "argued for a certain direction, then people would listen carefully" and usually follow his suggestion. The resulting atmosphere encouraged all group members, including the younger physicists, to comment freely while preserving a strong sense of team spirit.

By April 1973 the members of Lederman's group were prepared to conduct the single-arm experiment. Despite continuing accelerator problems, they obtained enough beam by the end of May 1973 to check the timing of the hodoscope counters and calibrate the lead glass array for the single-arm experiment. The group noted in its post-run analysis that electron events had been observed in the lead glass array for the first time.[62] News of the first real data meant a lot after months of fruitless attempts. Appel recalled working at around 3 a.m. one morning with Hans Paar, who was bent over a graph that showed "there was a real signal." That "was very exciting."[63]

The next step would be to measure electrons at various angles. Fortunately for the group, the accelerator operated more reliably in 1973 than it had in 1972. Except for some component problems in July, its performance continued to improve into the summer, and by September the machine was running "fairly well at 300 GeV," as reported in the monthly report.[64] With beam now streaming in at regular intervals, the group could turn its attention to such typical tasks as taking data and improving the experimental setup. The group realized that its highest priority was improving its "trigger," the crucial part of the experiment that sorted out the desired signal to their detector from the flood of other events.[65]

The group also began its data analysis. In October 1973 David Saxon wrote a series of FORTRAN routines to reconstruct tracks from the hodoscopes. In keeping with the Spartan conditions at NAL, data analysis for E-70 was done on computer equipment that was far from state of the art: the laboratory's second-hand CDC-6600, obtained from Lawrence

Berkeley Laboratory and a PDP-10.[66] The preliminary data showed an unexplained peak. The group took this with characteristic humor. In notes from an October 15 meeting, Appel reported that Lederman "buys champagne for everyone if . . . peak source found and removed. . . . Everyone buys champagne for LML if . . . peak is a quark." A note from a November 16 meeting explained that the spectra emerging from preliminary efforts would help the group get started on more sophisticated analysis programs and allow them "to see if junk triggers distort spectra unimaginably."[67]

In December, Appel faced the problem of high background swamping the primary electron signal. He identified three contributions to this background: pions that simulated electron events and two sources of secondary electrons converted from gamma rays (produced either in collisions with material upstream of the magnets or in collisions with the helium in the magnets). In February 1974, Paar and Repellin identified several more "uninteresting" sources of electrons, including the decays of neutral pions and neutral kaons and leptonic decays of kaons and hyperons. These observations resulted in various changes to the apparatus in the next several months.[68]

The Lederman group expected background difficulties in their single-arm experiment; "We were, after all, looking for the few little gems," as Bruce Brown noted. But the high background they experienced was not caused by limitations in the experimental apparatus, as they had assumed, but by two unexpected physical effects that were not understood in early 1974: parton scattering, which produced many more hadrons at large angles than previously expected, and the yet-undiscovered charmed-quark particles, which produce copious signals. These secrets would remain hidden to the group during the next six months, as its members continued to struggle against the high backgrounds in their data.[69]

The time had come to take the next major step, to design the two-arm spectrometer for Phase III. Some work on the spectrometer magnets had been completed between 1971 and 1973. Using computers at Argonne, Saxon tested various design parameters, including the aperture size and the position of coils. Preliminary work had also been done on the configuration of lead glass blocks, which were being continuously produced according to the group's specifications. They had to determine how many would be necessary for this phase of the experiment.[70]

Once the data taking began with the single-arm experiment in the fall of 1973, a major issue was whether they should begin Phase III with electron or muon pair detection. The group wanted to make both measurements, as mentioned in the 1970 proposal. In fact, ideas for a variety of schemes for using the apparatus were brewing. After a discussion with Lederman, the group decided to build on the techniques it had developed for suppressing the pion background with the single-arm apparatus and first study electron pairs.[71]

A number of other aspects of the original experimental plan were reevaluated. Bruce Brown's work to map out how particles would be tracked led to a more detailed design of the detecting system and, in turn, to a reevaluation of the use of the scintillation hodoscopes used in E-70. At this point, the group questioned whether it should instead use multiwire proportional chambers, a new technology then being used in a two-arm electron pair experiment at Brookhaven headed by Samuel Ting. Making the choice required understanding the advantages and disadvantages of each detector and guessing which would prove most relevant while making sure the components were well matched, complementary, and cost effective.[72] In a November 27 memo Bruce Brown compared using a scintillation hodoscope system and a multiwire proportional chamber system. He had "a strong bias toward using proportional chambers," because using a scintillation hodoscope system would cost almost twice as much. After discussion, the group members agreed that the lower cost allowed them more flexibility in deciding the number of chambers to build in the future. The decision to use proportional chambers ultimately led to a vast improvement in position resolution.[73]

The group made increasingly detailed plans. By November, the size of the array of lead glass blocks had been set. In December, Appel and Yoh considered various parameters of the overall configuration of the apparatus. Yoh, Hom, and Gaines outlined preliminary plans for the electronics hardware in a memo on Christmas day (obviously not a holiday for these young physicists).[74] Due to time constraints, the group did not make detailed calculations but used existing theory, which at this point did not give much confidence, merely as a rough check that their experimental work was preceding in the right direction.[75] Detailed plans for the electronics, most of which was to be designed by Sippach, were laid once the overall experimental configuration had been defined. At Lederman's suggestion, Bruce Brown sought technology for designing the multiwire proportional chambers. In May 1974, Brown reported that the design for the chambers, including their electronics, was completed. By this time, a prototype chamber was close to completion and chamber material and parts were on hand at Nevis or on-site at NAL. The two-arm experiment was now ready for construction.[76]

By then Wilson and Goldwasser had decided to clarify the terms of the laboratory's partnership with the Lederman group. Lederman drafted an agreement, to be signed by himself and Sanford, indicating that Phase III would cost $635,000, with $505,000 contributed by Columbia and $130,000 by Fermilab. The agreement was open ended: "Circumstances and needs will change as the design of the experiment and the plans for the experimental program develop." When the laboratory managers insisted that Phase III be assigned a new experiment number, Lederman agreed with characteristic humor.[77] Gently mocking the bureaucratic process of assigning a new number to this part of their experiment, the team wrote a proposal in February 1974 which was just one page in length.[78] It announced that after a series of grand discoveries, including the much-sought W boson, they would "publish these and become famous." The original proposal for Phase III had discussed measuring electron pairs. The plans set for the new experiment, E-288, were more ambitious, measuring muon, hadron, and neutral pion pairs as well. E-288 was accepted one month before the deadline. Wilson's ideal experimenter had fared well in the negotiations. Later, in 1976, Goldwasser insisted on a more formal explanation of the group's plans. Lederman outlined them in a six-page letter. He estimated that the group needed a little more than a year to complete the measurements now specified as Phase III. E-288 would in fact run until 1978.[79]

Direct Leptons—the Red Herring and the J/ψ, April, 1974–December 1974

In time the trail Lederman blazed would lead to a major discovery, but in 1974 one frustration after another lurked around every bend, even after exciting results. At the April meeting of the American Physical Society that year, the E-70 group announced that many more high transverse momentum electrons than models of the time predicted were being produced directly in its 300 GeV proton-beryllium collisions. In the next few months the group found the "marvelous result" that an unexpectedly large number of high-transverse-momentum muons was being produced directly. The next step was to discover the source of these "direct leptons"—those not accounted for by decays of longer-lived mesons or other backgrounds. They were coming neither from interactions in the detector nor from the decays of known particles. They identified five possible candidates for the parents of these direct leptons: massive virtual photons, large-transverse-momentum vector mesons, charmed particles, heavy leptons, and the coveted W particle. The Lederman group appeared poised to do some interesting new physics. By July 1974 they were ready to present preliminary conclusions in *Physical Review Letters*.[80]

The data taken in the one-arm experiment suggested that something unusual might be causing the high level of electrons and muons observed. But the data did not show any of the dramatic features hoped for: peaks associated with either the W particle or Lee-Wick boson, a heavy lepton, or a charge asymmetry. After accounting for the electrons and muons which would be expected from the decay of the lightest mesons (pions and kaons), there were still a lot of observed leptons to consider. The group speculated that the direct leptons could come "primarily from the phi," a known vector meson. These particles also have a decay mode involving pairs of leptons and antileptons, and the rate of φ meson production, which could have been responsible, had not been measured. They remarked at the time, "An overwhelming production of vector mesons in close hadron collisions, if confirmed, may have profound implications for the strong interaction." When plotting their data, members of the group noticed another intriguing phenomenon. In both the muon and electron data, the lepton signal had "a constant level of (approximately) $10^{-4}$of the π^0's." The unexplained constant relationship between the yields of pions and leptons seemed, as Lederman later noted, to be a constant of nature.[81]

Claiming the prerogative of using beam time without official approval for the sake of obtaining the best results, Lederman and his group quickly mounted a new experiment using the single-arm apparatus to see whether φ mesons were indeed a major source of the observed direct lepton signal. During this new study of single leptons, little attention went to preparing for the double-arm phase needed for the E-288 measurements. Because of the group's efforts at Columbia, the Phase III experiment was essentially ready for installation. But the laboratory caused a delay, because, as the laboratory's managers explained, there were no funds to rig the experiment as promised during the summer and fall of 1974. Lederman's response was to focus more attention on the one-arm spectrometer, which looked at this point as though it might yield exciting results. Lederman and his group were thus "lulled into a lack of urgency" for E-288 and they "did not push" the issue.[82]

The group continued to make measurements of direct leptons at various angles while waiting for the chance to install the two-arm apparatus for E-288. These failed to produce interesting results. They planned to continue their effort for another month, while also preparing for their φ

search, one of several short experiments the group performed that had not received a new number. As a double-arm apparatus was needed to do this special search, the group constructed what Chuck Brown later called "an ersatz, cheap double arm" on the same side as the existing single arm.[83] This new second arm, which was placed below the first, contained three scintillation counters, two counter hodoscopes, a proportional wire chamber, and a hadron shower calorimeter. Later, when their data were analyzed in 1975, the group found "no significant φ signal." Their data set an upper limit on φ production "ruling out the φ meson as a significant source of the high transverse momentum leptons observed to be produced directly in hadron collisions." What then was the source of the direct leptons?[84]

Speculation about the existence of a fourth quark known as "charm" was by then causing a stir. In the summer of 1974, theorists Mary K. Gaillard, Benjamin Lee, and Jonathan Rosner had issued a long review paper entitled "The Search for Charm."[85] Subsequently, on November 4, Ting's Brookhaven team and Burton Richter's SLAC team announced their discovery of a particle, which the Brookhaven team named "J" and the SLAC team named the ψ. Their findings provided compelling evidence for the existence of quarks.[86]

Yamanouchi remembered hearing the news of the J/ψ discovery during lunch at the Fermilab cafeteria. After talking to a number of colleagues about their competitor's success, he reported back to the E-70/288 experiment: "I am sorry, but I think it is true." The Lederman group realized immediately that they could have made the discovery had they proceeded more quickly to the double-arm experiment and not taken the time to so thoroughly investigate the direct single lepton production.[87] The realization that their direct lepton signal was itself largely due to the newly discovered family of particles was particularly galling. By Lederman's count, this was the *third* time that he had missed the J/ψ. The first was by failing to interpret "Leon's Shoulder" in the Brookhaven AGS data as a J/ψ signal; the second was choosing the wrong trigger in his CERN ISR experiment, mounted at the same time as E-70; and the third was because of E-70's preoccupation with measuring single-arm direct leptons. Also, they subsequently discovered that the 10^{-4} "constant" of leptons to pions was, in Lederman's words, just a "red herring . . . a dumb number, which turned out to be an accident. . . . It was 10^{-4} for no good reason."[88]

The physics community quickly recognized the J/ψ findings as decisive. Although many contradictory explanations for the new particle surfaced after the November 4 announcement, by December Harvard physicist Sheldon Glashow and others had argued persuasively that the J/ψ was composed of a charmed quark and its antiquark. "In one stroke, the discovery of the J/ψ proved the existence of quarks and of charm."[89] The December 6 issue of *Science* likened the discovery to that of strange particles in 1947, noting: "Many physicists think that the discoveries may have opened a whole new dimension in the world of subnuclear particles."[90] The discovery sparked what was later called "the November Revolution," a burst of progress in high-energy physics that would culminate in the classification of all elementary particles within a new theoretical framework known as the Standard Model.[91]

The fact that the J/ψ had been found at SLAC and Brookhaven caused some Fermilab physicists to sense they were losing ground in their pursuit of the physics frontier. Just a year earlier, in 1973, Fermilab physicists had missed the discovery of weak neutral currents, in which CERN triumphed. After the discovery of the J/ψ, many Fermilab experiments emerged from the proposals of university researchers who suggested clever ways to quickly and inexpensively address the new phenomenon of charm. As recalled by Fermilab physicist Lach, "It was a fad, if you didn't make a proposal with charm, you couldn't get beamtime." But, as Peoples explained, charm physics "was messy, and the techniques needed to do experiments with hadron beams simply didn't exist." Given the difficulties caused by Wilson's tendency to cut corners, life was especially difficult in this era for those struggling to make progress in high-energy physics.[92]

Even in this period of disappointment, Fermilab's leaders upheld Wilson's vision of the laboratory. In January 1975, in discussing the neutrino program, Goldwasser argued passionately for maintaining the focus of the laboratory squarely on outside users having numerous quickly done experiments rather than concentrating manpower, money, and beam time on a few large experiments. His rationale, like Wilson's, was that "the lifeblood of high-energy physics research has always been the new people who continually come to the fore with new ideas and new methods. If a given group, no matter how competent and imaginative the members, is permitted to become frozen into a given program and a given set of apparatus, the research may very well become stultified, and suffer accordingly."[93] It was therefore necessary for Peoples to argue forcefully with Wilson while discussing the possibility of authorizing E-516 (discussed in chapter 11), a large charm experiment with a strong core of laboratory physicists.[94]

After a brief attempt to observe the J/ψ signal with their makeshift double-arm apparatus, the Lederman group finished its scheduled direct lepton measurements. A fundamental paper by James Bjorken, Samuel Berman, and John Kogut eventually explained the plethora of leptons obtained, which had both intrigued and frustrated the Lederman group and many others at NAL and the ISR. The data taking for E-70 ended in December 1974, and attention shifted to E-288. As the effort to dismantle and rearrange the E-70 experimental apparatus began, graduate student David Hom reported on the status of E-288 data acquisition and online analysis planning, and Weiss and Yoh addressed the always-problematic issue of signal-to-background rates. Appel remembers that the group was disappointed, but not thrown off course, by the J/ψ announcement. With characteristic humor, they sent a photocopied invitation announcing a party on December 9 to celebrate the end of E-70 data taking. The invitation from Lederman featured a drawing of a gravestone with the epitaph "E-70 Run, 1970–1974, R.I.P." On the grave lay a wreath, labeled "from E-288." A lead glass block sculpture in Fermilab's art gallery commemorates their efforts. With E-70 and its disappointments laid to rest, the group proceeded, as planned, to the double-arm measurements, still hoping to discover other phenomena.[95]

The Road to the Upsilon

During the spring and summer of 1975, Lederman's group was gearing up for Phase III of their experiment, E-288, using the double-arm apparatus (fig. 7.5). Bruce Brown concentrated on developing the double arm and testing ideas for a water Cerenkov counter.[96] One arduous task was rearranging the apparatus. The group had to move magnets, rebuild the target hall area, rearrange massive shielding, and install proportional chambers, the trigger, lead glass, and Cerenkov counters. Chuck Brown remembers being impressed by the installation of the lead glass blocks. Whereas the single-arm apparatus had forty-five blocks, the double-arm apparatus had ninety-six. Lederman continued to borrow equipment—for example, amplifiers from Jim Christenson's group at New York University.[97] At this point the group had more than two years of experience working with their equipment, with each other, and with the Fermilab staff and facilities. Over the next two years, under the rubric of E-288, the group would measure electron pairs (in a subexperiment called eeI) and conducted two experiments that measured muon pairs (μμI, and μμII). Between the two muon pair experiments, the string of experiments following from the original proposal for E-70 also included an experiment, E-494, to measure hadron pairs.[98] While Lederman

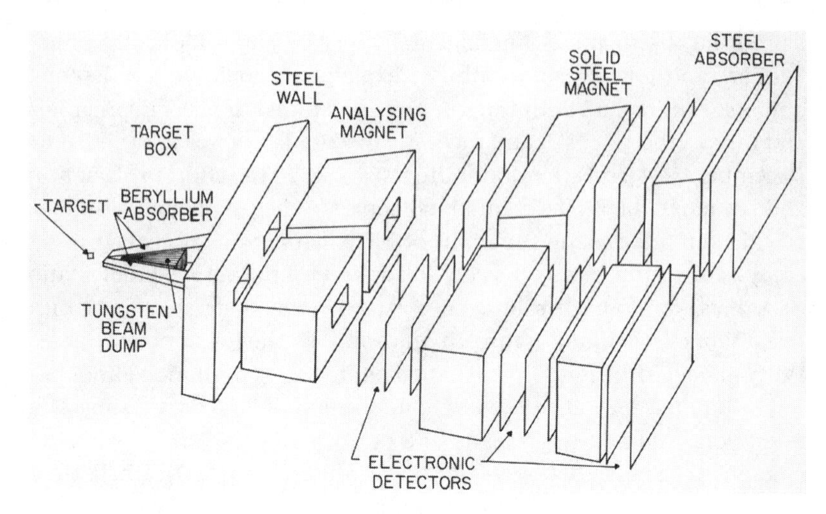

7.5 Apparatus for E-288, 1977. (Courtesy of FNAL Visual Media Services.)

had led the initial inquiry, the string of experiments that evolved included other spokespersons (including Bud Good, John Rutherfoord, Chuck Brown, Joel Moss, and Jen-Chieh Peng). Appel continued, now as a Fermilab staff member, as did Yamanouchi (fig. 7.6).[99]

By fall 1975, data taking with the new apparatus was underway, although modifications in the experimental setup continued throughout the experiment.[100] The group still relied heavily on the resources at Nevis. For instance, while it was constructing the double-arm device, Nevis engineer Milton Storch played a critical role in drafting and fabricating the frames of the proportional chamber. These frames were difficult to assemble because of their large size. Sippach, the fast-electronics expert, designed a clever track reconstruction box and a readout system that could take data from all the proportional chambers in the span of one millisecond.[101] But the Lederman group could not rely on Nevis for all its technical support because of the long distances involved. Nor was Fermilab structured to supply most of the needed services. Wilson's environment challenged individuals to find their own creative solutions, and Fermilab experimenters were often forced to do specialized tasks themselves. "Berkeley, SLAC, Nevis, any place else had a lot more support than Fermilab," said Bruce Brown.[102] Yoh was struck by the difference between Fermilab and CERN, where technical support was relatively abundant. "We had to understand computing, we had to understand mechanical engineering."[103]

Although the laboratory was far more developed than it had been in the first years of the experiment, conditions remained rather crude in 1975 at the time the double-arm spectrometer was being set up. In Daniel Kaplan's words, the "pit" was at this stage "basically a large hole in the ground rather than a building, with no crane, corrugated iron walls, and a makeshift roof which leaked whenever it rained. Until the Lab could be convinced to install numerous gutters (inside the pit!) to run the water off to the sides, we draped the detectors with plastic for protection." As Yoh remembers, such measures did not solve all their problems, and they had high voltages and a leaky roof, which made the working environment "somewhat dangerous." Humidity was another problem. They had trouble with phototube discharge in the first phase of the double-arm experiment. Because it was necessary to run the air conditioners in the electronics hut even in winter, the devices would often ice up and

7.6 Members of the E-288 collaboration, 1977. (Courtesy of FNAL Visual Media Services.)

cause the electronics and online computer to overheat, recalled Kaplan. The prospects for data analysis were also limited because the group had access only to "an assemblage of obsolete computers, which operated in batch processing mode and were programmed using punched cards," even though by this time, "most computing installations had moved on to timesharing." The conditions of work were also hampered by the absence of running water. In the first years, the group had to make do with only an outdoor portable toilet. Later, they had the luxury of an indoor restroom, but it was located about a hundred meters away, in the Proton Area Operations Center.[104] Overall, the Proton Area of the E-70 days was "extremely primitive. . . . We were literally in mud. We were . . . in clouds of mosquitoes. . . . There were times when you had to walk maybe 200 feet to get to your experiment. . . . There wasn't even a path." Postdoc Jeffrey Weiss said he "felt like a pioneer out there on the prairie," a vision the participants shared and sometimes even enjoyed at the times when their work gave them the sense that Wilson had glorified, of being engaged both with the elements and each other in a united mission to make grand discoveries at the frontier.[105]

eeI, μμI, Oops-Leon, and the Dihadron Interlude, 1975–1976

When Lederman's group finally started measuring pairs of electrons in the fall of 1975, the data were analyzed in two ways. One program, written by Innes, was complicated, precise, and time consuming to use, but, in the words of Jöstlein, "did everything right." Another program, written by Yoh, "was maybe only 10 percent as complicated and did 95 percent of the job," but gave "results within a few days," allowing them "to check if something was going wrong." Such fears proved unwarranted.[106]

Hom's oral presentation and the press conference that Appel and others held on E-288 at the American Physics Society meeting in New York told how their subsequent analysis showed "a clear J/ψ signal" that was "consistent with our estimated apparatus resolution." They also found "electron-positron pairs . . . with masses as large as 10 GeV! . . . In addition there is the clustering of up to 12 events within our 100 MeV mass resolution in the region near 6 GeV, 12 out of the 27 events above 5.5 GeV." Appel remembered that when these data were coming in the group "got very excited" about the evidence "for a new resonance at 6 GeV." One weekend we got three . . . pairs of electrons." The good-humored team expressed this excitement by proposing various schemes for posting how many new particles were seen each day.[107] Wilson suggested putting the information on the laboratory's "Channel 13" TV monitors distributed around the laboratory.[108]

Not all members of the group agreed on how to interpret the 6 GeV result. As Kaplan later explained, the "dilemma" was that "the statistical case for the new resonance was marginal . . . , and a conservative person would not claim it as a discovery." They also feared that if the effect were real, then once they published their data, members of SPEAR, the electron-positron collider at SLAC, might beat them to the discovery. This was a painful prospect for Lederman and his group as they continued to take data. By the end of 1975, Kaplan explained, the group was "the proud possessor of the world's largest sample of massive lepton pairs: no other experiment was even close."[109]

A meeting was called in the conference room known as the "Snake Pit" to consider a name for the possible new particle at 6 GeV. The night before, Weiss had had a brainstorm while working with Chuck Brown to take data on a 4 a.m. shift. They consulted the *Particle Data Book*. and in "going through the Greek alphabet trying to discover any Greek letters that hadn't been used," Weiss concluded, "it's got to be Upsilon."[110] Innes noted that it was a good name, because "should it prove a statistical fluctuation, it would be the "oops-Leon." Despite this cautionary note, the group remembers that naming the still-disputed particle rallied support for its existence. In Yoh's words, the "bandwagon" was fueled by the fact that they had "a very nice name for the particle." Bruce Brown found that despite his earlier skepticism, "all of a sudden, the plot . . . looked very much better." He explained, "I just had this gut feeling: well, I don't know . . . but this felt good to me.[111]

Outside the laboratory, reaction to the 6 GeV result was mixed. Some, like the eminent SLAC physicist Martin Perl, were supportive of the findings. Others were not. Appel, who represented E-288 at the APS press conference at the New York APS meeting in January 1976, remembers that after David Hom's talk a colleague approached him and said bluntly: "I don't believe it." When the experiment subsequently submitted the results to *Physical Review Letters*, "the reviewers had a terrible time with it." The paper went back and forth from the reviewers to the group. "One reviewer sat on it," saying "he didn't believe the statistical analysis."[112] In May *Physical Review Letters* published the article, which reported that the clustering of events from 5.8 to 6.2 GeV "suggests that the data contain a new resonance at 6 GeV." To strengthen their future claim to the desired discovery, Lederman devised a footnote stating: "We suggest the name Υ (Upsilon) be given to the resonance at 6 GeV if confirmed." In other words, if the discovery was not confirmed, the name Upsilon could be given to any other higher-mass object.[113]

SLAC physicists promptly searched the 6 GeV region and found no signal. The possibility still remained, however, that there existed a 6 GeV resonance which, in Kaplan's words had "a large hadronic production cross section but extremely small electronic decay width." To explore this possibility, the group decided "to switch to muon pairs." Doing this would involve placing shielding material between the analyzing magnets and the target to absorb most of the hadrons, which would then not even enter the detectors. They also switched to using beryllium to reduce scattering of the muons and enhance the mass resolution.[114] The running period for this double-muon experiment, referred to as $\mu\mu$I, was short but decisive. As the group reported in fall 1976, they observed 159 events in the mass range 5.5–11 GeV and concluded that "within limitations of resolution and continuum uncertainty, the dimuon mass spectrum provides no evidence for fine structure above 5 GeV." The group members recalled feeling very let down and embarrassed. Yoh summarized: "we had egg on our faces." The premature announcement of the Υ would become legend as the "oops-Leon."[115]

Yoh saw something else while plotting the results from both electron and muon pair measurements that fall of 1976. He noted in a November 17 memo: "Evidence for" a resonance at "9.6 comes from both ee and $\mu\mu$." He then speculated that the current run would be unlikely to yield enough data to confirm or deny the effect. In a postscript, he noted, "As if to make me eat my words—a new event at 9.44 was just found. What the hell is happening? 4 events near 9.6 within 12 days (11/5–11/16)." The evidence prompted Yoh to put a bottle of champagne in the group's refrigerator marked "9.5," so that group members would be able to toast the new resonance should further results confirm its existence. Yoh's hesitating memo on the 9.5 data bore the title: "From the People who Brought you the Υ, a Bigger (But Not Necessarily Better) Resonance." Other group members had even less confidence in the possibility of a 9.5 resonance and did little to publicize the results. As Yoh later explained: "We'd been burned once. Obviously there was no point embarrassing ourselves again."[116]

In the meantime, the group members expanded their work in a new direction. They decided to adapt their open spectrometer with lead glass to detect hadrons pairs. Bud Good's Stony Brook group joined the collaboration and contributed two Cerenkov counters for hadron particle identification. By early 1976 the group was working not only on its dilepton measurements but on planning a dihadron experiment as well.

In May, Goldwasser assigned the dihadron work the new experiment number E-494. Good was listed as the experiment's spokesman. Just a month earlier Goldwasser had asked Lederman to provide a more detailed accounting of plans for dilepton runs to be performed as part of E-288.[117] These moves reflected the penchant to implement more formal procedures at the laboratory, as well as the need to emphasize, at least on paper, the extent of outside user participation.

Besides one of the Stony Brook Cerenkov counters, a hadron calorimeter was added to each of the two spectrometer arms to use as a hadron trigger. Once again, Lederman demonstrated his borrowing ability by obtaining two multiwire proportional chambers that Ting's group had used in the discovery of the J (and thus called "J chambers").[118] Tensions rose in the group when it had to decide how long the dihadron run should last. Should it run as long as possible to ensure having sufficient data, or should they stop the run as quickly as possible and start their search for exotic phenomena? Some of the newcomers had little experience with the problems encountered in the long-shot search for exotic phenomena. Bruce Brown remembers that some of the Stony Brook novices "got very discouraged because . . . it looked like [they] could never get enough statistics to measure anything."[119]

As in all times of crisis, Lederman stepped in. R. L. McCarthy, one of the Stony Brook physicists who preferred to stay longer on the dihadron work, remembers that "Lederman himself made the decision" to give them "a week." The dihadron enthusiasts accepted this decision, took their data, and then left to do data analysis at Stony Brook as the experimental area was then rearranged to explore dimuons. Those interested in the dimuon work agree that they came out well in the collaborative effort to include a dihadron search, for that additional study did not, in their opinion, impede their search for exotic phenomena. Chuck Brown noted, "We needed the extra help and they [brought] valuable equipment and technical help." An added bonus was that Robert Kephart brought along a talented technician, his wife, Karen.[120]

The Discovery of the Upsilon, February 1977—June 1977

The dihadron experiment finished running in February 1977, and the group made radical adjustments to the setup. They changed the experimental solid angle by bringing detectors closer to the magnets, opening up the aperture-defining collimators upstream of the magnets and symmetrizing the spectrometer arms vertically so that each was centered on the neutral beam. Such changes were the product of more than a year of intense calculation, discussion, and debate. The group members often wrote lengthy memos to one another; Chuck Brown called it "memo warfare." Kaplan felt that this "highly collaborative" process demonstrated the open, freewheeling atmosphere inspired by Lederman's leadership style.[121] Lederman later commented on the issues discussed in the memo warfare: "We realized that in order to draw any conclusions about the rarer, higher masses we would have to observe many more events. At the same time we would have to improve the resolution to obtain interpretable data.[122] When acquiring the necessary beryllium, Lederman again demonstrated his talent for procuring difficult-to-obtain materials. After many hours on the telephone, he found surplus quantities of the precious metal at Oak Ridge National Laboratory. Unfortunately long-term exposure to the material was known to cause lung disease, so group members had to monitor their exposure.

Around April 1977 the team started seeing "some structure in our data near 9 GeV, . . . we had all kinds of grandiose speculations as to what this would mean," Lederman reflected much later.[123] They decided to improve their identification of muons by adding a solid steel magnet to each arm, as proposed by Bruce Brown. And to improve their measurement of the muons' momenta, they also added multiwire proportional chambers inside their air-gap analyzing magnets, an innovation designed by Innes.[124] The group obtained beam for test runs in late April 1977, and after a few weeks of accelerator maintenance, data taking began in mid-May. Kaplan remembered that because of the substantial improvements in the detector, now every "good day of running was comparable in sensitivity to the entire $\mu\mu$-I run." The group also benefited at this point from the energy of the Main Ring, which was then running at 400 GeV. As early as May 2, Yoh told the group members that based on his quick data analysis, he thought he saw something around 9.5 GeV, just as he had noticed the previous fall.[125]

This productive time was not without drama. In the midst of data taking on May 22, "a recently installed precision shunt used to monitor the current in the air-gap analyzing magnets overheated, spraying molten metal on its surroundings and starting a fire," Kaplan reported. When he arrived to take his shift, P-Center was surrounded by fire trucks and the experimental area was cordoned off. "Since burning PVC insulation releases chlorine, the application of water by the firemen created an acidic mist which would eventually have destroyed the thousands of amplifier/discriminators mounted on the wire chambers."[126]

Once again, Lederman's extensive contacts paid off. Through a European colleague, Jean-Marc Gaillard, who had dealt with a similar fire at CERN, Lederman tracked down M. Gerard Jesse, a Dutch salvage expert then living in Spain, who was working with a Swiss firm. Arrangements were made to immediately fly Jesse and his essential chemicals to Fermilab. Chuck Brown remembered that "he came in with two 55 gallon drums of his secret cleaning fluid, which felt, looked, and seemed for all the world just like detergent." Cleaning of the experiment began within seventy-two hours on the tenth floor of the High-Rise, where the experimenters had their offices; the experimental hall itself was too dirty. All the electronic equipment was brought from the experimental area to be processed in a "production line" cleaning effort, with a long line of people who sat at stools and dipped the pieces of electronics in "baths of this special, secret fluid." These workers, who included anyone the group members could convince to help—wives, girlfriends, secretaries—put each piece "in the first bath, scrubbed, then rinsed, then put it in the second bath, repeated the process, then put it in a third bath." After this thorough cleaning, the equipment was returned to the experimental area, which had in the meantime been cleaned by janitors. In the end, the fire cost the experiment only a week of running time. By June 6, Yoh had checked both prefire and postfire data runs and determined that the 9.5 GeV enhancement was significant in both runs. This time, he calculated that the effect was in the range of at least eight standard deviations.[127]

Based on the predictions of the Standard Model, the group began to feel confident that they had indeed found the resonance that indicated the bottom quark. In an effort to be sure the finding did not result from some effect in the programming or the detector, Yoh carefully rechecked his work. Steve Herb made an independent check using a different data analysis program. Chuck Brown remembers that Lederman, who was attending the annual PAC meeting in Aspen, called and asked: "How's it looking?" Since the results from the two programs "were agreeing with each other" they told him: "We believe it, we're writing a paper. . . . It's there Leon, definitely" (fig. 7.7). After obtaining permission from other group members, Lederman gave Wilson the exciting news of their discovery. The name Upsilon was resurrected for the new particle.[128]

As Lederman reflected some years later, "The Upsilon represented an exotic 'atom' of a bound state of a new quark, the bottom quark and its antiparticle twin." The mass of the "b" (short for bottom quark) proved to be about five times the mass of a proton. Together with the new "tau"

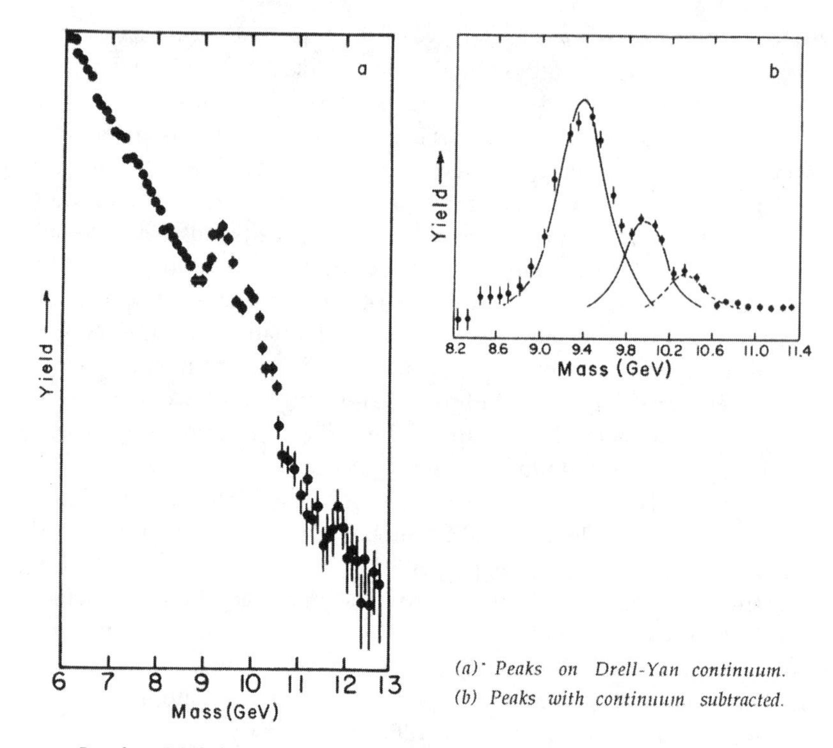

7.7 Data from E-288 showing the discovery of the bottom quark, 1977. (Courtesy of FNAL Visual Media Services.)

lepton, discovered by Martin Perl, it was now very clear that a third generation of quarks and leptons exists in the Standard Model. Lederman recalled that "all the powerful-enough e^+e^- machines tuned to this energy," and before long, the DORIS group at DESY in Hamburg found upsilon, upsilon-prime, double prime and triple prime. It was confirmed that indeed this was a new quark having a mass of approximately 5 GeV as well as a complex spectrum of excited states.[129] The only undiscovered quark in the Standard Model was now the postulated "top" quark. That would also be found at Fermilab, seventeen years later.[130]

On June 22, soon after Lederman returned from Aspen, the group held a meeting at which they decided to announce and publish their discovery of the bottom quark. These goals were accomplished quickly. On June 30 Herb gave a talk about the discovery to an enthusiastic Fermilab crowd. The next day, following a Columbia tradition, Lederman flew to New York to hand-deliver the paper to *Physical Review Letters*. The tradition was meant to expedite the publication of major discoveries. On

August 1, Goldwasser granted Lederman's request to publish the paper without a formal laboratory review.[131] Lederman later recalled, "We had finally realized our 20-year-old dream of a new, high mass resonance. The excitement at the discovery was palpable. As the data kept accumulating, we made out what first appeared to be two closely spaced mass peaks and later, it looked like three!"[132] It happened that on the day that Lederman announced the discovery of the bottom quark, at a conference in Budapest, the city's electricity supply failed, in fact during his talk! "Could be a sign from God that he thinks we're getting too near his secrets," he quipped.[133]

The group had finally broken through. Lederman had proved that even with the limitations of Wilson's laboratory, a clever experimenter could perform his part of the bargain, that is, assemble the resources to perform an experiment capable of making a major discovery. And Fermilab played its part by offering an accelerator which by then was regularly operating at 400 GeV. The strong energy dependence of the production rate might have been why they missed making the discovery at 200 GeV. "Everyone who works at Fermilab had a piece of the Upsilon action," Wilson announced triumphantly to his staff in July 1977. "In some way everyone had contributed and we all have the right to feel proud of what they have done and what we have done."[134]

Wilson's Vision and the "Higher Obligation"

The success of the Lederman group called certain aspects of Wilson's vision into question. Their results came partly because they were allowed to remain on the P-Center experimental floor for a long time. "You need some real estate," reflected Bruce Brown, "where your successes can be retained and your failures replaced."[135] Despite Wilson's and Goldwasser's policy of sponsoring large numbers of relatively simple experiments that could be quickly done, Lederman had managed to convince the laboratory to make an exception. His group was allowed to stay in place for seven years, during which they could create a relatively sophisticated experiment.

Other experimental groups at Wilson's laboratory experienced difficulties similar to those endured by Lederman's group. Nevertheless, most NAL experimenters recall the early days of Main Ring research fondly, as a time of adventure and excitement, as well as a time of hardship. The interviews conducted with Fermilab researchers for this study reveal few complaints about the machine's performance, despite the initial delayed start of experimentation. But did Wilson's laboratory meet the "higher obligation" he defined of facilitating great discoveries?

The most celebrated particle physics discoveries of the decade were ones that NAL missed in 1973 and 1974: neutral currents and the J/ψ. Experimenters at Wilson's laboratory came close to seeing both, but as Peoples noted in 1974, Fermilab's construction "wasn't quite finished."[136] In 1974, NAL experimenters had access for the first time to all four experimental areas. Beam intensities were still below the 5×10^{13} that had been specified in the 1968 design, although the accelerator could produce beams of 1.4×10^{13} within about a factor of 3 of this specification. Accelerator reliability had improved, but experimenters still did not receive beam as scheduled. By 1975, the accelerator was "operating regularly at 400 GeV."[137] Wilson had invested most of the laboratory's resources in providing an accelerator with optimal capabilities. The result was that the laboratory had little money left to invest in experiments. This strategy paid off in the case of the Upsilon's discovery, which required a 400 GeV beam. It can be argued that if Wilson had diverted considerable resources to experiments in the early 1970s, the experimenters in Lederman's group might not have had an accelerator to use in time to beat the 1974 experiments at SLAC and Brookhaven. But what of Fermilab's overall record of experimentation, once the accelerator began operating above 400 GeV in the mid-1970s?

In 1979, a year after Lederman became Fermilab's second director, he and theorist Quigg issued a report that placed Fermilab's major experiments of the previous year into the context of the crucial questions of that time: "How many quark species are there? How do the quarks behave within hadronic matter? What is the nature of the weak neutral current?"[138] In helping to answer these questions, experimenters at the laboratory found the third generation of quarks and the unanticipated excess of particles produced with high energy at large angles, helped refine quark and scaling theories of elementary particles, produced the first charmed baryon (although the first one may well have been a false signal), and created and studied the J/ψ particle using photons, neutrons, pions, and protons. Despite these achievements, many physicists, including Wilson, felt that the laboratory had achieved only mixed success in meeting the higher obligation that he had assigned. Years later, Wilson concluded that NAL's experiments in the 1970s "were not . . . leading edge work."[139]

In separate interviews, Fermilab's third director, John Peoples, and SLAC director Burton Richter reflected on the source of the difficulties:

that Wilson and many of those proposing experiments during his tenure failed to appreciate the implications of not mounting the largest, most sophisticated experiments. These larger, higher-priority experiments proved necessary for exploring quarks in the 1970s.[140] Wilson admitted opposing the increase in scale: "I didn't want any part of it."[141] SLAC, in contrast, concentrated its resources on a few large detectors built by inside groups who were in a better position to manage massive projects than were groups composed of geographically scattered visitors. By serendipity, SLAC reaped another competitive advantage; in crucial experiments SPEAR produced stronger signal-to-background effects than Fermilab's proton synchrotron. For this reason, Fermilab's experiments required more sophisticated apparatus to detect similar phenomena.

Wilson's insistence on pushing the limits and reaching 500 GeV with his authorized 200 GeV Main Ring without extra funding had kept Fermilab in the game, enabling Lederman's discovery of the bottom quark. In Lederman Wilson found a physicist capable of making the university-laboratory partnership pay dividends. For in the end, the frontier setting offered at Fermilab brought the best out of this "city slicker," who managed to succeed because he found inventive ways to alter the terms of the agreement. Yet Richter, Peoples, and many others felt that even Lederman suffered in the NAL environment in its first decade, and they noted that Lederman and his team might have made more discoveries, including the J/ψ, had they been allotted more equipment money.[142] The fact that Wilson's celebration of small, quickly done, clever experiments failed to prepare researchers for the much larger, high-energy experiments required in particle physics in its next decades was another of the many ironies of Wilson's legacy.[143]

EIGHT

Beyond the Horizon: The Energy Doubler, 1967–1978

Between the fixed-target experiments that can be made with the highest luminosities, and the colliding-beam experiments that will reach the highest energies, we can anticipate a leap forward in our knowledge of particles and forces—all through the magic of superconductivity. ROBERT R. WILSON[1]

The energy of the accelerator was among many frontiers that Wilson explored as director of Fermilab. He pursued ways to extend this frontier well before completing the Main Ring. Always eyeing the limitations constraining this pursuit, he conceived a daring plan that employed the phenomenon of superconductivity. He viewed this mysterious state of matter, in which certain metals and alloys lose all their electrical resistance at low temperatures, as an "elixir to rejuvenate old accelerators and open new vistas for the future."[2]

The plan involved doubling the accelerator's energy to a trillion electron volts (1 TeV), while at the same time saving power, by adding a second ring to the existing machine with roughly a thousand superconducting magnets. When operating close to 1 TeV, this "Energy Doubler" (later known as the "Energy Saver," or "Saver," and ultimately as the "Tevatron") would consume only half the energy of the

This chapter is based partly on Hoddeson 1987.

Main Ring. This was a daring promise to make during the period's national energy crisis. As Wilson told Congress in March 1977, "Our electrical bill is over $7 million and would be over $10 million if we ran as much as we could, were there not economic restrictions. If we could replace the copper wire by the superconducting wire that is now becoming practical, we should be able to save as much as $5 million per year of electrical energy that is now being lost—not to mention our extreme need to use the $5 million to help run the laboratory."[3]

First Attempts to Build Superconducting Magnets

The hope of building economical high-field superconducting magnets that dissipate little or no heat dates back to 1911, when Heike Kamerlingh Onnes first observed superconductivity in Leiden. To a few physicists this discovery suggested a way to create magnets that could inexpensively operate at much higher fields. That seemed possible because of the development of a helium liquefier allowing systems to be cooled down to very low temperatures. Kamerlingh Onnes himself recognized in 1913 that superconducting materials might in time allow economically created high-field magnets.[4] But while Kamerlingh Onnes speculated about the promise of the superconducting magnets, he and his associates noticed a troubling aspect. In experiments using superconducting lead above a critical magnetic field of a few hundred gauss, the superconductors often suddenly would "quench," returning to their normal state while releasing a large store of energy. They usually melted in the process. In the 1970s, this strange and dangerous phenomenon of the quench was still challenging those who were developing superconducting magnets.

The road to actually building high-field magnets out of superconductors like niobium nitride, niobium tin, or niobium titanium was opened after the Second World War.[5] Of particular importance were experiments conducted between 1949 and 1954 by Berndt Matthias and John Hulm. Working together at the University of Chicago, where Hulm was a postdoctoral fellow and Matthias was visiting from Bell Labs, the two followed Fermi's suggestion to undertake an experimental search for new superconductors. Developing a set of empirical rules, they found that the factor most crucial in determining superconductivity was a material's atomic valence number: superconductivity is possible if there are between two and eight valence electrons, and it occurs more readily if this number is odd. They also concluded that superconductivity is favored by certain kinds of crystal structures and by the amount of empty space in the crystal. With these findings, they were able to produce a large number of superconducting materials.

When Matthias returned to Bell Labs in 1951, he set up a program to find even more of these materials.[6] Around 1960 Hulm, Matthias, and J. Eugene Kunzler started to apply their results to building actual high-field superconducting magnets. In 1961 Kuntzler reported a critical field of 88 kilogauss for niobium tin at 18 K, with average current exceeding 100,000 amperes.[7] Embedding the superconducting alloy in highly pure copper made it possible for the solenoids to hold the high currents.[8]

This line of research began to impress particle accelerator builders as work began in the late 1960s on building NAL's Main Ring and CERN's Super Proton Synchrotron (SPS). A few smaller superconducting magnets for bubble chambers had already been built successfully at both Argonne and Brookhaven.[9] By the time the Main Ring was completed, a few enterprising accelerator builders (including John Adams at CERN and Wilson at NAL) saw superconductivity as a road to the next energy frontier.

Before long, seven groups were studying the problems of building high-energy superconducting accelerators—three in Europe, three in the United States, and one in Japan. Adams hoped that part of the SPS would include superconducting magnets.[10] For this purpose, the Rutherford High Energy Laboratory in Great Britain, the Center for Nuclear Studies at Saclay in France, and the Karlsruhe Institute for Experimental Physics in West Germany formed the Group for European Superconducting Synchrotron Studies (GESSS). Each GESSS member group built several short (1–2-meter) prototype dipole magnets. But as Adams's idea was not strongly supported at CERN, and because GESSS progressed more slowly than anticipated, the idea of making part of the SPS superconducting was dropped and the GESSS group was dissolved.[11] At Berkeley, the project known as the Experimental Superconducting Accelerator Ring (ESCAR) began in July 1974 as a pilot project for researching superconducting accelerators. Like GESSS, it contributed valuable research, especially on dipole magnets and refrigeration. ESCAR ended abruptly in June 1978, when a serious quench burned much of the equipment.[12] The Japanese effort never entered the mainstream of research.

By 1978, therefore, only two groups remained in the race to build the first superconducting magnet particle accelerator: the groups building Fermilab's Energy Doubler and ISABELLE, Brookhaven's intersecting storage accelerator, which evolved from the effort to build the 1000 GeV proton machine recommended by the Ramsey Panel.[13] For over a decade, ISABELLE and the Doubler stood in competition with each other.

Early Discussion of a Superconducting Accelerator for NAL, 1967–1972

Wilson's idea of using superconducting magnets to double the energy of NAL's accelerator had come to him in the early years of designing NAL. When he announced it informally during his Oak Brook design conference in the summer of 1967, he impressed some of the participants with the advantages. Such a superconducting ring could be used to lengthen the time over which beam stored accelerated particles or was fed out to experiments. It could be used in colliding the beam from such a machine against an opposing beam that emerged from the Main Ring to achieve even higher energy.[14] But while most of the physicists at Oak Brook considered these concepts interesting, a superconducting accelerator appeared well ahead of the current feasibility. They felt that designing and building the Main Ring was the job at hand. Wilson did not disagree. After the Oak Brook meetings he issued an informal edict prohibiting active work on a superconducting accelerator until the Main Ring was functioning.[15]

Wilson nevertheless insisted that adequate space be left "free in the tunnel of the Main Ring so that a second magnet system can be placed just above (or beside or below) the original magnets." He wrote in his notes of September 1970: "The idea then is to place [in the tunnel] a second ring of superconducting magnets . . . the proton energy could be raised to 1000 GeV."[16] Richard Lundy, who had worked on the Main Ring and later on the Doubler magnets, recalled that "Bob [Wilson] did enforce the idea that space be left clear . . . [although] it was never exactly obvious what would go in there." Claus Rode, who worked on the Doubler's refrigeration system, remembered considering this space "sacred territory."[17] The artifactual evidence of this stage of planning for the Doubler included a set of now vestigial spheres on each conventional magnet stand in the original ring. These orbs, installed in 1971 to serve as feet for the stands of the superconducting magnets, were never used, because (as we will see in what follows) the Doubler was ultimately built below the ring of conventional magnets.[18] Wilson's earliest sketch of a Doubler magnet, circa 1971 (fig. 8.1), hung on his door for a number of years.

In March 1971, shortly before the crisis of the Main Ring magnets erupted, Wilson believed that his accelerator was about to operate. It seemed the right time to inform the Joint Committee on Atomic Energy (JCAE) about his idea of doubling the energy of the Main Ring. "The

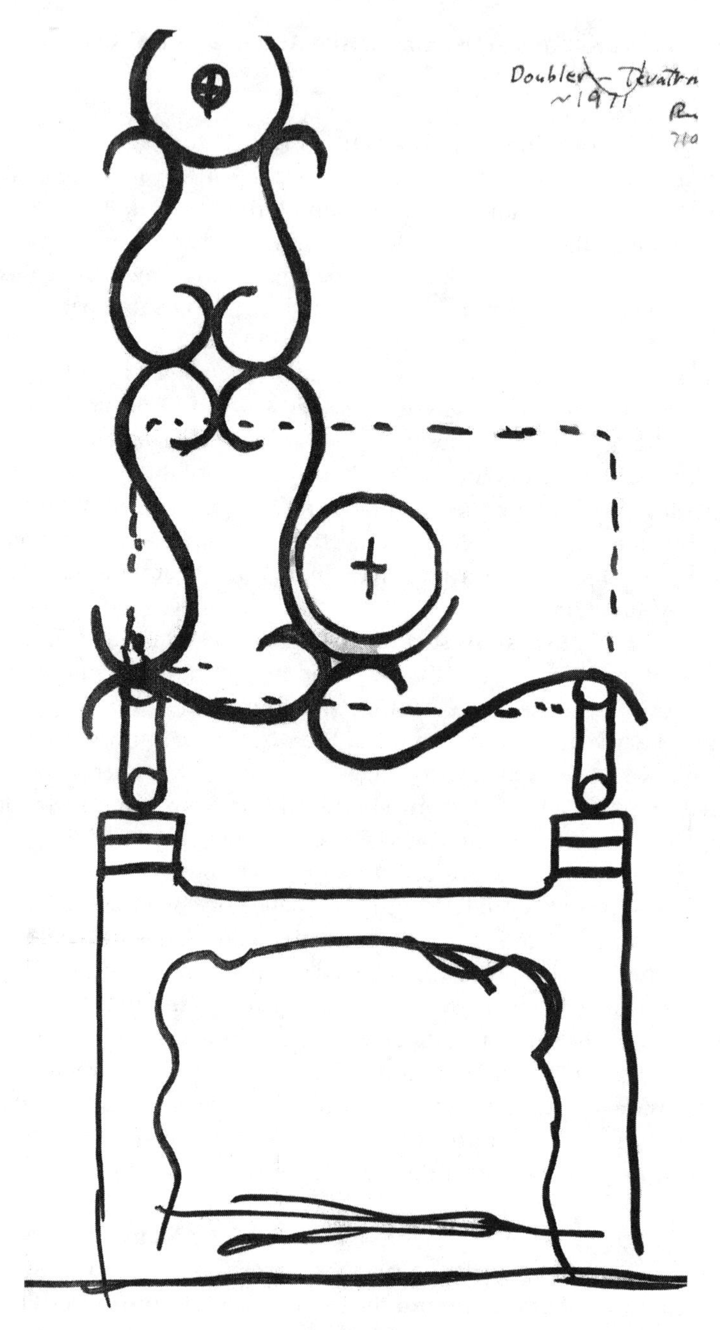

8.1 Drawing of a magnet design for the Energy Doubler by Robert R. Wilson, 1971. (Courtesy of FNAL Visual Media Services.)

idea," he stated, "is to take the protons out of the present magnet ring and then inject them into the new ring of superconducting magnets . . . piggy back upon the other." With the added ring, he continued, "we could then double the energy. . . . If the protons were transferred at 500 BeV, the energy could become 1,000 BeV." He fantasized about multiplying the idea: "You do not see all of the tunnel in the picture, but it is completely empty, and one could install one of these rings after another, taking the beam from one to the next, doubling the energy each time." He estimated that "because the bore of the new magnets would be so small, because no new tunnel or buildings would have to be constructed, we can hope to be able to build such a device for less than $20 million, possibly even for less than $10 million. All of these considerations, it must be emphasized, are based only on the most preliminary of studies."[19] Seven years later, this optimistic estimate of the Doubler's cost—but not Wilson's reminder of its "most preliminary" nature—would be recalled as Washington DC officials reviewed proposals requesting twice this amount for the Doubler.[20]

The AEC responded positively. On July 19, 1971, the director of the AEC Division of Research, Paul McDaniel, asked Wilson and his staff to "perform the necessary work in the coming fiscal year to clearly define the scope of this undertaking and to ascertain whether the inclusion of energy doublers can be achieved within the $250 million authorized for this project [the NAL]."[21] That summer, Ron Oram, Robert Sheldon, and Bruce Strauss built the first prototype Doubler magnet, a half-meter-long beam-focusing quadrupole. They concluded that "a suitable doubler magnet can be produced at a not prohibitive cost, and that further development work in this area would probably be rewarding."[22]

Start of the Doubler Effort, 1972–1975

The initial wave of optimism about the Doubler continued for seven years. On March 1, 1972, the Doubler received informal authorization when Wilson lifted his edict barring work on the Doubler to help celebrate the achievement of NAL's first 200 GeV beam on that day. A week earlier, two NAL physicists, William B. Fowler and Paul Reardon, had prepared a short proposal outlining a preliminary organizational plan. Fowler had previously worked on superconducting magnets used with bubble chambers; Reardon had led the Booster as well as the business office. The informal document included a several-year management and testing plan, with proposed magnet tests performed in the Protomain, the experimental area in the Fermilab Village replicating some 200 feet of the Main Ring tunnel.[23] While this proposal underestimated the complexity of building the Doubler, it began the project.

Six months later, on September 1, Wilson established an informal working group under Paul Reardon to consider the technical questions involved in building the Doubler. "I referred to it as the rag-tag outfit," recalled Reardon. "I essentially looked for people who were interested in new technology who didn't fit in where they were in a nice way, and formed a group of them who really wanted to work hard together." Taking possession of the blue building in the Village that had been used earlier for the Main Ring as well as the Protomain, the new group met twice a week, its meetings open to anyone interested.[24] Special invitations were extended to a number of experts (Fowler, Reardon, Strauss, Don Edwards, Hinterberger, Malamud, Donald Miller, Rae Stiening, Teng, Young, and McDaniel). Wilson's handwritten notes for their first meeting included a sketch of the Doubler's placement at four possible locations.[25] In this period the group still expected that the Doubler would be built above the Main Ring's trajectory, probably hung from the tunnel's ceiling, or resting on the spherical orbs erected on the existing magnets.

In its first year of work, the Doubler group produced a small number of superconducting magnets—two dipoles (the 3-foot Mark I, of "pancake" design, and the 1-foot Mark I, of "shell" design), both tested successfully in January 1973.[26] That month construction began on a 1-meter-long pancake model (P6C) with a 5-centimeter bore.[27] Refrigeration tests in the Protomain suggested that it would be practical to cool long strings of magnets to liquid helium temperatures.[28] Relying on makeshift equipment, including an old liquefier from an industrial plant in Hightstown, New Jersey, Peter VanderArend, Stanley Stoy, and Donald Richied designed a long helium transfer line with a 400-foot helium pump loop.[29]

The project was still without formal authorization in January 1973 when the Doubler group moved into the Accelerator Division, then led by Reardon. The group's participants had other primary responsibilities and worked on the Doubler in their spare time. As Reardon reported in May 1973, "The design effort is proceeding at NAL with the full understanding that the Doubler is not a critical project and has a low priority and that other more important activities take precedence whenever the needs arise."[30] The work was still limited to the development of models, both for magnets and for refrigeration systems; the group did not yet concern itself with arranging components into a working accelerator.

At this stage Wilson did not see the Doubler as a construction project, but rather as an experiment in how to build a superconducting accelerator. He estimated that the Doubler would cost less than $20 million, about the amount left over from building the Main Ring in Wilson's frugal way. This was not a lot of money for an accelerator of the envisioned scope. In December 1973, Goldwasser asked the AEC Chicago area manager, Frederick Mattmueller, whether he would authorize the Doubler "within the construction funds currently available to the Laboratory."[31] Within weeks the Chicago office of the AEC granted this approval.[32] The URA board of trustees encouraged NAL to proceed with the Doubler and prepare a financial plan for the disposition of the balance of NAL's construction funds. But the plan encountered a roadblock when the AEC rejected Wilson's request to spend the money left from the Main Ring's construction for the Doubler. In October 1974, to insure compliance, the AEC removed $6.5 million from the original authorization, stating that these funds were needed elsewhere.[33]

Despite this setback, Wilson's Doubler group, which by this time had grown to roughly thirty members, proceeded to make important technical decisions concerning the material of the beam tube and cryostat and the best methods of producing components. For the magnets, Wilson chose a warm-iron design in which the iron was placed outside the cryostat (rather than inside, as in the cold-iron design selected for the ISABELLE magnets). Although delivering a somewhat lower field, the warm-iron design cost less because, the magnets were physically smaller and therefore easier to cool and support. The team obtained a 1,500-watt helium refrigerator for testing each magnet before installation and purchased about 15,000 pounds of niobium-titanium for the magnet's wire. Fermilab provided this material along with detailed specifications to a number of outside firms who were eager to help with processing the superconducting wire.[34]

Another decision concerned how to arrange the wire in making the superconducting cable for the magnets. The group chose a twisted multifilament-strand cable design, developed in the early 1970s at the Rutherford Laboratory for the GESSS effort. This Rutherford cable replaced the monolithic cable (the so-called Danby wire) that the Doubler group initially had tried. Although the monolithic cable was mechanically stronger, the Rutherford cable reduced through geometric cancellation some magnetic problems interfering with magnet strength, such as hysteresis and eddy current loss.[35] Other advantages of the Rutherford cable included its ratio of superconductor to copper and its good packing fractions. This choice of multifilament-strand cable differentiated the cable used in the Doubler magnets from the braided cable being used by ISABELLE for its superconducting magnets since the summer of

1971. (Later observers would point to ISABELLE's braided cable as the fatal flaw of that project.)

The Doubler group was intensely concerned with the fact that superconducting magnets quenched at lower than the design field. To minimize its "training," and for greater economy, the Doubler group decided to wind its magnet coils into the three-dimensional saddle (shell) configuration favored by Hank Hinterberger, rather than into the flat pancake design favored by Reardon and by Willard Hanson, the inventive engineer then head of the laboratory's magnet facility. The pancake offered better wire placement and easier winding of the coils, but the saddle-wound magnets offered better field quality.[36] In August 1973, the saddle-wound magnet program advanced to the "dual dipoles" project, in which two identical 29-inch, saddle-wound, dipole magnets were successfully built and tested. The experimental question was: could two magnets be made sufficiently close in their characteristics to allow them to be used together in an accelerator?[37]

In February 1974, the first full-length prototype 20-foot saddle-style dipole magnet was ready to test. Composed of four shells of Rutherford-like cable, the early saddle magnet design showed poor training; many quenches were needed before the magnets attained full field. To track down the problem, Wilson had a model made with a single shell. It produced quite a low field but trained in a few quenches. On that basis he gambled: he authorized going from four shells to two, at the same time expanding the width of the cable from $1/4''$ to $3/8''$ by increasing the number of strands in the cable from seventeen to twenty-three. This gamble paid off. By the end of March 1975, Fermilab's new "D series" magnets were working well. A large effort was made to design circuits to detect and protect against quenches.[38]

Further progress was made as well on the refrigeration system for cooling the magnets to temperatures low enough for them to turn superconducting. Early in 1973, VanderArend and Fowler developed a "counter flow" refrigeration scheme in which single-phase liquid helium from a reservoir would be compressed, pass through the magnets for some 120 meters, and flow through a valve that would divide the helium into two phases (through partial boiling into vapor and cooling of the helium steam). This two-phase system of liquid helium and vapor would then be returned. The idea was tested in the Protomain; as there were not yet any magnets, transfer lines containing heaters simulated magnet loads. The next step was to implement the system with two dozen independent and constantly running refrigerators installed around the ring. Testing of this concept was delayed, however, by litigation between VanderArend's company, Cryogenics Consultants, Inc. (CCI), and the Airco cryogenics company over the contract for a prototype refrigerator.[39]

In line with Wilson's preference that individuals and groups take a particular project from beginning to end, the original Doubler effort was not "naturally disposed," as Appel put it, toward developing vendors. But over time, the Doubler group learned to work with vendors and produce "bid packages" that included performance specifications, the supply expectations of the vendor and laboratory, quality testing procedures, responsibilities regarding failures and procurement, and the steps for transferring the responsibility for completing their technology to industry.[40] In November 1975, to help streamline the business side, James Finks, who had been working in Fermilab's business office, was moved directly into the Doubler group to handle a variety of administrative tasks such as arranging contracts. This move was dictated by Wilson's more general policy of decentralizing business dealings in the laboratory. By allowing scientists to make their own agreements with outside vendors, rather than delegating this chore to a central business office, Wilson hoped to circumvent conflicts between projects having different priorities.[41]

Even as the Doubler took steps toward becoming a large-scale project, the group working on the effort retained the character of a club of engineers and physicists led by Wilson, who dominated the discussions and controlled its decisions. Fowler was Wilson's associate director. Expertise was not a quality that Wilson particularly valued; only Fowler and Strauss had any previous experience with superconducting magnets.[42] Engineer George Biallas recalled that Wilson's "spirit of 'go out and push and throw together' . . . really worked at the beginning." He claimed that what saved the group "was Fowler's insistence on quality control, almost hiding it from Wilson at those early meetings."[43]

Within the larger laboratory, the Doubler effort still had low priority relative to the Main Ring. Only gradually did it become apparent that the Doubler's challenges were of such magnitude that they could not be overcome without experts and with a total commitment from the laboratory.

All-Out Research and Development, 1975–1978

By the spring of 1975, the Doubler group was in need of new scientific staff. In the previous year, in an effort to inject more scientific discipline into the group, Wilson had added Darrell Drickey, but Drickey took ill and died. Two other physicists in the group, Reardon and David Sutter, planned to leave Fermilab. Tollestrup, the soft-spoken physicist from Caltech, would provide what the Doubler needed. He arrived at Fermilab on April 1, 1975. Already active in the experimental program, Tollestrup was then starting a year of sabbatical. He planned to work on the Doubler magnets for nine months and spend the rest of his sabbatical at CERN. He "brought a whole new technical light to the project, which absolutely saved it," Biallas recalled.[44] According to Livdahl, Tollestrup "had a remarkable ability . . . to break every problem down to the sophomore [physics] level."[45]

By the end of 1975, Tollestrup had already made two pivotal contributions to the Doubler project. The first related to the strong mechanical support that superconducting magnets need in order to provide field uniformity and prevent movement caused by the strong electromagnetic forces that act on current-carrying conductors. Even slight motion generates heat, which can cause a quench. At the time Tollestrup joined the group, the Doubler magnets were supported by an endoskeletal porcelain ring supplemented on the outside by a series of steel bands. These spiraled in the opposite direction of the coil windings to balance the torque causing the magnet to unwind. The Doubler group, especially Hanson, had developed this structure over a period of several years and thus had a considerable investment in it. They had discarded an earlier proposal for a bolted "horseshoe collar" developed by Reardon, Hanson, and Hinterberger, because Wilson considered it too expensive.[46]

As a newcomer, Tollestrup was in a position to inject fresh perspectives. Taking a cold look at Hanson's banding scheme, he noticed that the entire bending movement was taken up by the rings inside the magnets. Calculating the stress on the rings in November 1975, he predicted that in the case of full-length magnets all the rings would break. Experimental tests subsequently performed on magnets D-17 and E-17 supported Tollestrup's calculation; the porcelain rings cracked either at 90° or 180°K, just as Tollestrup predicted.[47] In addition, the magnet itself unwound by 10° or 15°K. After this discovery, Tollestrup decided to spend a longer time at Fermilab and canceled his plan to visit CERN.

He spent Christmas vacation that year at Brookhaven, During this time he worked further on the support problems of Fermilab's superconducting magnets and after a few weeks proposed a new kind of steel support, which the group started calling Tollestrup's "collar."[48] The magnet windings were held by an external clamp. Then Hinterberger and Hanson redesigned Tollestrup's collar into a better-engineered concept, and the group showed that it was possible to "keystone" the magnet wire (i.e., form its cross section into trapezoidal sections that made better contact and reduced shorts without degrading it), so it could pack densely when wound into a Roman arch–like structure, supported and prestressed from the outside.[49] This new Tollestrup-Hinterberger-Hanson collar design evolved into the final "clamshell" design, in which collars are placed loosely on the magnets at room temperature and then squeezed under 3,000 pounds per inch of length, so that the windings remain compressed even after cooling, despite the strong electromagnetic forces they experience. Such "prestressing" offered good results for all values of the magnetic field. Because of the flexibility with which Wilson led the Doubler project at this stage, the group was able to complete the first full-scale (22-foot) collared magnet by December 1976. According to Livdahl, Wilson "was willing to try things that didn't have a high probability of working but were worth some effort." And if an idea then looked poor, he was "perfectly willing to drop it."[50]

Tollestrup's second crucial contribution to the Doubler project during his first year at Fermilab was his suggestion to wrap the magnet wire with Kapton insulation. Concerned about electrical shorts arising from the solder used to attach coils, or from metallic chips produced during wire preparation, Tollestrup and his coworkers developed a way to clean the wire ultrasonically. When the shorts continued, he tried to insulate the wire with Mylar or Kapton material. To his surprise, the Kapton not only removed the shorts but also solved a more important problem: the magnets made of wire which had been wound in Kapton required significantly less training. The Kapton seemed to form a cocoon around the wire that helped to insulate it against heat generated by friction as the wire moved. Because of its slipperiness, the Kapton also reduced scraping.[51]

During 1975 and 1976, Fermilab received a number of proposals for colliding-beam experiments but Wilson rejected them all, because he believed they would delay progress on the Doubler. But he did ask Tollestrup to organize a two-day meeting to explore the whole question of colliding beams within the context of finding uses for the new Doubler.[52] Tollestrup's "Modest Colliding-Beams Meeting," in January 1976, launched Fermilab's colliding-beams program (see chapter 12). As part of their discussions the attendees raised the question of placing the Doubler below, rather than above, the Main Ring, so that the Doubler could be used more effectively in colliding-beams work. But to discuss this question seriously it was important to know whether enough space had been left for building the Doubler below the Main Ring. A group of

8.2 Cover of the design report for the Energy Doubler, 1976. (Courtesy of Angela Gonzales.)

the participants visited the Main Ring tunnel to determine this. Riding around the ring on bicycles, they found plenty of space.[53] Wilson had issued an order years earlier to keep the space clear, and, as Tollestrup realized, "he had forgotten this. But Bill Fowler reminded him." (See fig. 8.2.)

Now the group proceeded to weigh the pros and cons of building the Doubler below the Main Ring. They decided that it would make good sense. As Tollestrup explained, "The original position was at the top of the tunnel. . . . That is where Helen had installed a first superconducting magnet test." The problem with that position was that "the radius was different and the two beams were separated by a large distance, making them very hard to get into collisions." That would have caused the luminosity to be too low. Another issue concerned the space in the Main Ring just above the original magnets. It was crowded, and it would be difficult to install the Doubler there.[54] Noting Lee Teng's demonstration that magnets with circular apertures, rather than elliptically shaped ones, would facilitate using the new superconducting ring as both a storage ring and an accelerator, the group began to adjust the design of the Doubler magnets to provide for colliding-beams experiments.[55]

Then an intensive program of research on superconducting magnets began. Some of the research was conducted by a group led by Yamada based outside the Doubler group. At times Yamada's group competed with the Doubler group. For example, while the Doubler group was testing isolated magnet coils in vertical dewars, Yamada's group was testing short prototype magnets in their cryostats, and it also studied various magnetic properties (including magnetization, hysteresis, and alternating current loss) of short samples of superconducting cable and wire using a computerized data bank that tracked correlations between the maximum quench current properties of short samples, such as composition, resistance, temperature, and pressure. This work demonstrated conclusively that Rutherford cable was superior to the other alternatives and that the addition of solder greatly increased eddy currents, one source of quenching. As a result, the use of solder in bonding cable strands of Doubler magnets was discontinued.[56] The work of Yamada's group also showed that the alternating current (ac) loss of short magnets, an energy loss due to induced currents that can heat the strands of superconductor cable and cause a quench, increases abruptly due to mechanical deformation of the magnets. The results led to the building of a "zebra" conductor, a 50-50 mixture of strands coated with "ebonal" (oxidized copper) and strands coated with "stabrite" (silver-tin), a conductor favorable for stabilization and having low ac loss. Later Yamada's group constructed the Magnet Test Facility where the final Doubler magnets would be tested.

A novel component of the Doubler's program was its in-house magnet factory, an assembly-line construction shop manned by a staff of about ten reporting to Hanson. Wilson had built this factory in the period when he was building the Main Ring magnets, believing it could serve as a research tool for structuring magnet technologies and to allow better-controlled and more economical assembling of magnets on-site. This factory proved crucial in the development of the Doubler magnets. Being able to produce magnets in-house allowed rapid production of more than one hundred sample magnets. By building prototypes one at a time, it was possible to observe their behavior and use that in designing subsequent magnets, even large ones. This empirical approach to designing magnets resembled the "cut-and-try" methodology used in developing the atomic bomb at Los Alamos.[57] Karl Koepke recalled that "we built almost a hundred one-foot magnets" in the years 1975–1978, "because we knew so little about what we were doing."[58] Although "people may think that we went in a logical progression from the first magnet to the last magnet," in fact "we were picking the best of the designs that followed from the one-foot experience, and those we built concurrently in longer lengths, like five-foot, ten-foot, and finally 22-foot."[59] He went on to explain: "It took us close to a couple of months to build a 22-foot magnet in the early days, whereas the one-footers would be completely done in a week. So what we did was we built the short magnets, tested different cable, different insulation schemes, different geometry, different pre-load, to see what might work, no matter how wild the idea was, since it didn't take much effort to test it.... And then whenever a short magnet tested well... we would then try to incorporate those design features in the long magnets."

Some of the testing involved full-scale magnets. Tollestrup observed that "in a quantitative sense, you can't walk up to a magnet and predict that this will take three quenches to train and predict very well where the quenches will be.... The short ones would train in a few quenches and the question with the 22-foot ones was: Is it going to take 22 times as many quenches to train? It turned out remarkably enough that it didn't, that they all trained in just a few quenches. So that was an exciting discovery. But it had to come from full-scale magnets."[60] The Doubler's first 20-foot dipole magnet was wound at the magnet facility in January 1976. Over the next year, the dipoles grew to 22 feet and remained at that length until 1979, when Lederman, after succeeding Wilson as director, accepted the recommendation to trim them down to 21 feet (see chapter 10).

The Doubler group's approach to magnet development expressed Wilson's persona as a craftsman who worked with real materials toward practical ends. ISABELLE's typical approach was to focus research on a relatively small number of prototypes in the hope of finding the perfect design to guide industrial production. In contrast, during 1977–1978 alone, the Doubler group built hundreds of prototypes, roughly ten times as many magnets as the ISABELLE group, in an effort to empirically develop a workable design. Lundy compared the ISABELLE approach to classical procurement, in which one develops a set of blueprint specifications with a contract to bind the manufacturing corporation. The idea behind the approach, Lundy explained, was "private enterprise is ingenious and they will find cost-cutting ways to build it that you would never have dreamed of." The flaw in this reasoning was that it assumed one understood the complex behavior of the superconducting magnets.[61] Moreover, in the standard approach it was often difficult, or impossible, to communicate to the manufacturer any changes that came from new understanding. Waiting for industry to respond to such changes was usually slow and costly.

An additional advantage of Fermilab's approach to magnet development was that in fixing the magnet design, one also determined the optimal tooling for making many identical magnets. Wilson knew that developing the tooling for creating one thousand identical magnets had to be an integral part of the Doubler program. He recognized that only if the magnets were built exactly alike could strings of magnets function together satisfactorily. As Lundy reflected, the "properties of the magnet were determined in complicated ways by the kind of tooling it was built with, and how it was built. We didn't understand the inner workings of that process." He elucidated, "The magnet was a complicated enough device that apparently harmless changes in the tooling often had disastrous results." It was therefore important "to develop tooling to make a good magnet and then not to change the tooling; in fact, in an almost superstitious way, not to tamper with it until we understood the effects."[62] However obvious this argument might appear in hindsight, until the 1980s no other physics laboratory took these arguments into consideration in building superconducting magnets.

Fully aware of how little was understood about the complex behavior of the superconducting magnets, Tollestrup insisted that only one factor should be varied at a time in this process. In this way the cause of any new behavior could be isolated. The ISABELLE team could not afford such control because it was studying too few magnets. When the ISABELLE group's famous Mark V prototype reached a 5 Tesla field in 1977, well above its specification of 4 Tesla, the Brookhaven group decided to double ISABELLE's design energy.[63] But ISABELLE fell on hard times when the rest of its magnets failed to perform as well as Mark V. And by 1982, when Brookhaven finally understood the reasons behind its magnet problems and redesigned the magnets, it was too late. Events were unfolding (see chapter 13) that would bring DOE to terminate ISABELLE, by now renamed the CBA, Colliding Beams Accelerator.

Tollestrup's group addressed a variety of magnet design problems, such as field quality, mechanical constraints, reproducibility, and training, often with new analytical tools. For instance, Robert Flora devised the "scissometer" for making measurements on magnet coils immersed in dewars filled with liquid helium. Crossed glass fibers (the scissors) impregnated with epoxy were attached to the immersed coil. From the displacement of the fibers an indicator outside the dewar would reveal how far the coil moved when the magnets were excited.

The most challenging problem concerned magnet training: how could a magnet be brought to full field before it quenched? The team members became experts in measuring the properties of the superconductor and calculating from these measurements how much current and field a magnet could support before quenching. Continuing studies using circuits built by technician Wally Abrilowitz and others explored four main causes of quench: mechanical motion, eddy currents, wire quality, and cryogenic effects. Stiening, who had played a crucial role in commissioning the Main Ring in a period in which there had been little diagnostic instrumentation, now initiated the use of microprocessors in dealing with quench protection. Flora developed these devices further. The critical issue was how to detect the onset of a quench early enough to treat it, for example, by placing a sensing coil in the "throat" of the magnet. Sutter developed the use of microphones to amplify the sound of the quench to help locate its origin. A related problem was to design special circuits that quickly shut the system down when necessary.[64]

In developing a system based on large numbers of magnets working in concert, strings of magnets must be studied together. For this purpose, a special aboveground area was built simulating the Main Ring tunnel; it was referred to as B12, because of its location near the B12 station of the Main Ring.[65] The motivation for creating this B12 area arose in a May 1975 experiment in which Helen Edwards and Claus Rode were studying beam transport through two superconducting magnets (one 3 feet and the other 10 feet) hung from the tunnel ceiling.[66] The cooling system, located outside the tunnel in two trailers, had been connected to the magnets by an 80-foot-long helium transfer line.[67] To house the area, William Reay built what was called the "awning," because Wilson's first idea for protecting the area had been to use an awning. Among the problems studied in this environment were installation and vacuum problems, cryogenic operation, quench protection in magnet strings, power systems, refrigeration, and prototyping of control systems. As the test continued into winter, the awning had to be replaced by a regular building.

The tedious quench tests conducted at B12 wracked the nerves of the physicists involved. The tense, isolated, and continuous effort burned out researchers unused to "working where the wolves are howling and the blizzards are blowing, and . . . [with] a long string of magnets which at any moment might do terrible things."[68] Peter Limon, the exuberant experimentalist who conducted the first B12 tests in February 1976 on a 10-foot magnet, remembered one early study in which he suddenly heard a "tremendous hissing noise . . . like a 747 taking off." The area soon "filled with fog, with little pieces of super-insulation floating by." He is reported to have run out so fast that he left a footprint on the back of the white lab coat of a technician he passed over. Koepke recalled: "As soon as a quench occurred we'd hear this bang and then . . . this roar for about 3 or 4 minutes as the helium exhausted through these vents. Anyone standing outside would see this white vapor cloud coming through the cracks in the building and the doors as if the building were on fire. . . . I got used to walking through that vapor cloud by crawling against the wall to keep my bearings, because you can't see anything."[69] Initially the tests at B12 involved only a small number of magnets, but eventually they tested strings of up to sixteen 22-foot Doubler magnets. The tests continued for five years, until late spring of 1981, when magnet installation in the Main Ring tunnel took priority.

Meanwhile, substantial progress was being made on the Doubler's refrigeration system. Early in 1975, while litigation between the CCI and Airco companies held up work on the original counterflow refrigeration scheme, VanderArend and Fowler had a new idea, which VanderArend subsequently patented.[70] They conceived of building a central helium liquefier with a system of dewars arranged around the ring, and with smaller satellite refrigerators that increased the number of liters of liquid helium produced. The satellites would consume liquid helium (supplied by the Linde Company) and turn this into vapor. The helium vapor would then be returned to gas storage tanks at the central liquefier to be reliquefied. And every twenty-four hours liquid helium trucked around the ring from the central liquefier would refill the satellite dewars. They saved $8 million by using a recycling scheme in which the group turned surplus air compressors found at the Defense Department's Santa Susanna rocket engine test station near Los Angeles into helium compressors for the central liquefier.[71] The trucking to the dewars was later replaced by a four-mile-long helium transfer line, then managed by experimentalist Ken Stanfield. The central helium liquefier building was completed in 1978. To further cut costs, the construction of numerous components, including heat exchangers and distribution dewars, was turned over to a local machine shop.[72]

The Doubler's Funding Crisis

By 1978, research on the magnets and cooling system had solved most of the technological problems of the Doubler's individual components, but it was by no means yet clear that the Doubler would succeed as an accelerator. For as Livdahl, who had replaced Reardon as head of the accelerator, reflected, the components of the Doubler were then still far from able to work together in an accelerator system.[73] And the R & D funding needed to continue work on the Doubler was as uncertain as it had been throughout the previous decade. Meanwhile, a number of planned design changes, such as an increase of the magnet aperture proposed by Teng, were threatening to increase the Doubler's cost.

Wilson's attempts to secure funds for the Doubler suffered rejection. Throughout the 1960s and early 1970s, Wilson had dealt comfortably with the JCAE and with the AEC itself, an agency in which his Berkeley and Los Alamos connections gave him an insider's access to AEC chairman Glenn Seaborg, another of Lawrence's boys. But the JCAE's authority vanished in January 1975 when President Gerald Ford's administration replaced the AEC with two new organizations: the Energy Research and Development Administration (ERDA) and the Nuclear Regulatory Commission (NRC). Whereas the AEC, led by a civilian board, had been a special-purpose agency, with the JCAE as its congressional oversight committee, ERDA and the NRC were overseen by a number of congressional committees. Further reorganization in October 1977 would produce a new cabinet-level Department of Energy (DOE), replacing ERDA, to address the myriad of energy issues confronting the country. And while the AEC had concentrated on basic nuclear-related technology, including accelerators, ERDA's and DOE's mission covered the full range of the nation's energy needs. Without the special status the AEC had enjoyed, ERDA and DOE had to fend for themselves in the competition for funds that were also claimed by poverty, highway, and defense programs.[74] The difficulties of finding adequate funding were magnified during the energy crisis of the mid-1970s.

In 1976 Wilson's dilemma of finding construction funds for the Doubler was frustrated by ERDA's commitment to ISABELLE.[75] In 1977 a HEPAP subpanel advised ERDA to authorize ISABELLE as a construction project; its groundbreaking was set for October 1978, and $23 million was authorized for construction in fiscal year 1979 despite ISABELLE's failure to build reliably working superconducting magnets. Although the HEPAP panel recognized extensive progress in the Doubler effort, it viewed ISABELLE, not the Doubler, as the "next natural step beyond the only other existing device of its kind, the ISR (Intersecting Storage Ring) at CERN." ISABELLE researchers believed they had triumphed when the panel recommended that only $12.8 million be authorized for the Doubler in fiscal year 1979.[76]

The Doubler project thus faltered. In October 1977, Wilson wrote to James R. Schlesinger, President Jimmy Carter's secretary of energy, to complain about "the critical lack of support of the Fermi National Accelerator Laboratory." Pointing out that CERN, with a budget 2.5 times the size of Fermilab's, was threatening to "overwhelm us," he explained that the Doubler offered America a chance to "regain the advantages of uniqueness" while reducing Fermilab's annual electric power bill by $5 million. The letter closed with these fateful words: "My own continued participation as Director will depend on a change in the laboratory's present dreary expectations for the future."[77] For Wilson the lack of funding for the project in which he had invested much of his creative energy in the previous six years meant merely administering a laboratory whose frontier had closed.[78]

Wilson's plea to the secretary of energy had three strikes against it. First, by insisting earlier that there was only a 50-50 chance that the Doubler would work, Wilson had conveyed to Washington the impression that the Doubler's technology was not yet ripe.[79] Second, Wilson's habit of proposing low budgets, which had helped him in the early days of Fermilab, now made Washington wonder whether Fermilab's Doubler effort was well conceived. Finally, and perhaps most important, Wilson's power base in Washington had seriously diminished with the change from the AEC to ERDA and from ERDA to DOE in less than three years.

As Wilson's relations with DOE grew more difficult, he lost heart as the director of Fermilab. "The goddamned place was a great big bureaucracy and I could no longer have any effect," he explained. Worst of all, the bureaucracy could be traced to *him*! "To see this place built into a bureaucracy just used to pain me, just pained me like somebody sticking a sword in me."[80] And to make matters even worse, John Deutch, DOE's new director of energy research, interpreted Wilson's request for support of the Doubler as inappropriate criticism of another DOE project, Brookhaven's ISABELLE.[81]

Wilson's Farewell

From January 25 to 27, 1978, a severe blizzard struck the laboratory. Record snowfalls and winds gusting to more than one hundred miles per hour canceled an on-site conference and sent employees home early. As the temperature fell below 10°, the security staff and the roads and grounds crews took charge of the site. Stranded visitors to the laboratory took refuge from the storm; a few camped in the lobbies of the modest dormitories and other visitor housing in structures from the former village of Weston.

While the storm raged across the Fermilab site, Wilson resolved his own personal storm. On February 9, 1978, he submitted his resignation to Ramsey.[82] On the same day, he circulated a letter of clarification to the members of his staff, his "fellow-workers at Fermilab." He explained that for some time he had "been deeply concerned that the future of Fermilab is seriously jeopardized by the low rate of funding." He had offered a warning in his "fight for better funding" that he "would resign unless the rate was substantially increased." As President Carter's budget did not provide such an increase, "regretfully, I have submitted my resignation to the Trustees of the URA through its president, Norman Ramsey. This, of course, makes me very sad, but I will not have done right by you, or by myself, without having pushed this fight to the bitter end."[83]

The news of Wilson's resignation was reported a week later, on February 16, 1978, in the *Village Crier*, the laboratory's newsletter, which dated back to the early days of working in the former village of Weston. The lead article included passages from Wilson's letter to URA. "Presently we are operating at about half of our capacity to do physics experiments." Referring to the financial resources of CERN, "which are considerably more than double our own," he explained, "our scheme to leap-frog their financial advantage by increasing the Fermilab proton energy to 1,000 GeV through the application of superconductivity has been confounded by indecisive and subminimal support, as have been our modest proposals for intersecting beams."[84]

As soon as Tollestrup learned of Wilson's resignation he made his way to Wilson's on-site home bearing a bottle of inexpensive wine that he had hastily purchased at a nearby store. Tollestrup sat down with Bob and Jane Wilson at their kitchen table. The three drank the wine and spoke at length. Then Bob descended to the cellar and returned with an expensive bottle of wine that he had put away for the ten-year anniversary of the laboratory. Over Jane's complaints about drinking the special wine just then, they consumed that too, while recalling events of the previous decade.[85] A week later, Harold K. Ticho, chair of the URA search committee for a new director, informed Fermilab's staff that the committee was developing criteria to guide the search. He asked for "an expression of your views" and explained that "Dr. Wilson has kindly agreed to continue to serve as Director until a suitable replacement can be found."[86]

During the following months, Wilson's deep frustration sometimes erupted. On May 5, 1978, he distributed copies of a handwritten letter to all employees at the laboratory:

> Dear Colleagues,
> An all too common failing of large institutions is to fall into the bureaucratic morass—complicated procedures, red tape, and all that. That's Terrible. Let's try hard to keep the good old, can-do, informal spirit of Fermilab alive! I ask each of you to be intolerant of creeping bureaucracy.[87]

To help him adjust to the approaching end of his time as director, Wilson worked on a metal sculpture intended for the reflecting pond in front of the High-Rise. "If you're going to leave, it seems to me you have to leave a kind of a signature," he explained in an interview.[88] His parting gift to Fermilab was a striking 32-foot-high steel plate hyperbolic obelisk. When he learned that professional welders would charge $20,000, he said, "That's ridiculous. It's too expensive. . . . Oh hell, I'll do the welding." When the welders told him, "You can't weld here," he asked, "Why can't I? I'm the Director, I can do anything I please." They replied. "You can do anything you please, but we'll walk out. This is a union shop and you're not a union welder." So Wilson joined the union and became an apprentice to James Forester, the machine shop's master welder. "Learning welding from him is like learning physics from Fermi," Wilson later remarked, "He'd hold my hand and tell me how to do it." Wilson worked on the sculpture at every possible opportunity, both before and after working hours.[89]

He named the sculpture *Acqua alle Funi*. The Italian expression had been used at NAL as a rallying cry in the final months of constructing the Main Ring. During those earlier difficult times, Wilson had told his staff about an incident said to have occurred in Rome during the sixteenth century when an Egyptian obelisk was being raised to the vertical in St. Peter's Square in the Vatican. As hundreds of horses and nearly one thousand men tugged on ropes attached to the structure, the chief erector, Domenico Fontana, issued his commands while the assembled crowd of citizens and dignitaries watched under an order for complete silence.

When that earlier obelisk had been raised to about 45°, the midday sun began to heat the ropes, causing them to lengthen and slide on the capstans. As the obelisk started to sag, an old Genoese sailor named Bresca di Bordighera suddenly began to shout "acqua alle funi!" (water to the ropes!). The frightened crowd expected Fontana to order Bresca executed for breaking the silence. Instead he ordered the erectors to throw water on the ropes. That caused the ropes to contract and stick to the capstans. After this, the erection of the obelisk proceeded successfully.[90]

The evening of May 9, 1978, found Ramsey and Wilson sitting in a skiff in the reflecting pond in front of the High-Rise. The small boat also bore a bottle of champagne. Ramsey slowly rowed the boat across the pond in the dark for a small ceremony to accompany the erection and christening of Wilson's obelisk. When a slight problem occurred, Wilson cried out to a cheering group: "Water to the ropes!"[91] (See fig. 8.3.)

Wilson's anguish enlivened one of his early interviews with Hoddeson conducted for the Fermilab History and Archives Project, which Wilson established during his last year as director. From her point of view the interview was not going well, for Wilson was merely retelling well-known stories about the start of NAL. A knock on the door interrupted the interview.[92] Wilson was called to an urgent phone call from URA. He politely excused himself.

Whenever possible, Wilson would avoid answering his phone. His administrative assistants recalled that he would disconnect his own phone and sometimes bury it in a large plant that stood on the floor of his office.[93] "Bad news usually comes by telephone," Wilson told Hoddeson.[94] When he returned to the History Room, he was visibly upset, but he wanted to proceed with the interview. His tone, which had earlier been rather formal, was now animated and relaxed. He spoke of his passions and ambitions, and how he and his staff boldly overcame the countless obstacles that they had faced in building NAL. Hoddeson never learned exactly what the call had been about, but there was talk around the laboratory that URA had formally accepted Wilson's resignation that day, a result he had not expected.[95]

While the laboratory adjusted to Wilson's resignation, Goldwasser made his decision to return to the University of Illinois, accepting the position of vice chancellor for academic affairs. Like Wilson, Goldwasser felt he had grown out of touch with experimental physics. He wanted to try university administration. Wilson informed his staff of Goldwasser's departure in a letter to them on May 24, 1978: "Although our best wishes go with him in this important new endeavor, along with our congratulations to the U.I., we cannot help but give expression to deep anguish at

8.3 Robert R. Wilson's obelisk sculpture *Acqua alle Funi*, 1978. (Courtesy of FNAL Visual Media Services.)

the loss of a founder and creator of Fermilab." On a more personal note, he added: "I have been honored and privileged to be associated with this great physicist and lovable man in the adventure of Fermilab."[96] After his retirement tribute in August 1978 a small park of trees, "Goldwasser Grove," was planted onsite at the end of Wilson Street.

Two weeks later Goldwasser explained in his own letter to the laboratory that he had been on leave from the University of Illinois for eleven years. Although "asked several times to return to Urbana," he had resisted until now, "because it seemed to me that my experience and skills qualified me best for the job that was to be done here."[97] He said that a major change had occurred during 1977–1978, raising "all sorts of new uncertainties" about the laboratory's direction. The change was "a result of the furor generated by the funding crisis at Fermilab, by Bob's meeting that crisis with a tendered resignation, and by his eventual decision to make the resignation final." Given these uncertainties, Goldwasser felt he "could no longer responsibly gamble on mere conjectures."[98]

When Wilson formally stepped down on July 17, 1978, he circulated yet another letter to his staff:

My dear Colleagues;
With great pride in the privilege of having served with you in creating Fermilab, but in anguish and pain at the parting, I now step down as your Director. I am deeply grateful.
Affectionately,
Robert Wilson

He added a postscript:

As Tennyson about the aged Ulysses said:
"Tho' much is taken, much abides; and tho'
We are not now that strength which in old days
Moved earth and heaven; that which we are, we are;
One equal temper of heroic hearts,
Made weak by time and fate, but strong in will
To strive, to seek, to find, and not to yield."[99]

In an attached letter Ramsey wrote about Wilson's "brilliant personal contributions to the accelerator, the research program and the beauty of the structures." These, he wrote, "can be found everywhere and will remain as monuments to his genius." Ramsey also conveyed that the

URA Trustees had accepted Wilson's suggestion to name Philip Livdahl "Acting Director until a new Director is appointed."[100] Livdahl later described his activities as "directed toward keeping the ship upright and putting out the little fires that kept cropping up."[101]

Sixteen years later, in a written "Toast to Bob Wilson on his 80th Birthday," Goldwasser reminded his audience of a few of Wilson's courageous departures from "then-current practice" in accelerator building: "separated function for such an accelerator, small magnets, small tunnel, localized rf, no pylons to bedrock, and a single extracted beam feeding multiple experimental areas." He applauded "Bob's commitment to build this kind of a facility responsibly," and he noted that Wilson always kept "a watchful eye on broad societal problems and using what relevant muscle such a large project might have in a way that could address some of those problems." Goldwasser also referred to Wilson's concern that "every feature of the design should be held closely to the boundary of predictable failure," allowing Fermilab to build its Doubler in a time when dollars were scarce. "But," reflected Goldwasser, "he always appreciated that a laboratory would atrophy and die if it were not building for the future while it tended to the present."[102]

By fixing his gaze on the frontier, even in the face of DOE opposition, Wilson had struggled to build the Doubler his way and thus brought his time as the director of Fermilab to an abrupt end. After his retirement as director, he continued working on Fermilab's programs, especially the Doubler, whose group within the Accelerator Division he continued to lead.[103] Holding a joint appointment in the Department of Physics and the College of the University of Chicago from 1967 until 1980, he continued to lecture and consult on many topics ranging from magnet design to architectural and artistic design. In an article for *Reviews of Modern Physics*, he rhapsodized about the future of Fermilab. The article begins:

> Reflecting about plans far in the future can be futile. Fancies can be fantasized for fabricating future facilities at Fermilab, but fulfillment will finally depend upon the fickle unfolding of physics—and on finding funds, on the focus of other facilities, on forceful friends and fierce fights; but finally most of it will depend on new facts, new findings, new fancies. Thus at Fermilab we might find our way to five TeV; or we might find it preferable to fill in facts about physics as revealed at fifty GeV CM: or we might find more felicitous the flowering of photon physics at 500 GeV. In the following phantasmata, let me feel free to first figure on the most fruited fulfillment, let me flounder in a veritable fantasia of physics facilities; for realistic factors finally, "little by little will subtract faith and fallacy from fact."

Yesterday's fancy is today's fact, or is forgotten. The fulfillment of any fantasy is infinitesimal, or it being forgotten is finite, it fades to a figment of futile fabulation if its forecast into the future exceeds even a few years.

And finally let me quote from the ferocious Figaro of Fakespeare: "Oh fancie that might be, oh, facts that are!"[104]

After leaving Fermilab, Wilson became the Peter B. Ritzma Professor at the University of Chicago in 1978. In the fall of 1979, he was appointed the I. I. Rabi Visiting Professor of Science and Human Values at Columbia University, and in 1980, the Michael Pupin Professor of Physics. In 1982 he would return to Cornell as a professor emeritus and live there with Jane until his passing in Ithaca on January 16, 2000. He had suffered a stroke three years earlier from which he never recovered. According to Wilson's request, he was laid to rest in the nineteenth-century Pioneer Cemetery on Fermilab's site on April 28, 2000. Wilson had received many honors during his life, among them election to the National Academy of Sciences in 1957, the National Medal of Science in 1973, the Enrico Fermi Award in 1984, and election in 1985 as president of the American Physical Society. None of these honors meant as much to him as the "privilege" of founding the National Accelerator Laboratory in a suburb of Chicago.

A Change in Command: The Interregnum

The first candidate that Ticho's search committee identified for the position of Fermilab's director was Burton Richter, later the director of SLAC. Along with Samuel Ting of Brookhaven, Richter had won the 1976 Nobel Prize for the discovery of the charm quark. Richter visited the laboratory to look it over and declined. In August, Ramsey asked Leon Lederman, Fermilab's highest-profile user, whether he would be willing to be nominated for the position. Lederman had an outstanding record of physics research and extensive experience in science leadership. But he hesitated when Ramsey reached him with his offer in Corsica, where Lederman was attending a physics workshop.

Lederman recognized that the responsibilities of leading Fermilab would cut deeply into his career as a physicist and probably end it. Wilson intervened. "The community has been good to you," Wilson said, encouraging Lederman to take the job. "It's your turn to be good to the community."[105] Some years later Wilson reflected: "A new person was the main thing that was needed. I felt really that we weren't doing

8.4 The triumvirate during the interregnum: Robert R. Wilson, Philip V. Livdahl, and Leon M. Lederman. (Courtesy of FNAL Visual Media Services.)

the proper experiments. We weren't getting the funds to do the right experiments. That took someone who knew and was interested in the physics. And clearly Leon was that kind of person."[106]

Lederman spent a week in Washington DC surveying the prospects for Fermilab and the future of particle physics. Late in August he met with key staff at Fermilab and asked if they would work with him.[107] By month's end, he had agreed to serve as director designate and assume the position of director in June 1979. One reason for Lederman's delay in accepting was that his Columbia-based group was setting up a new experiment at Cornell's electron synchrotron and needed his direction. Another was that he wanted first to learn the ropes and examine the "skeletons-in-the-closet."[108]

Ramsey, as president of URA, made the official announcement of Lederman's acceptance on October 19, 1978, during a two-day workshop of ICFA, the International Committee on Future Accelerators (later named the International Committee for Future Accelerators), held at Fermilab. At DOE, John Deutch announced the news. As Fermilab's acting director, Livdahl conveyed the announcement to Fermilab employees, who gathered to attend a ceremony held in the auditorium of the High-Rise. Livdahl presented Wilson with three parting gifts: a lab ID with his original payroll number, 1, a pair of welding gloves, and a yellow hard hat labeled "Boss." Wilson told the group, "I can't think of a stronger team to administer this laboratory than Phil and Leon." (See fig. 8.4.) After his usual jokes, Lederman graciously thanked Wilson on behalf of the past, present, and future users, for creating "a singular laboratory in which physics can be carried out at the forefront of knowledge."[109] Livdahl then served as Lederman's deputy director for eight more years. As for Wilson's fight for the Doubler, Lederman knew he could not postpone resolving this issue until he formally became director. As director designate he had to act immediately.

PART THREE

The Road to Megascience

NINE

Lederman's Vision

This is, in real life, the way basic research is done. . . . There is an important element of humor in basic research. . . . It has something to do with astonishment at the outrageous simplicity of mechanisms in nature when . . . they are finally comprehended. LEWIS THOMAS[1]

When Lederman stepped into Wilson's shoes he found them rather large. "I was a good physicist," he admitted, but "Bob was a great man."[2] Moreover, while the Illinois laboratory that Wilson had founded with limited funds held the promise for world-class discovery, it was wavering dangerously just then. Lederman need not have been so unsure of his ability to lead Fermilab. Soon he would stabilize the laboratory, improve its funding environment, and nurture a vigorous megascience there, in a "truly national laboratory" whose frontier science was integrated within the cultural matrix of the American Midwest.

Like Wilson, Lederman structured his vision for Fermilab on his self-image. But while Wilson's image was based on his personae as pioneer, engineer, and Renaissance man, Lederman's rested on his all-embracing passion for physics. He was unabashedly romantic, often admitting in his public talks and popular writings that "the life of a physicist is filled with anxiety, pain, hardship, tension, attacks of hopelessness, depression, and discouragement." He insisted that the supreme pleasures of physics, especially experiencing rare "epiphanies," made the research worth all the pain.

This chapter draws partly on material published in Hoddeson and Kolb 2003.

These "Eureka!" moments of discovery that punctuate the life of the physicist with "flashes of exhilaration, laughter, joy, and exultation," he loved to exclaim, derive from "the sudden understanding of something new and important, something beautiful," especially when those moments occurred "in the small hours of the morning, when most people are asleep, when there are no disturbances and the mind becomes contemplative." The experimenter typically sits off in a lonely laboratory staring at numbers. But then, "You look and look, and suddenly you see some numbers that aren't like the rest—a spike in the data. You apply some statistical tests and look for errors, but no matter what you do, the spike's still there. It's real. You've found something. There's just no feeling like it in the world."[3]

Lederman found the everyday tasks of experimental physics almost as exciting as the epiphanies. In one anecdote of the 1950s, he and Gilberto Bernardini, a visiting experimentalist from Rome, were building a particle counter at Columbia University. Working together into the night, long after the machinists had gone home, the two finished insulating, soldering, and flushing gas through a charged system to which an oscilloscope was attached. Suddenly Bernardini peered at the green oscilloscope trace displaying the readout and "went stark, raving wild," reported Lederman. "Mamma mia! Regardo incredibilo! Primo secourso. . . . He shouted, pointed, lifted me up in the air—even though I was six inches taller and fifty pounds heavier than he—and danced me around the room." When Lederman asked Bernardini what had happened, the older physicist's eyes danced as he replied, "Izza counting! Izza counting!" Lederman shared Bernardini's excitement of building equipment that could detect cosmic-ray particles "with our hands, eyes, and brains."[4]

Addressing the multiple needs of Wilson's fragile laboratory required far more than passion, however. The feat drew heavily on Lederman's practical experience in high-energy physics research, and on his considerable social and political skills, especially his ability to motivate large teams to work together for decade-long periods. As an experimenter, he knew he had to turn the Fermilab facility into a users paradisio (in contrast to Wilson's purgatorio)—a total environment supporting the joyful day-to-day teamwork conducted over periods of months or, more often, years (fig. 9.1). Lederman's image of a user-friendly environment for physics research drew from his own recent work in Wilson's leaky trenches and in the experimental halls of other institutions. It matched the needs of the researchers much better than Wilson's. He was one of them.

9.1 Fermiland, 1979: artist's rendering of Fermilab's place in the physics world. (Courtesy of Angela Gonzales.)

Biographical Sketch

Leon Max Lederman was born in New York City on July 15, 1922, to Ukranian-Jewish immigrants, Morris Lederman from Odessa and Minna (Rosenberg) from Kiev. His initial inspiration for the science came from parents who "venerated learning," according to Lederman, even though they owned and operated a hand laundry. His older brother Paul, "a tinkerer of unusual skill," gave Lederman a "lifelong fascination for the kind of mechanical relationships that are a part of science." As for his sense of the cultural support needed for science, he claimed that the city of New York "provided the streets, schools, entertainment, culture and ethnic diversity," which nurtured "many future scientists." After graduating from James Monroe High School in the Bronx, Lederman entered New York's City College, where in 1943 he took his bachelor's of science degree in chemistry. There he "fell under the influence" of several future physicists, including Isaac Halpern and Martin J. Klein, a high school friend. Lederman decided to switch to physics for his graduate study, in part due to the influence of Klein, who later became a distinguished historian of physics.[5]

During the Second World War Lederman spent three years in the U.S. Army Signal Corps, rising to the rank of first lieutenant. His training for experimental physics began before he was assigned abroad, when he was sent to MIT to learn the new technology of Doppler radar. Stationed in France and Germany, Lederman was assigned to head the New Equipment Introductory Detachment to examine captured enemy equipment. As it happened, among those working under Lederman in this detachment were two of Lederman's future Columbia University colleagues, Joaquin Luttinger and Jack Steinberger. Both would become internationally renowned physicists, Luttinger in the field of quantum solid-state physics and Steinberger in high-energy physics.

After the war, like millions of other veterans who under normal circumstances would not have been able to attend college or graduate school, Lederman took advantage of the GI Bill. He enrolled in Columbia University's Graduate School of Physics; its already famous department chair, Isidore I. Rabi, became one of Lederman's lifelong physics heroes. Rabi had recently won the 1944 Nobel Prize in physics for developing a method of measuring the magnetic properties of atoms, molecules, and nuclei. He was also a crucial force in the founding of Brookhaven in 1947 and of CERN in 1954. In later years, Lederman claimed that Rabi taught him the distinction between observation and measurement in experimental physics. Rabi had said, "First comes the observation, then comes the measurement." Lederman came to understand that this meant: "Observations are experiments which open new fields. Measurements are subsequently needed to advance these."[6]

Lederman's physics training benefited from government funds in another way. His doctoral thesis research was aided by the Office of Naval Research (ONR), one of the most important physics funding agencies of the early postwar era. ONR supported the postwar construction of Columbia's 385–400 MeV synchrocyclotron on the beautiful, sprawling Nevis estate in Irvington-on-Hudson. There, in 1948, Lederman began work on his doctoral thesis under Eugene T. Booth, the director of Nevis. Lederman built a Wilson cloud chamber for "seeing" elementary particles as tracks of condensed water droplets. After completing his Ph.D. in 1951, Lederman was invited to stay on at Columbia, where as a member of the physics faculty he taught and conducted research for the next twenty-eight years. In 1958 he was promoted to professor; in 1972 he became the Eugene Higgins Professor of Physics.

Lederman became known for his good taste in selecting and solving fertile physics problems. In 1956, while working at Brookhaven National Laboratory, he found the long-lived neutral K meson, whose existence had been predicted on the basis of fundamental symmetry properties. With Richard Garwin and Marcel Weinrich, he collaborated in several experiments dealing with the static and decay properties of the μ meson (later called the muon). One result was the first measurement of the magnetic moment of the muon using magnetic resonance techniques. Another was the observation in 1957 of parity violation in the muon's beta decay.[7]

Lederman spent his first sabbatical from teaching, in 1958–1959, as a John Simon Guggenheim Memorial Fellow. He worked at CERN, where he organized the g-2 series of experiments and measured the spin of the muon. These experiments would continue for almost twenty years and motivate several important CERN collaborations.[8] Lederman's study in 1961–1962 of the neutrinos arising from muon decay, performed with his Columbia colleagues Melvin Schwartz and Jack Steinberger, proved the existence of two different kinds of neutrinos, one related to the muon and the other to the electron.[9] For this work, which triggered much interest in research employing neutrino beams at many laboratories, the three would share the 1988 Nobel Prize. While pursuing studies in many of the world's high-energy laboratories, Lederman considered how to improve the conditions in which experimenters worked. This concern prompted him to write his 1963 paper (chapter 3) on the "truly national laboratory," a notion that became permanently associated with Fermilab when Wilson named it the National Accelerator Laboratory.

Over the years Lederman's experiences brought him into service on national science advisory panels; for example, in 1959 he joined Jason, the Institute for Defense Analyses, which advises the Department of Defense on the physics underlying national security issues.[10] In 1961, at the age of thirty-nine, he became the director of a major high-energy physics center, Columbia's Nevis Laboratory. This position increased his opportunities to contribute to national program recommendations and science policy. Further distinction followed in 1965, when he was elected to the National Academy of Sciences and received the National Medal of Science. In 1967, he became a founding member of HEPAP, a panel advising the AEC and later ERDA and the DOE. He and Wilson were also among the early organizers of the International Committee for Future Accelerators (ICFA), whose work by 1975 would lay the basis for the Superconducting Super Collider (chapter 13) and the International Linear Collider.[11]

Increased participation in positions of scientific leadership did not slow the pace of Lederman's research. In 1968 he directed a Brookhaven team that measured the mass spectrum of muon pairs produced in interactions with the primary beam of the AGS. Submitting one of the first experiment proposals at NAL, Lederman studied the pair production of electrons and muons in large transverse-momentum hadron interactions. In the course of this work (chapter 7), he and his group saw the "shoulder" that gave the first evidence of "charm," but they missed identifying this shoulder as the J/ψ particle. It was later identified in 1974 by Ting and Richter, who received the 1976 Nobel Prize for this major discovery. Lederman's work did, however, launch the study of the associated production of lepton-antilepton pairs in high-energy hadron interactions. The research culminated in the 1977 discovery of a family of Upsilon particles offering the first evidence for the bottom (or b) quark, also called "beauty." The discovery was soon confirmed by experiments at DESY and Cornell.[12] In the same period, Lederman also collaborated at CERN with Luigi DiLella and others to discover an abundant pion production at large transverse momentum, which demonstrated an incisive new means for exploring hadron structure.[13] His impressive body of experience in both the research and leadership of physics made Lederman an ideal choice as Fermilab's second director.

Lederman's Leadership

When Lederman took the reins, he "found good and bad news." It was good, as he explained, that "the group that Bob Wilson had assembled was really magnificent," and that Wilson's "tradition of architecture and ecology" was in place. It was good also that Wilson had established "the style which somehow strove to preserve individuality within the need for impressive collaborations," and that the "vision of a superconducting accelerator (Energy Doubler/Saver) was very clear." The bad news was that Fermilab's mission was no longer clear; for "in the transition, the focus had been lost, the Lab was impoverished, and the physics program was hostage to an under-funded Saver project: a project whose physics goals were ill-defined."[14] Lederman knew he had to decide quickly whether to terminate Wilson's Doubler and attempt to create a facility for colliding particles in the problematic Main Ring, a possibility that few considered

9.2 Leon M. Lederman, director of Fermilab, 1978–1989. (Courtesy of Leon M. Lederman Collection, Fermilab Archives.)

feasible. His decision, as well as his decision-making process, is discussed in chapter 10.

Lederman's leadership style, like Wilson's, had a strong rhetorical component. But while Wilson would rally his staff using folktales, legends, and great works of literature, Lederman (fig. 9.2) held his audience by tempering his impassioned rhetoric with jokes.[15] And, unlike Wilson, who had been unwilling to play "the political game" with the DOE, Lederman resigned himself to his role in Washington's corrdors of power. He proved highly effective there.[16] "I'm constantly shuttling to Washington to cajole federal and congressional officials into keeping budget cuts to a minimum," he told an interviewer.[17] Confidently assessing each audience, he seemed to know instinctively when to draw on science, knowledge, emotion, or humor. In later years, he joked about how Wilson prepared him to steer the laboratory through crises by providing him with three sealed and numbered envelopes, not to be opened until a major crisis arose. When the first crisis arose, Lederman said, he opened the first envelope. It directed him to "blame his predecessor." When he opened the second envelope to deal with the second crisis, the message instructed him to "reorganize and blame the Department of Energy." And in 1988 when he turned to the third envelope, he was told, "Prepare three envelopes!"[18]

Even without the pipeline to power that Wilson had enjoyed in the early years of NAL, Lederman succeeded in attracting funds to the laboratory. He learned of opportunities through the networks he had cultivated as director of Nevis and through his work on advisory committees. Lederman negotiated with a more relaxed and softer style than Wilson did, a style he adapted to different situations.

Lederman and Wilson both hated bureaucracy. But while Wilson's reasons typically hinged on philosophy or aesthetics, Lederman's were usually pragmatic: bureaucracy bled resources from the laboratory. In distancing himself from bureaucratic practices Wilson created tension between himself and those who could have helped him achieve his goals. Lederman endured as much as seemed necessary to secure funding, often with the skill of a poker player. Fermilab's Bruce Chrisman recalled Lederman's cool performance during a visit to Washington in the early days of his tenure as director. The purpose of the trip was to review the budget of the Accelerator Division. Lederman hoped to secure several million dollars promised for the Doubler. The Fermilab group and some DOE officials sat waiting around a conference table in DOE's Forrestal Building because John Deutch, the director of energy research, was late. When Deutch finally arrived and sat down he leaned back and casually stretched his legs out across the top of the table. During the subsequent interplay, which was "not exactly friendly," Deutch frequently interrupted Lederman's prepared remarks. Others might have lost their composure, but Lederman ignored the jibes, recalled Chrisman, and calmly continued his presentation. Deutch released the funds.[19]

"The Wilson-Deutch problem was an unnecessary tragedy," reflected Lederman some years later. "They were both charismatic personalities who could have gotten along very well." He suspected that their problem related to DOE's embarrassment over the magnet failures recently encountered by ISABELLE, a project in which DOE was heavily invested. Lederman guessed that DOE may have thought, "How could these guys from Illinois do something that the wizards on Long Island couldn't do?"[20]

Wilson's philosophy had called for underdesign in building apparatus and facilities. If a component did not work properly, it could, he felt, be fixed or modified.[21] Unfortunately, the time for repairs grew more precious during the 1980s as the competition with European laboratories intensified. Lederman therefore broke from Wilson's philosophy of building frugally for the short term. Instead he authorized the building of sounder facilities to help Fermilab enter and compete successfully in the era of megascience. The many lasting constructions under Lederman as director included a new magnet factory, a muon laboratory, a computing center, an antiproton source, and two gigantic colliding-beam detectors and collision halls (chapter 12).[22]

Following in Wilson's footsteps, Lederman promoted the advancement of Fermilab's energy frontier. As soon as the Doubler reached 512 GeV on July 5, 1983, he turned his attention to what he saw as Fermilab's next machine, the Superconducting Super Collider (SSC) (chapter 13). In 1987 he endorsed a substantial upgrade of the Fermilab accelerator, the Tevatron, intended to increase its luminosity and double its intensity. This upgrade, later called the Main Injector, would be completed during the administration of Lederman's successor, John Peoples.

Lederman continued Wilson's practice of letting physicists run the laboratory, but he introduced a fresh outlook by frequently reorganizing his upper management staff. Many changes appeared overnight. Thus, one day early in 1979, Russ Huson, the head of the Accelerator Division, surprised his staff by announcing that Lederman had decided that Rolland Johnson would head the 400 GeV program and Don Young would steer the new colliding-beams program. Lederman put Rich Orr and Helen Edwards jointly in charge of the Doubler (fig. 9.3). Later in 1979, the Physics Department in the Research Division was reorganized to report directly to the director. That year, Richard A. Lundy, then serving as business manager, became head of the Doubler's Magnet Division, while Bruce Chrisman replaced Lundy.

Wilson had mistrusted experts, but Lederman drew heavily on them, attracting many to the laboratory. In the end Lederman made the laboratory's major decisions, but he typically went out of his way to seek advice from others, such as the appointed members of his Science Advisory Group (SAG). In his first year as director, Lederman empowered the Underground Parameters Committee (UPC), a group of experienced accelerator physicists that had formed under Wilson's tenure, to evaluate the design of the Doubler (see chapter 10). Similarly, in 1982, Lederman

9.3 Accelerator physicists J. Richie Orr and Helen Edwards outside the Main Ring, 1979. (Courtesy of FNAL Visual Media Services.)

called on Maury Tigner of Cornell to advise Fermilab on its new antiproton source (chapter 12). In 1983, Lederman asked Joseph Ballam of SLAC to head a committee to make recommendations on Fermilab's future computing needs.[23] Unlike Wilson, Lederman felt no need to manage details. He would typically set a direction and delegate the work while he moved on to set other goals. With this style of leadership, Fermilab physicist Lincoln Read explained, Lederman "would let a thousand flowers bloom" and encourage people to stretch their imaginations.[24]

Lederman brought Georges Charpak from CERN so the laboratory could benefit from Charpak's creative detector inventions. To encourage more interplay between theory and experiment, Lederman expanded the theory staff in September 1979, hiring James "Bj" Bjorken, a leading theorist from Stanford. To recruit Bjorken, Lederman offered him and his wife, Joanie, the director's house, formerly occupied by the Wilsons. While Bj stimulated the physics environment, Joanie Bjorken led the Fermilab Guest Office and NAL's Women's Organization, NALWO. She cultivated numerous family and community events for Fermilab until her premature death in 1983. As the Bjorkens were living in the director's house, Lederman and his wife, photographer Ellen Carr, settled into a renovated 1865 farmhouse on the site. There, in the informal yet gracious tradition of settlers at the frontier, they entertained countless visitors, including dignitaries, agency officials, politicians, and international scientific leaders. Many enjoyed riding the Ledermans' horses.

The expanded theory group pursued the most exciting particle physics theory of the time: quantum chromodynamics, strong interactions, lattice gauge theories, grand unification, quantum field theory, wormholes, superstrings, and supersymmetry. Chris Quigg continued as head of Theory from 1977 until he moved to Berkeley in 1987 to join the Central Design Group for the Superconducting Super Collider (SSC) (chapter 13). In 1987, William Bardeen succeeded Quigg, serving as the head of Theory until late 1992, when he moved to Texas to serve as head of Theory at the SSC and Keith Ellis took the helm.

Lederman also invigorated Fermilab by adding a theoretical astrophysics group. After discussions in 1979 with David N. Schramm at the University of Chicago about building a NASA-supported center for astrophysics at Fermilab, and spearheading URA's unsuccessful 1980 bid to host at Fermilab what became the Hubble Space Telescope Science Institute, Lederman hired Edward W. "Rocky" Kolb and Michael S. Turner in 1983 to create the Fermilab/NASA Theoretical Astrophysics Group. This collaboration, the first of its kind at a national laboratory, soon achieved an international reputation for its work on such topics linking the

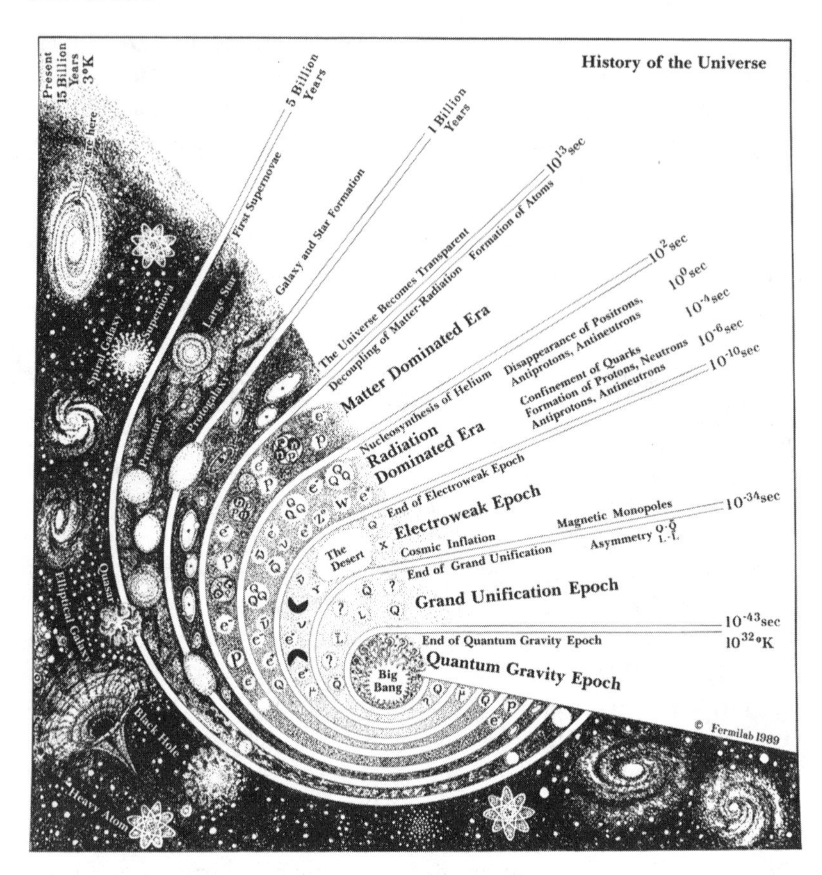

9.4 The history of the universe. (Courtesy of Angela Gonzales.)

subnuclear world (inner space) with studies of the universe (outer space), such as the evolution of the universe (fig. 9.4), cosmological phase transitions, quantum cosmology, cosmic strings, wormholes, and dark matter.

As for realizing the users' paradise, Lederman came closer than Wilson to this ideal. In 1980, ICFA adopted Fermilab's policy of merit-based user access for laboratories in all countries, while this committee worked quietly but openly to create a "world laboratory" (discussed in chapter 13).[25]

Neither Wilson nor Lederman had the resources to meet all of Fermilab's users' demands. But whereas Wilson tried to address them by elevating frugality to a virtue, Lederman simply asked whether an experiment really needed all the requested resources. He compared these requests with what his own needs would be under similar circumstances. The result was often improved conditions. Under Wilson many experimenters, including Lederman, had worked in pits where leaking rain or ground water dripped into dark, unbearably hot or cold (depending on the season) concrete rooms resembling wartime bunkers. The offices of experimenters were typically found in crowded trailers. Lederman upgraded the experimental areas so they could serve physicists working in the areas for months, or years. In the new Muon Laboratory, constructed in 1984 and 1985, the users worked in roomy, airy, environmentally controlled, and properly illuminated offices with convenient bathrooms and telephones.

Most of the Wilson-era experiments had been relatively small, of the size found at Brookhaven's AGS. But as Roy Rubinstein reflected, "All the easy experiments were done by 1977." As the scale, cost, and complexity of experiments increased, and as their lead times grew, the flow of the experimental program clogged up with many experiments that had been approved earlier under Wilson's tenure. To open the flow Lederman imposed tighter schedules; his PAC rejected large numbers of proposals for relatively small experiments. The resulting program included fewer experiments, most of them larger in scale than those of the Wilson era. Even so, Lederman is thought to have approved too many experiments. Rubinstein claimed they put a lien on the Fermilab budget that extended far into the Peoples era.[26]

The greater scale and complexity of the experiments planned for the colliding-beams program pushed the frontiers of computing. Wilson had opposed the expansion of the laboratory's computing because this violated his sense of what it meant to be a physicist, Lederman, however, confronted the growing computational needs of experimenters, especially when, with the advent of silicon microstrip detectors (see chapter 11), the computing problem threatened to go out of control. Lederman supported the Advanced Computer Program and its MAPS (Multi-array Processor System) for lattice gauge theory calculations, and he endorsed building a new computing center. Dedicated on December 2, 1988, and named after physics icon Richard P. Feynman, the new center housed the "central computing upgrade project" initiated by Al Brenner and completed under Jeff Appel. The architectural consultant for the new building was Robert R. Wilson.[27]

Cultivating Community

Lederman's stabilizing efforts extended to Fermilab's community. Addressing more than seven hundred employees at his first director's meeting in mid-July, 1979, he proclaimed: "We are carrying out an activity at

9.5 Lederman with Drasko Jovanovic and colleagues at the daily Director's Coffee Break, 1984. (Courtesy of FNAL Visual Media Services.)

the frontier of civilization." Aiming "to be at the top" he called for improving the laboratory's quality of life. "We have to see that our communication remains free and open; that the surroundings—the trees, lakes, wildlife—remain pleasing to the eye. In a sense, that's related to what we are doing at Fermilab: studying nature at its most basic level."[28] To help preserve the natural habitats on the site, Lederman supported the development of a long-term environmental research plan, extending Wilson's Prairie Restoration Project. He asked the DOE to institute a policy for all national labs to plant ten thousand trees each year to help absorb greenhouse gasses. Along with Fermilab's site managers, John Paulk and Rudy Dorner, Lederman pursued the formal designation of Fermilab as a National Environmental Research Park, a status granted in April of 1989.

Lederman created many opportunities to have contact with his staff. Moving Wilson's Director's Coffee Break from noon to 3:30 p.m., Lederman established a convenient regular time when he could mingle with colleagues in the art gallery adjacent to his office (fig. 9.5). Once a year he hosted his "Run with the Director," in which employees jogged with him around the Main Ring and then joined him for bagels and juice. He took part in canoe races for staff and users in the Main Ring's Cooling pond, added a gymnasium, a restaurant (called Chez Leon; fig. 9.6), and the Children's Center (fig. 9.7), the first employee childcare facility at a DOE laboratory.[29] He fostered community by visiting users at their experiments, usually late at night, and by celebrating achievements with labwide parties (fig. 9.8) for every milestone.

How did Lederman manage to create so many opportunities in periods of austerity? Andy Mravca, an official who had worked with the laboratory in its early days alongside AEC officials K. C. Brooks and Fred Mattmueller, returned to the laboratory during Lederman's administration to serve as DOE's in-house liaison at Fermilab. Mravca noticed that when Lederman wanted to change something he just did it.[30] Later, when DOE would judge a certain change impossible, it had already been made. Thus, for example, by the time Mravca reported back to Lederman in 1980 that the DOE would not authorize the Children's Center, Lederman had already established it. In the early 1990s under Secretary of the DOE Admiral James Watkins, the department mandated that all its laboratories offer childcare.

Lederman helped to solidify Fermilab's place in Illinois with a legacy of science education outreach. Among his achievements was Saturday Morning Physics. Missing the contact with students he had enjoyed at Columbia, especially in his Physics for Poets course aimed at nonscientists,

9.6 Chef Tita Jensen with her staff at Chez Leon in the Users' Center, 1985. (Courtesy of FNAL Visual Media Services.)

9.7 Lederman visited by children from the Fermilab Children's Center, 1982. (Courtesy of FNAL Visual Media Services.)

Lederman authorized Drasco Jovanovic to initiate in October 1980 a program of physics classes taught by staff physicists for gifted local high school students (fig. 9.9). This led to the Friends of Fermilab, which Stanka Jovanovic and Marge Bardeen founded in 1982 to raise support for innovative national math and science education programs. Fermilab's Summer Institute for Science Teachers was held for the first time during 1983. Fermilab joined the Valley Industrial Association (VIA), a local organization dedicated to the advancement of worthwhile programs in mathematics, science, and computer education. The VIA in turn created the Corridor Partnership for Excellence in Education, a group of local corporate and education leaders dedicated to creating the means for students to develop stronger skills in mathematics and science. This Corridor Partnership then worked with the Friends of Fermilab to found the Illinois Mathematics and Science Academy (IMSA), an innovative public residence high school for gifted Illinois students; IMSA opened its doors in 1987. Later Lederman, Bardeen, Jovanovic, and the Friends of Fermilab worked with educators to found the Teachers Academy for Math and Science (TAMS), which worked from offices at the Illinois Institute of

Technology to improve public education in Chicago. Lederman served as chairman of the TAMS board of trustees, as well as chairman of the Governor's Science Advisory Committee, and vice-chairman of the Illinois Coalition to help Illinois invest in science and technology enterprises.

Such Fermilab work in educational outreach was but part of Lederman's effort to strengthen Fermilab as an Illinois cultural and scientific center. He increased the number of programs in the performing arts series that Wilson had established. He planned physics exchange programs to help forge alliances and exchange scientific information with Latin American countries, including Argentina, Brazil, Colombia, and Mexico.[31] He organized lectures to educate the public about the dire impacts of Ronald Reagan's Strategic Defense Initiative (SDI), known as "Star Wars." He instituted fellowships for outstanding minority students. And in an effort to compete with universities for assistant-professor-level physicists, he established a program to award three-year Wilson fellowships to exceptionally talented young particle physicists. To help address the

9.8 Drawing for the end-of-the-run party, 1987. (Courtesy of Angela Gonzales.)

9.9 Lederman teaching the Saturday Morning Physics class, 1980. (Courtesy of FNAL Visual Media Services.)

shortage of young physicists who were entering the field of accelerator science, he created a program of cooperative studies with universities to grant Ph.D. degrees in accelerator science.

To facilitate the transfer of Fermilab's research and technology into the larger social and economic culture, Lederman in 1980 appointed Richard Carrigan to organize Fermilab's Industrial Affiliates, a group of corporate partners interested in applying Fermilab's basic research ideas in the marketplace. One result was the Loma Linda Medical Accelerator, based on Wilson's early idea of treating cancer patients with protons. Steered by Philip Livdahl, this effort extended the work of the Proton Therapy Cooperative Group, which had met at Fermilab in 1985. An agreement made in the following year, between Loma Linda University in California, Science Applications International Corporation in San Diego, and Fermilab, led to the development and construction at Fermilab in 1988 of a 70–250 MeV proton synchrotron for use at the Loma Linda University Medical Center in 1990. One of the treatment rooms was dedicated to Robert R. Wilson.[32]

Lederman also expanded the scope of Wilson's program in the history of accelerators to include studies in the history of particle physics. This book grew partly out of that initiative. To stimulate awareness of "our heritage," as Lederman called it, he also instituted a series of international symposia on the history of particle physics. The first two of these were held at Fermilab—in 1980, on the birth of particle physics, and in 1985, on developments in the field during the 1950s. The third and last in the series, hosted by SLAC in 1992, explored the emergence of the Standard Model from 1964 to 1979.

These nonphysics as well as physics contributions allowed Lederman to stabilize and professionalize the laboratory that Wilson had built quickly and frugally for the short term. Under Lederman Fermilab became a long-term, adequately funded research facility whose programs were integrated within the culture of its many relevant communities. Wilson's fast-paced, minimalist approach had been necessary in the late 1960s. During the 1980s Lederman's efforts made Fermilab an apt home for a megascience that would establish Fermilab at the frontiers of physics for decades to come.

TEN

Completing the Doubler, 1978–1984

To look in a place where no one has ever looked before, to observe deeper into the core of the atomic nucleus than has ever been probed; to measure in a domain more remote than human experience—much more remote—than the surface of the moon and Venus—to be able to recreate, in microcosm, the conditions which existed in the earliest instant after creation—this is the exalted privilege that has been given to us. LEON M. LEDERMAN[1]

Lederman knew that he could not postpone settling the Doubler's fate.[2] Handling the issue in his first month after becoming director designate on October 19, 1978, he had to know first whether the Doubler was feasible technically. He appointed "three wise men" as his expert review committee: Boyce McDaniel of Cornell, acting as chair; Matthew Sands, then at the University of California at Santa Cruz; and Burton Richter of SLAC. After meeting extensively during October 21–23 and November 6–7 this committee of experts judged the Doubler feasible.[3]

The next round of advice came out of a "shoot-out" that Lederman scheduled for Armistice Day, November 11, 1978. Here opponents and supporters of the Doubler faced off against each other in a heated debate that began at 9 a.m. and ended close to 3 a.m. the next day.[4] "The ground rule," Philip Livdahl recalled, "was that everybody was to have all the time necessary to state their cases."[5] Lederman

This chapter is based partly on Hoddeson 1987.

invited accelerator experts from inside and outside the laboratory to act as judges: "The role of the judges is to embarrass the advocates as much as possible with penetrating, incisive questions," Lederman wrote in his announcement of the shoot-out. "The efforts, money, time scales must be realistically appraised in the light of our heavy commitment in the 400 GeV and Doubler construction program. The format of the review will call for prepared presentations and will permit (almost) unlimited discussions from the proponents."[6]

Most of those who attended the shoot-out were convinced that the race with CERN to discover the W and Z particles could not be won with a collider based on the Main Ring, an accelerator probably limited to 200 GeV when working in collider mode. Wilson's frugality had allowed Fermilab to win the battle for existence, but the Main Ring's "underconstruction" now threatened to lose the war for physics leadership. Another disadvantage of employing the Main Ring in a colliding-beams machine was that its design prevented functioning in the traditional fixed-target mode while it was operating as a collider. Many experts pointed out that any of the suggested schemes to achieve a colliding-beams machine using the original Main Ring accelerator would require a major design effort over a period of time comparable to that needed to complete the superconducting Doubler. Once the three wise men declared the Doubler feasible, Fermilab's best strategy was clear: complete the Doubler and forgo the race for the W and Z. A proton-antiproton collider based on the Doubler would raise Fermilab's energy to almost 2 TeV and secure Fermilab's frontline position at the frontier for decades by offering roughly a fourfold advantage over CERN in the center-of-mass system. CERN's 270 GeV in each beam amounted to only 540 GeV in the center-of-mass system.[7] Offering additional support to the Doubler was the fact that a superconducting ring would consume less power, even when functioning at its top energy.[8] When Lederman discussed his decision the next morning with his wise men over bagels and coffee, they offered their stamp of approval.

In the meantime, during October 1978, while Lederman was pondering the fate of the Doubler, Fermilab and DOE achieved a partial compromise on funding the Doubler. It was to be a phased agreement based on the completion of milestones. DOE agreed first to fund the construction of one-sixth of the machine, its "A-sector," as an R & D project. If a test of this A-sector succeeded, $38.9 million of construction support would follow on Fermilab's submission of a realistic budget for completing the remaining five-sixths.[9] Congress approved this funding plan. But the newly formed DOE was skeptical of the Doubler, perhaps because of the many previous failures that had occurred in building superconducting magnets, or because of mistrust associated with Fermilab's unconventional magnet facility; it was customary to subcontract such manufacturing work.[10] The result was that DOE specified yet another condition to be met before it would authorize construction of the A-sector: the Doubler would have to demonstrate that a string of ten out of twelve full-scale Doubler magnets tested in the aboveground area at B-12 met all the design specifications. Only on passing this B-12 test could construction of the A-sector proceed. If the A-sector then tested successfully, the rest of the construction funding for the Doubler would follow. In his characteristic style of working with, rather than conflicting with, the DOE bureaucracy, Lederman accepted this working plan. And then the Doubler group immediately turned to preparing for its critical review at B-12.

As these events unfolded, a powerful internal force at the laboratory was gathering momentum.[11] In the early years of building the laboratory, Wilson had excluded many members of the accelerator community at his laboratory from the Doubler project, perhaps to maintain their focus and avoid delay in completing the Main Ring, or possibly simply to avoid interference from these experts. The result was that many who wanted to be more directly involved in the Doubler work felt marginalized. "Wilson's style was not one of soliciting the opinions and ideas of other people in an open forum," Livdahl recalled. "He would tend to make design decisions on his own and then expect the people that were going to carry these decisions out to react to them either with better ideas or reasons why the decisions weren't appropriate."[12] This style did not mesh well with everyone on the staff.

In the year before Wilson stepped down, a small group of Fermilab's accelerator physicists constituted themselves as an informal committee, the Underground Parameters Committee, or UPC. Tollestrup chose the name jokingly to emphasize the fact that Wilson had not officially formed this committee. Its members considered themselves a vital part of Fermilab and wanted to be included in planning the laboratory's future. The Doubler was by now "the only show in town," according to one prominent UPC member, Helen Edwards. The UPC members felt alienated. They were concerned, in particular, that while much progress had been made in developing the Doubler's magnets, Wilson had discouraged his staff from thinking about the Doubler as a whole system with actual parameters.

Edwards remembered the UPC as "pretty much an informal group . . . of people discussing topics that they were worried about."[13] One worry focused on how best to extract the Doubler's beam. Would the accelerator be laid out in a way that would account for the fact that the beam extraction process would not be perfectly efficient?[14] Wilson supported the UPC's work, according to Orr, but he did not attend its meetings. "Bob and I had a private deal. He was not supposed to go because he was one of the reasons we called it the UPC."[15] After Wilson stepped down, Livdahl, as acting director, followed Wilson's practice of supporting the UPC's activities from a distance,[16] but Lederman would occasionally attend the UPC's meetings.

While the work toward the B-12 test proceeded, the UPC met almost daily to prepare for the installation of the A-sector, which would follow this test at B-12, and for the important A-sector test that would be made after this sector's installation. The many tasks included listing, outlining, locating, and describing every element to be installed. At times, experts were brought in to advise the group on detailed issues, such as the controls, the power supplies, or quench protection. Later the UPC would help Helen Edwards prepare the official Doubler design report for DOE.[17]

The basic plan for the A-sector test was to extract a 90 GeV beam at the Main Ring's AZero straight section, send it through the string of Doubler magnets arranged in the sector and cooled by the first satellite refrigerator, and finally steer the beam into a beam "dump." This test was designed to answer several questions that could be answered only with magnets actually placed in the ring. For example, is the quality of the magnets good enough to store and transport a beam? Can a beam pass through a string of magnets in the ring without causing the magnets to quench?[18]

To install magnets appropriately into the A-sector required understanding all the relationships arising from the placement of the superconducting and conventional magnets. To simply duplicate the configuration of the conventional magnets would not work, because the superconducting magnets were somewhat longer. Solving this problem of the so-called magnet lattice obviously hinged on developing a consistent plan for extracting beam from the Main Ring at one point and sending it through the Doubler magnets. Rich Orr, who at that time was serving as Fermilab's business manager, was asked at one UPC meeting to produce the plan for installing the Doubler magnets into the tunnel. He called for expert advice on Tom Collins, "the guru of that sort of thing." In an effort to cajole Collins into helping, Orr showed him the lattice design that he had developed with Stanley Pruss. As Orr expected, Collins went "through the ceiling," and said, "You did it wrong!" Then Collins redesigned the lattice. When Orr and Pruss found that they could not make the magnet-to-magnet connections that Collins had specified, they urged Collins to once again redesign the lattice, which he promptly did on his handheld calculator. He called his new lattice design "The Great Doubler Shift." Orr confessed, "I never did figure out how he did that so fast."[19] Helen Edwards could then take on the problem of extracting the beam. She enlisted Lee Teng as well as her husband, Don Edwards, who had worked for many years on the Doubler.[20] But actually installing the A-sector before conducting the crucial A-sector test still hinged on passing the critical aboveground pretest at B-12 demonstrating that ten out of twelve magnets in a string operated successfully. That hurdle was overcome on June 1, 1979, making it possible to install the A-sector.[21]

The most controversial suggestion by the UPC was to upgrade the "good-field" region for colliding-beams work by adding a set of correction coils to the Doubler's dipole magnets. These coils would control the alignment of the dipoles and correct any undesirable shifts, but the only way to make room for these coils was to shorten each 22-foot dipole by 1 foot. Wilson had firmly opposed this suggestion from the time it was first made in 1972, because implementing it would reduce the energy of the Doubler by about 5%. As no decisive argument had been made for shortening the dipoles, some 130 22-foot magnets had been built in 1978.[22]

The UPC continued to press for the correction coils. The matter was finally settled at a UPC meeting that happened to be chaired by Lederman. As Don Edwards recalled, "Tom [Collins] presented arguments for making the magnets shorter. Alvin presented arguments for not making them shorter." In a vote everyone at the meeting except for Tollestrup, who remained torn, favored shortening the dipoles. Then Lederman simply ruled that the shortening would occur, and the 21-foot magnet program began in April 1979.[23] Lederman's decision on this "circumcision" proved to be a turning point in the history of the Doubler. By changing a long-standing feature of the project to one favored by the group that would oversee the assembling of the Doubler's components into a working accelerator, Lederman brought the entire laboratory behind the Doubler. By the time he officially became the director of Fermilab in June, the Doubler was the laboratory's first priority.

The design for the Doubler was now set, but its funding was not yet official. There was still no authorization or construction funding other than the partial compromise with DOE achieved in October 1978. In this tense period, newspaper reporters presented Fermilab's case to the American public. On June 26, 1979, the journalist Malcolm Browne, whom Lederman had invited to visit the laboratory for several days, published a prominent article in the *New York Times* echoing the argument that Wilson had made over a year earlier in his letter to James Schlesinger. Referring to the scientific competition abroad, Browne wrote that the United States was losing the race in high-energy physics to its wealthy European competitors at CERN, who might soon discover the W particle. "Hobbled by chronic money shortages, scientists at the Fermi National Accelerator Laboratory here are in a race against a wealthy European scientific consortium to find a hypothetical but vitally important nuclear particle called the W." While that characterization was not exactly true, given that Fermilab had by then forgone that race in favor of completing the Doubler, Browne's public pitch was poignant and clear. He explained that if Fermilab could develop its superconducting accelerator by 1983, it would have four times the energy of CERN in center-of-mass collisions. America's high-energy physics program would then be the first to enter this new highest-energy frontier.[24] Browne's account alerted the public to the state of American high-energy physics and thus helped pave the way for DOE's subsequent decision to fund the Doubler according to the compromise negotiated in October 1978. When the DOE review committee subsequently met at Fermilab in June 1979, it in fact raised the earlier estimate of $38.9 million for the Doubler to $46.6 million (in effect adding contingency funds to avoid the need to reauthorize the project should the earlier estimate be exceeded by 25%).[25]

Fermilab received a huge boost of morale in the first days of July 1979, roughly a month after Lederman's title changed from director designate to director, when it was finally learned that the Doubler would be officially authorized as a construction project with funding to follow.[26] At a hastily called celebration at the magnet facility, Lederman announced the happy news, referring to the new superconducting accelerator as "an improved spaceship that will carry us deeper and deeper, farther and farther into the atom."[27] Later that month he met with Mattmueller and James E. Leiss at DOE headquarters in Germantown, Maryland, to give the official imprimatur to the Doubler, signing the official "Energy Saver Project Management Plan."[28] The press emphasized that in this period of "energy crisis" the "Saver," as the Doubler was now often called, would cut power usage at Fermilab by more than half, saving $5 million of taxpayers' money every year, just as Wilson had declared in his March 1977 testimony to Congress.[29]

A successful working relationship replaced the earlier tensions between Fermilab and the DOE, which had been exacerbated by Wilson's criticism of the DOE-supported ISABELLE project at Brookhaven. An important factor in dissolving these tensions was Lederman's close working relationship with Andrew Mravca, who returned to Fermilab in July 1979 as DOE's in-house liaison to the laboratory. In his earlier period of work at the laboratory during Wilson's tenure, Mravca had been told by Kenneth Dunbar, manager of the Chicago office of the AEC, "Andy, we need a success in the AEC, and we need to get that accelerator built. The country needs to look good.... Let's do something big that'll have national recognition." Those were days, according to Mravca, when bureaucratic process had been regarded as an obstacle to overcome. K. C. Brooks had also stressed, "We're going to do whatever it takes to get this job to be a success.... We'll roll over anybody who is an unnecessary or bureaucratic stumbling block." This attitude had been useful to Wilson as he went about building NAL.[30] But as Mravca helped Lederman understand, the DOE of 1979 differed greatly from its grandparent agency, the AEC.[31] Some of DOE's basic research projects had already initiated formal project management procedures analogous to those of weapons labs. For example, Lawrence Berkeley Laboratory had borrowed schemes from Livermore to build the Bevatron in the early 1950s. In the words of Daniel Lehman, one of several officials who were brought to DOE to establish project management oversight, "There wasn't good project management on basic research projects as a whole," so DOE "began emphasizing the importance of better planning and record keeping." In that way, the department could satisfy its obligation to bring projects in "on time and on schedule."[32]

Wilson had flatly refused to cooperate with the bureaucratic expectations of the DOE, but Lederman and Mravca, the two "new boys on the block," decided that it would be better for the laboratory if they fell in step. Mravca knew that the memory of Wilson's conflict with DOE needed to be rewritten to reflect the sea change he noticed. They worked together to establish friendly relations between the laboratory and DOE, meeting frequently with DOE officials, keeping them informed of progress and of the laboratory's compliance with official requirements. In this "instant image change" in the Fermilab-DOE relations, Mravca explained, "my perception is that there was a mellowing with respect to relationships in all of DOE with respect to Fermilab."[33] Some members of the Fermilab staff felt that Lederman's concessions introduced an unwanted level of DOE micromanagement to the project, but the cooperation with DOE helped relieve Fermilab's funding pressures.

DOE wished to implement a project management plan based on a cost-estimating procedure which was then widely used in reactor development, but it was not yet standard practice in high-energy physics. When Fermilab physicists initially opposed the additional bureaucracy,

Mravca and Lederman pushed for a compromise. The Doubler formed its Project Management Group (PMG), which included everyone who bore major responsibility in the Doubler project, and also many others who had worked with the Doubler in its earlier phases (e.g., Helen Edwards, Fowler, Lederman, Limon, Livdahl, Lundy, Paul Mantsch, Frank Nezrick, Orr, and Tollestrup and a representative from the business office). The PMG became a mechanism not only for cost estimating but for discussing problems and finding solutions. Meetings were held regularly on Tuesday afternoons in "the dungeon," a room near the accelerator control room that came to be viewed as a crisis center. During the Main Ring magnet crisis of 1971 and 1972, this space had housed the historic meetings of Wilson with his commissioners and his "three managers" (Livdahl, Lundy, and Orr), all of whom were now members of the PMG. Acting as secretary, Tollestrup prepared detailed minutes for DOE.[34]

By the end of the summer of 1979, the transition that began with Lederman's decision to shorten the dipole magnets was complete. Fermilab's accelerator professionals now controlled the Doubler, and the Doubler's jobs had gained priority over those of the Main Ring. In July 1981 a significant administrative move dissolved any lingering friction between the Doubler and the laboratory's Accelerator Division. Orr, who since early 1980 had served as director of the Doubler, also became head of the Accelerator Division, with Helen Edwards as his deputy head.

Another important step in the evolution of megascience in the Doubler program under Lederman occurred in the magnet facility. On taking office Lederman enlarged this facility and turned it into a more professional operation. He replaced Hanson, whose health had become poor, with Richard Lundy, a strong manager as well as a good physicist.[35] Lundy had been working with Fowler on cryogenics and magnet measurements. Those working at the magnet facility noticed a clear distinction between Lundy's professional approach and Hanson's less formal style. Another obvious constrast was between Lundy's keen analytical style and Wilson's almost "anti-analytic feeling."[36] Lundy would remain head of the Doubler's magnet facility until September 1984, when he became Fermilab's associate laboratory director for technology.

In his first year as head of the magnet facility, Lundy solved the last major technological problem of the magnets, that of inadequate stability in the orientation of the vertical magnetic field of the dipoles. The first twelve magnets in the 21-foot dipole series (magnets TA 200 to TA 212) had been used in the B-12 ten-out-of-twelve test for DOE, passed in June 1979. It was hoped, as Lundy recalled, that now "everything was known about the magnets, and all that remained was to build approximately

1,000 of them with no change."[37] However, certain tests had not been carried out to the required precision and had to be redone. Later that summer of 1979, Helen Edwards read the reports made on the more precise second measurements and noted to her dismay that the orientation of the main magnetic field component in the so-called vertical plane inside the magnet yokes was not always the same from cooldown to cooldown.

Edwards, Collins, Sho Ohnuma, and others had been concerned for some time that no one was measuring whether the dipole field held its direction. In September 1978, a small rotation had been noticed in the 22-foot magnets, but it had been ignored.[38] When physicist Dan Gross discovered that his quartz blocks, used for positioning the wires, were grooved in two places, it appeared that the source of the problem had been found, but this proved false. Edwards kept rechecking the magnets and by mid-1979 had gathered enough evidence to convince Lundy that there was a real problem with the dipole field being able to hold its direction. Magnet production then stopped for several months. The news of Fermilab's "vertical plane problem" reached CERN. On a trip there in this period, Peter Limon recalled, "people would come, and with enormous grins on their faces . . . say, 'Isn't it terrible what's happening at Fermilab?'"[39]

Lundy's group eventually traced the variations in the orientation of the vertical magnetic field to a change that had been made in the cryostat design at the time the magnets were shortened. This change made it impossible to center the magnet coil accurately inside the warm iron. Diagnosis is not a solution, however, and on February 28, 1980, more than seven months after the discovery of the vertical plane problem, it had not been fixed. Tollestrup reported "the vertical plane problem is still with us. A special effort will be made over the next week to collect together all of the data on this particular problem and to formulate a couple of possible solutions to the difficulty."[40]

Lundy finally solved the problem using brute force. He decreed that all existing cryostats be scrapped and that new ones be built having four anchors instead of one. Approximately one hundred magnets had to be remade.[41] Lundy's approach worked, but it caused new problems.[42] When a magnet is warm, the prestress that is placed on it creates great force on its supports. To keep the heat leak low, these supports were composed of a glass-epoxy composite referred to as G-10. Unfortunately, this material "creeps" (changes shape) at room temperature, so that a magnet maintained at room temperature would lose its prestress. Lundy solved this second problem with "smart bolts" (support mechanisms with springs in them) that regulated the load on all bolts to allow centering and decrease creeping.[43] By July 1980 the magnet factory had determined that for magnets bolted in this way the vertical field measurements were satisfactory. By early September the design of the dipole magnets was finally fixed, and the problems settled down. Tollestrup reported, "Everyone agrees the magnets that are now being measured are acceptable for installation." Mravca reflected that had the DOE known about the problem of the rotating vertical plane field at the time they reviewed the project, "there would have been a big question mark as to whether the project would have started. . . . I doubt it would have."[44]

After installation of the A-sector between May and December of 1981, preparations for the critical A-sector test could be made. These continued from January to June of 1982, for the actual test had to wait until the 400 GeV fixed-target run ended in June 1982. At this time Fermilab's 400 GeV facility was permanently turned off. The test checked the cryogenics, power supplies, voltage to ground, quench protection, pressure piping, and reliability of operation. It was a rip-roaring success, reaching 4,200 amperes! Helen Edwards reported: "The successful carrying out of the [A-sector] test was the first step in the development of an overall system that would be reliable enough for operation when beam commissioning started about one year later."[45] Those involved in this test tried in every way to "break the machine," Orr recalled. "We ran it to the highest possible energies." But the machine did not break. While the B-12 strings could not run with currents higher than 3,000 amps, the A-sector test ran strings above 4,000 amps, corresponding to 900 GeV. "We quenched the whole thing at once with the valves closed so we could get to the highest possible full ring quenches," recalled Orr. "We blew covers off the top of the relief stacks, the legend is as high as the High-Rise. We quenched three-quarters of the sector with high voltage riding on top of the normally induced voltage on the magnets to see if we could cause electrical arcs. . . . We couldn't break it."[46] The machine appeared reliable enough for operation.

Now installation of the full ring of Doubler magnets could finally begin. In the final year of building the Doubler the focus turned to meeting serious schedules in completing every aspect of the accelerator. Some installation had been performed during a machine shutdown in the summer of 1981 and for a month at the end of 1981. The full installation, led by Peter Limon, Thornton Murphy, and Larry Sauer, required numerous "tunnel rats" to be working around the clock in the underground ring on many different jobs (such as electrical contracting, pipe fitting, leak checking, drilling, aligning, and connecting).

Murphy was responsible for coordinating the many tightly interlaced activities of the installation so that all components advanced in step, without forward or reverse salients that would slow down the evolution of the whole system.[47] Touring the tunnel several times a week in a golf cart while speaking into a tape recorder, he noted the progress of pipe fitters, leak checkers, electricians, riggers, millwrights, and other workers. He estimated that at the peak of activity as many as two hundred people were at work in the four-mile tunnel on any given day. "Their activities and travel patterns had to be understood to keep from getting a total traffic jam." He recalled that during the first nine months of this massive effort, starting in June 1982, the installers were hampered by not knowing the magnet production schedule, which was slowed by certain problems. "We never knew exactly how many magnets would be available on what date to install." In time these problems abated and magnet production actually got ahead of installation.[48]

As a large facility, Fermilab was able to shift its staff to meet the temporary personnel crisis created by Doubler installation. Drawing staff from elsewhere in the laboratory, Limon's installation group received roughly fifteen additional people. Having the flexibility to regroup people "who would have been waiting and twiddling their thumbs otherwise, and to do it in very short order . . . within the order of days," was a strength of the project, reflected Murphy. Had decisions been acted on methodically, those affected in the group might have been left "polishing their desks or cleaning their tool kits" and thus have been unable to meet the unexpected technological problem efficiently.[49] Only minor technical problems—for example, a high failure rate in vacuum seals (quickly corrected)—were encountered during the Doubler's subsequent installation phase. The last magnet was installed in the ring on March 18, 1983 (fig. 10.1).[50]

The small superconducting magnet effort that Wilson established as an informal back-burner project was now a professional, large-scale group effort that included all the oft-mentioned leaders (such as Helen and Don Edwards, Fowler, Limon, Lundy, Murphy, Nezrick, Orr, Rode, and Tollestrup) and many others as well (fig. 10.2). Among the others were Gerry Tool, head of electrical engineering, who developed control and power systems as well as a quench protection system. Another unsung hero was the physicist and Jesuit priest Father Tim Toohig, who coordinated construction. Toohig saw that service buildings, long runs of pipe and cable, utilities, cryogenics, and other structures, were built or modified by the time they were needed. He described his philosophy as "You think far enough ahead to know what they should be doing next,

10.1 Interior of the Main Ring with the ring of superconducting Energy Doubler magnets below the original ring of conventional magnets, 1983. (Courtesy of FNAL Visual Media Services.)

and then you get the matrix ready for that." He encountered Herculean problems in modifying the tunnel to accept the Doubler. At one point some power lines were cut accidentally because drawings had not been updated. To minimize such incidents, Toohig kept color-coded progress charts on all aspects of the construction—the magnets, the central helium liquefier, refrigeration, power supplies, controls. By circulating updates of his charts and by constantly asking questions—such as "Where do you want the piping?" "Where do you want the outlets?" "Where do you want the ducts?"—Toohig made construction jobs drive the solving of technical problems urging the army of workers to meet a schedule that converged in the spring of 1983.[51]

The Doubler was ready for commissioning in June 1983, after the first magnet cooldown in May. Commissioning "went so smoothly as to be hardly worth mentioning," recalled Limon. "We had a pleasant surprise," recalled Orr. "The machine was a lot better than we thought it would be."[52] It was an unexpected contrast with the Main Ring, which had taken more than six months to be commissioned. What was the reason the Doubler simply worked?

10.2 Richard A. Lundy, J. Richie Orr, Helen Edwards, and Alvin V. Tollestrup, recipients of the 1989 National Medal of Technology for their achievement of the first large-scale superconducting accelerator, in the Red Room of the White House. (Courtesy of Janine Tollestrup.)

Clearly, the approach taken to creating this novel machine had much to do with its success. The fact that the Doubler budget was less pinched and that many accelerator experts were involved was important. The absence of commissioning time also derived from the important principle used in building the Doubler: to develop and individually test each small component (magnets, cryogenics, power supplies, control systems) before moving on to use it in the evolving machine. This principle had to some extent also been used in building the Main Ring, but it had been applied far more rigorously in the case of the Doubler. Lundy ascribed the success of the Doubler to its built-in correction capability. He contrasted the Doubler "to the Main Ring, where a very Spartan arrangement was tried. And when it wouldn't function, we were driven to add additional corrections and diagnostics, and then finally it worked." In the case of the Doubler, "the needed corrections were available from the beginning and were just brought into play without any fuss or bother."[53]

On June 2, 1983, the first beam injected into the Doubler made a full turn at 100 GeV. Two weeks later, the entire ring was ramped to over

500 GeV. Then, on June 26, a coasting (nonaccelerated) beam was achieved. And on July 3 the first acceleration of beam to 512 GeV occurred (fig. 10.3). Newspapers across the country heralded the success of the Tevatron.[54] Over the next months the Doubler achieved a few more milestones, such as when a resonant extracted beam reached the switchyard area on August 12 at 700 GeV. Experiments using the Doubler at 400 GeV began in October 1983. And shortly after a further record of 800 GeV was set on February 16, 1984, experiments began near the new 1 TeV (trillion electron volts) energy frontier.

Once the Doubler became a permanent part of Fermilab's facility, the 800 GeV Fermilab accelerator was renamed the Tevatron. The research program was divided into two complementary initiatives corresponding to the two modes in which the Doubler and Main Ring could function together. The "collider mode," a proton-antiproton colliding-beams program with energies close to 2 TeV, would be referred to as the Tevatron I program; the enhanced "fixed-target" mode, supplying protons of roughly 1 TeV to a variety of experiments, would be called the Tevatron II program.

10.3 Control room celebration of the achievement of 512 GeV, July 3, 1983. (Courtesy of FNAL Visual Media Services.)

At the official dedication of the Doubler-Energy Saver on April 28, 1984, H. Guyford Stever, former presidential science adviser and director of the National Science Foundation, and then president of URA (1982–1985), recalled some of the changes that had taken place since the dedication of the Main Ring: "At the dedication a decade ago, though we did not know it at the time, we would soon enter an era in which the operating electrical power for accelerators would soon become much more costly; already the oil embargo of October 1973 had occurred. Today, with our Saver, our accelerator is operating successfully in a power-saving mode and, as is becoming habitual, the decade has brought another doubling of the proton particle energy for our high-energy physics experiments." Lederman spoke next, thanking Wilson, "whose vision provided our essential blueprint . . . [and who] worked well with Alvin Tollestrup to solve many of the SC magnet problems which, over and over again, threatened progress in our mastery of this difficult technology." Danny Boggs, deputy secretary of the DOE, expressed confidence that the new machine "will open new doors and explore a new world," adding, "We all look forward to the new scientific territory that will be explored by the Energy Saver/Doubler."[55]

The Doubler project had succeeded despite its long odds. It was born of Wilson's imagination and developed through his infectuous enthusiasm and willingness to employ empirical methods, especially needed in this case because scientific understanding of superconducting magnetic materials was lacking. Tollestrup's insistence on varying just one factor at a time allowed systematic tracking of the effects of every change. Fermilab's unconventional on-site magnet factory enabled rapid feedback in the R & D and control of the tooling for constructing of many duplicate magnets.

From an institutional point of view, the Doubler benefited from the laboratory's readiness to adjust the project's administration, from cooperation between groups, and from their frequent informal meetings, which sustained coordination of the complex effort within the laboratory. The fact that leadership and support by the laboratory director and his deputies were directly available, under both Wilson and Lederman, allowed the project to become the laboratory's first priority, thus making a large workforce available for solving the many problems that arose. In having that much laboratory support, the Doubler project was unique among the attempts to build a superconducting accelerator in the same period.

Fermilab's bold step of converting its accelerator into the Tevatron between the late 1970s and the early 1980s, a period in which superconducting magnet accelerator technology was only in its infancy, allowed the laboratory to advance rapidly into a new frontier. From this vantage, it witnessed and nurtured, as the following chapters will illustrate, the emergence of a megascience filled with new problems, some of them yet to be solved.

ELEVEN

Bigger Science: Experiment Strings, 1970–1988

In these times of fiscal crises we appreciate the difficulty of seeking balance as between the needs of science and the needs of other activities; in science as between the obvious benefits of short-term technological advances against the vision of what particle physics, abstract as it is, will mean to the future culture and well-being of our society. LEON M. LEDERMAN[1]

Fermilab's path to megascience followed two tracks, one through its fixed-target program and the other through its colliding-beams program. The first, explored in this chapter, led to "experiment strings"—large experiments and their follow-ups, corresponding to long-lived (typically more than twenty years) research traditions within the laboratory's experimental program. Each subsequent experiment in a string differed from the previous one in but a small number of aspects, for example, a change in part of its detector.[2]

Historical probing of these strings reveals new insights into the phenomenon of megascience, defined in the introduction: a bigger big science shaped by restricted funding for physics after about 1970. The philosopher of science Mark Bodnarczuk has interpreted the practices of megascience as the response to a struggle for limited resources that caused

This chapter is based on the undated and unpublished draft by Mark Bodnarczuk "A Case Study in the Emergence of Big Science from the Fixed-Target Experimental Program" and the following three published articles: Bodnarczuk 1990; Bodnarczuk 1997; Bodnarczuk and Hoddeson 2008.

the physicists to behave in new and unusual ways. In an effort to hold on to their investments, for example, their apparatus or place in line for beam, the experimenters now did everything possible to extend their experiments and prevent others from using their experimental space. It made sense to conduct follow-up experiments in which resources were recycled. In addition, the increasing cost, size, and complexity of the apparatus extended the time needed for assembling and disassembling experiments. A kind of sluggishness set in resembling the bureaucracy of any large system. Wilson was saddened by this trend at his laboratory.

As Bodnarczuk explained, the resources in question circulate in overlapping economies.[3] For high-energy physics laboratories, these resources include particle beam, "experimental real estate" in the form of apparatus or experiment halls, physicists with appropriate training and expertise, computing power, and journal space. Survival at the frontier depends on how skillfully the researchers can extract such resources from these economies. As the researchers strove to maintain their precious investments, the experimental strings formed invisible mini-institutions corresponding to particular experiment programs, such as the study of charmed particles. While such strings would surely have evolved at Fermilab whether or not the Doubler had been built, the competition from other high-energy physics laboratories heightened as the energy of the accelerator increased, intensifying the struggle for access to protons, costly detectors, experiment halls, experienced and well-trained experimenters, and computing strength contributing to eventual publication of data.[4]

The new megascience was riddled with ironies. For example, as the detectors designed for specialized particle searches grew larger and more costly, convincing the laboratory to invest in building them required arguing that they would be long-lived features that numerous groups could employ. Yet once a group had expended the time and effort to develop a complex element, it could not afford to give it up to competing physicists. Soon the detectors originally designed to meet specific experimental goals were shaping the direction of physics explorations, as we will illustrate. The social structure of experimental groups (including the physicists, postdocs, graduate students, engineers, and technical support staff) began to match itself to the structure of the detector's subsystems. Physicists became typed according to the kind of detector they had learned to work with. Proposals were judged, at least in part, by whether they could attract physicists or students having the right expertise for doing a particular experiment. This process, in turn, affected the rate of production of Ph.D. candidates with the background to work on such specialized aspects of experiments.

The larger detectors affected other economies as well. The advent of the silicon microstrip detector (SMD), a device for measuring tracks to an accuracy of 15 microns using fine-grained detectors of silicon mounted to form *x*, *y*, and *z* coordinates, allowed more electronic detector channels to be used simultaneously. These numerous channels, in turn, created the need for more online and off-line computing, if the data was to be recorded on tape for later analysis.[5] Among the more striking ironies of megascience as it developed at Fermilab is that Wilson's mistrust of computers combined with his expectation that experimenters would do the bulk of their computing at their home institutions forced a greater reliance on triggers to reduce the amount of data gathered. But the triggering programs that specify the conditions under which the detectors record events are always based on the *prevailing* theory of particle behavior. If the theory is incomplete, the triggering assumptions may be incorrect and insufficient data will be gathered, as in the experiment discussed in this chapter.

Prominent examples of such strings at Fermilab include[6]

kaon decays (E-82, 226, 425, 486, 584, 617, 731, 773, 799), with spokespersons Valentine Telegdi, Hans Kobrak, Bruce Winstein, and George Gollin

neutral hyperon beam (E-8, 361, 415, 440, 495, 505, 555, 619, 620, 756, 800), with spokespersons Lee Pondrom, Gerry Bunce, Ken Heller, Tom Devlin, and Kam-Biu Luk

neutrino studies (E-21A, 262, 320, 356, 482, 616, 652, 701, 744, 770), with spokespersons Barry Barish, Frank Sciulli, Michael Shaevitz, Frank Merritt, and Wesley Smith

neutrino scattering (E-1A, 310), with spokespersons Dave Cline, Al Mann, and Carlo Rubbia

P-west high intensity (537, 705, 721, 771), with spokesperson Brad Cox

gas jet internal target in elastic and inelastic scattering of protons on protons, deuterons and He nuclei (E-36a, 186, 289, 317, 381), with spokespersons Ernie Malamud, Vladimir Nikitin, Adrian Melissinos, Rodney Cool, and Konstantin "Dino" Goulianos

lepton and hadron pairs (E-70, 288, 494, 605, 608, 772, 789), with spokespersons: Leon Lederman, Myron "Bud" Good, John Rutherfoord, Charles "Chuck" Brown, Joel Moss, Dan Kaplan, and Jen-Chieh Peng[7]

charmed particles with photon and neutron beams (E-87, 358, 400, 401, 687, 831), with spokespersons Wonyong Lee, Tom O'Halloran, James Wiss, Mike Gormley, and Joel Butler

charmed particles with photon and charged hadron beams (E-516, 691, 769, 791), with spokespersons Tom Nash, Mike Witherell, Jeff Appel, and Milind Purohit

While it would be valuable and interesting to write about all these experiment strings, space limitations restrict us to discussing the details of only one. We chose the string generated by experiment E-516, a study of charmed particles, because it highlights many of the general features of megascience and because Bodnarczuk studied this particular string in some detail. Our discussion is based largely on his work. In E-516 and all the experiments following from it, the beam of high-energy protons emerging from Fermilab's accelerator struck a target and produced new particles (photons in the case of E-516 and its immediate successor, E-691, and hadrons in the subsequent follow-ups E-769 and E-791). Beams of these particles then produced charmed states in another target and the data were recorded by the complex and costly detector known as the Tagged Photon Magnetic Spectrometer (TPMS).[8]

The Early Days of the Tagged Photon Laboratory

The story of the E-516 string began in the period when Wilson's Main Ring was being completed. A small group of physicists became interested in using the new machine to create a secondary or tertiary beam of photons whose energy would be "tagged" by scattered electrons. Their idea was to use this tagged beam in experiments conducted in the Proton East Area.[9] Tagging the photons offered a way to select the interactions of interest for the particular experiment's goals. By the summer of 1971, three competing groups had proposed experiments for a facility that would became known as the Tagged Photon Laboratory (TPL): P-25, a study of photon total cross sections; P-144, a study of general features of photon-induced reactions in the 50–500 GeV region; and P-152, a study of photoproduction.[10] Fermilab's management told the spokespersons for the proposals—David Caldwell (P-25), Sam Ting (P-144), and Clem Heusch (P-152)—that for budgetary reasons these groups would have to collaborate on a single design that would meet the needs of all three groups. When Ting then withdrew his proposal, E. Thomas Nash, who had worked with Ting at DESY and whom Ting had sent to Fermilab to help design P-144, was out of a job. He stayed on at the laboratory and joined P-25 as well as NAL's staff. Both P-25 and P-152 were approved. When E-25 made a commitment to build the Proton East beamline, it was awarded first use of the TPL from 1971 to 1976. Heusch's E-152 was second in line at the TPL.

While E-25 ran, Nash and a few collaborators (George Luste from the University of Toronto, Rolly Morrison from the University of California at Santa Barbara, Jeff Appel from Columbia, and Paul Mantsch from Fermilab) began pushing to build a sophisticated spectrometer that could fully utilize the TPL's photon beam in studying the production of charm particles. They formed a collaboration with Nash as spokesperson. In June 1976, they submitted a "Letter of Intent" to the Fermilab directorate; their "Proposal to Study Photoproduction of Final States of Mass Above 2.5 GeV with a Magnetic Spectrometer in the Tagged Photon Lab" was assigned the number P-516.[11] The plan was to study charmed particles by shining the photon beam on a liquid hydrogen target and produce secondary particles, including charmed particles. The electronic trigger would then select the charmed particles by seeking their signature, as specified by the prevailing "diffractive model." According to this theory, the charm particles would be accompanied by a recoil proton of particular momentum and angular properties. Observing a recoil proton with that signature in the spectrometer's recoil detector would trigger the recording of data out of which computer programs could "reconstruct" the charmed particles.[12]

The P-516 collaboration then set out to recruit physicists to help plan their new spectrometer, the TPMS, to detect charmed-particle decay products. Its computing architecture allowed it to select, record, reconstruct, and analyze events to elucidate such properties as charm particle production, cross section, and particle mass. The experiment planned was unusually large, costly, and complex by the standards of that day. To justify the large expense of the detector, the report argued that the TPMS would be a general purpose "facility" designed for a variety of experiments to be done by different groups.[13] The E-516 string would in fact hold on to its experimental real estate for fourteen years (except for one interlude when E-612 ran parasitically during E-516's scheduled operation).[14] The eventual approval of this and other long-lasting experimental efforts would bring Fermilab to break from Wilson's philosophy of doing small, "quick and dirty" experiments.

Aware of Wilson's preferences, P-516 did its best to portray its spectrometer as inexpensive. By this time, developing a proposal that appeared frugal had become a common means for extracting resources from the laboratory. Once the laboratory made its commitment to support a proposal, collaborations would proceed to "ratchet up" their request for resources, adding magnets, computing power, and other features not specified in the original agreement.

The proposal for the TPMS also discussed the experiment's "people design," demonstrating that enough physicists with appropriate expertise would be committed to carrying the experiment through to publication.

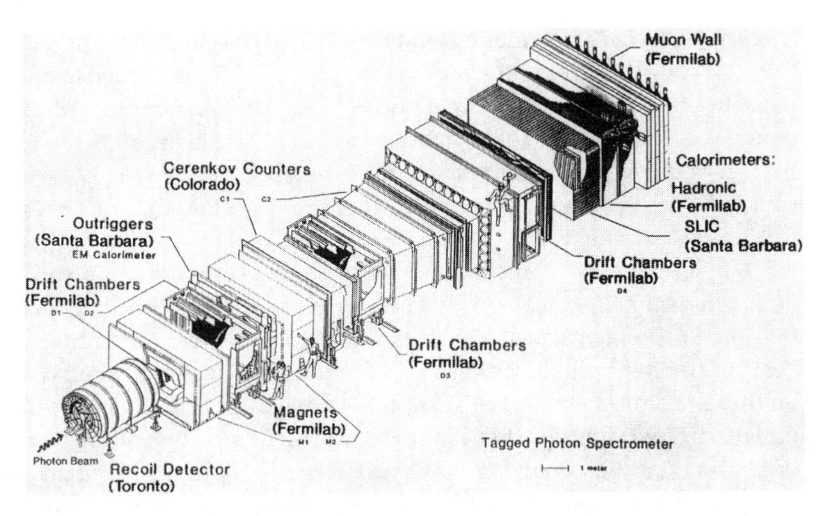

11.1 The Tagged Photon Magnetic Spectrometer of E-516, 1977. (Courtesy of FNAL Visual Media Services.)

As particular institutions took responsibility for the TPMS's six subsystems, the detector itself became a kind of map of the collaboration; its social organization became "detector driven." The experimental target and recoil detector for the trigger were the responsibility of the Toronto group, which also built and maintained the experiment's state-of-the-art trigger processor in collaboration with Fermilab. The electromagnetic calorimeter for measuring the energy of photons and electrons, the segmented liquid ionization calorimeter (SLIC), and the outrigger were the domain of the Santa Barbara group. The Cerenkov counters were associated with the University of Colorado. And the tracking system to analyze trajectories and momenta of charged particles, the hadronic calorimeter to measure the energy of uncharged hadrons, and the muon systems for measuring μ mesons penetrating the most forward chamber were Fermilab's responsibility. At this early stage, following the advice of the PAC on November 19, 1976, Wilson deferred the large and complex P-516, but he advised the collaboration, fourteen physicists from four institutions, to proceed with its spectrometer design (fig. 11.1).[15]

While P-516 waited in line for the TPL and worked on its spectrometer design, E-25 completed its run at the TPL and the next experiment in line, E-152, moved in. Represented by Heusch of the University of California at Santa Cruz, E-152 was an experiment mounted by only one institution. It was thus clearly vulnerable. Heusch had been promised only one run, but he realized that once E-516 and its enormous TPMS was installed, that experiment would not only displace E-152 but negate E-152's opportunity for follow-up experiments. Heusch proceeded to oppose this displacement. In early June 1977, some weeks after the TPMS design report had been submitted, he informed Fermilab's directorate that he would request additional running time after completing his first experiment.[16] If Heusch were granted this request before the end of his first run, that would delay E-516 for several years, and he could hope that P-516 might go away. Anticipating Wilson's question of what measurements he planned to make with the additional running time, Heusch told the directorate that he could not clearly define his future physics goals until after he had completed his initial run. Heusch also played on the laboratory's (and P-516's) promise to make the TPMS available to other users by claiming "that as much equipment as is technically feasible would be shared among different efforts."[17] Nash quickly offered many technical reasons why E-152's detectors would not meet P-516's needs, but for the moment E-152 was victorious; the experiment was approved on June 15 with 1,800 hours of beam time. P-516 could expect a delay.[18]

P-516 received another jolt when Wilson deferred it a second time after the June 1977 PAC meeting, for funding reasons. Despite the experiment's attempt to appear inexpensive, the spectrometer's size and complexity pushed the limits of the laboratory's resources. Already starting the process of ratcheting up their request for additional resources, the members of the collaboration claimed it would have to design and construct new analysis magnets at a cost of about $1 million. Concerned that P-516 may have underestimated the magnitude of its project, Wilson and Goldwasser demanded "a few specific examples of the kinds of measurements which you anticipate making during the initial operating period of the facility," both prototypes and tests to offer reassurance that P-516's detector could actually be constructed and operated to the specifications in the TPMS design report. They also asked for more evidence of the experiment's people design, in an effort to confirm that the collaboration had the technical expertise and manpower to complete the investigation and publish its results.[19]

The E-516 experiment was approved on October 3, 1977, a month after Nash and his team addressed these issues (fig. 11.2). In the summer of 1978, Wilson apportioned 1,000 hours of beam time to the experiment, specifying that the first 400 hours should go to an initial shakedown and engineering run, to make certain it was possible to deliver the design flux of 10^6 photons to the experimental target. The second run was to begin in November 1979; the third and final one in October 1980.[20]

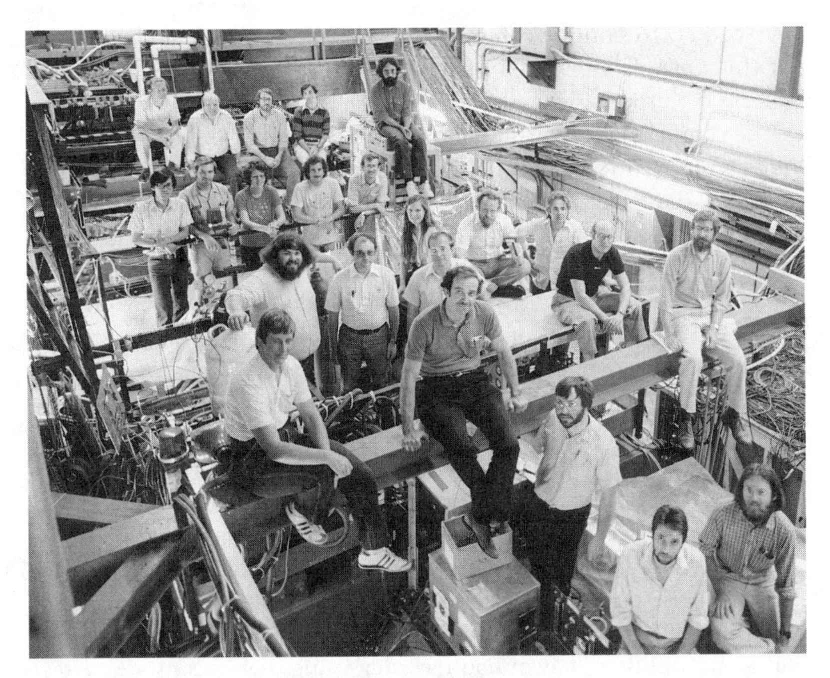

11.2 E. Thomas Nash (*in dark shirt, center right*) with E-516 collaborators, 1979. (Courtesy of FNAL Visual Media Services.)

Wilson's approval of E-516 was bad news for E-152, then occupying the TPL. E-152 soon lost its place in the TPL to E-516.

E-516's first major worry concerned the essential data-gathering and analysis resource of computing. The collaboration initially requested a relatively modest online computer, a PDP11-55, for Nash believed that the experiment's trigger processor would substantially reduce its data.[21] Predictably, in November 1977, after the approval of E-516, but before the start of the experiment's run, Nash began pressing John Peoples, then head of the Research Division, for more computing resources.

Working under Peoples and controlling Fermilab's computing resources was Al Brenner, the longtime head of Fermilab's Computing Department. Brenner supported Wilson's philosophy of frugality. He also understood, as did Peoples, that experiments with large computing needs often appealed to frugality in their proposals but then requested more support after approval. When Steve Bracker of Toronto, who besides Nash made the greatest contribution to the design of the trigger processor, began worrying that the requested online computer might not be able to handle both the acquisition and monitoring of the data, Brenner upgraded E-516's online computer from the PDP11-55 to a PDP11-55T, but this upgrade amounted merely to adding a tape drive; it did not increase the processing power.

The computing issue would blow up in December 1979, shortly after the start of E-516's second run. By this time Lederman had become Fermilab's second director and the experiment had completed its engineering run and installed its segmented hadronic calorimeter and drift chambers for the tracking system. Two senior members of the experiment, Morrison and Uriel Nauenberg of Colorado, warned Lederman that E-516 was in danger of failing because of its inadequate off-line computing system.[22]

The Reconstruction Shoot-Out and the Search for Charm

Anxiety rose within the collaboration during 1979. The members of E-516 had not expected to see immediate evidence of charmed particles, but they did not imagine that they would not see *any* such particles. By the time the second run ended, early in 1980, the group had already used two-thirds of its allotted beam time. Where were the charmed particles that SLAC and Brookhaven had recorded using their beams and detectors? Only gradually did Nash and his colleagues realize that charmed particles were penetrating their detector without registering evidence of their presence.

E-516's failure to detect charmed particles during its first two runs was grave, yet in the social context of the high-energy physics laboratory, it was prudent not to discuss such a matter publicly, not even at collaboration meetings. Because the approval of additional funds is based largely on performance, and because competing experiments always lie waiting to displace experiments that are not making good on their promises, Nash wrote a memo warning the collaboration about the dangers of "loose talk." "Credibility is destroyed by the public statement of incorrect or, more commonly, variable results," he explained. "It takes judgment and a thorough, open and continuing discussion within the group of any work to determine when a result has stabilized and is dependable. Until this point is reached it is essential that results—and difficulties in the analysis—not be discussed idly." Nash gave "a second reason to avoid loose talk." Referring directly to the resource economies, he reminded his team that the "funding and beam time . . . are in such short supply that administrators are looking for any excuse to cut back resources—even if already granted. Under these conditions a slight comment about [trouble with] a couple of bad channels in D1 returns as 'Their drift chambers don't even work. What do they need all that intensity for?'"[23]

Fermilab's experimenters had recently learned a hard lesson about announcing results before their findings had stabilized. Many at Fermilab recalled the embarrassing time when results from the E-1A neutral-current experiment had varied publicly (see chapter 7). Physicists referred to the episode as "alternating" currents, for in its race with CERN, E-1A had first claimed to discover neutral currents, subsequently retracted this claim, and eventually, after the discovery was announced by CERN, retracted its second claim asserting that neutral currents did exist.[24] Caution can be outweighed by pressure from the laboratory to announce and publish results, especially when laboratories are competing. That E-516 considered itself in crisis by the end of 1980 is clear from the correspondence exchanged between members of the collaboration. On December 9, 1980, some weeks into the experiment's final run (which began on October 15), Nash sent Morrison a letter expressing his concern: "Our experiment is in a very real crisis. There is more than a zero chance that we will come up empty handed."[25] This letter followed more than five years of preparation and an immense financial investment.

The pressure on E-516 to announce results heightened when Lederman announced the deadline of February 1, 1981, for submitting experiment proposals for the new Tevatron program, scheduled to begin as soon as the Doubler began operating in two years.[26] Thus, in the midst of what appeared to be a failing experiment, Nash and his collaborators had less than two months to design a follow-up experiment using the TPMS—or lose their place in line. In the same letter to Morrison in which Nash had described the crisis of E-516, he added: "I would like to assure you that I (with your and others' help and advice) can put together a respectable document that will serve to hold our place in line."[27] Proposing a follow-up experiment before completing one that is running is a common approach in holding on to experimental real estate. Such a proposal is typically made only when a collaboration has preliminary data to support its request. E-516 lacked such data, but the group could not afford to lose its place in line.

By this time, the cause of E-516's problems was clear to at least some members of the collaboration.[28] E-516's triggering specifications for turning on the detector (i.e., when it saw a single recoil proton along with a large missing mass in the forward region of the spectrometer) had been codified before the theory of quantum chromodynamics (QCD) had become prominent. It was based on the earlier diffractive model for charmed-particle production. Within the framework of QCD, charmed particles could be produced by incident photon beams, but the signature of a recoil proton would not dominate.[29] As the collaboration began to draft its proposal for a follow-up experiment, Morrison, Appel, and others in the group began to push for eliminating E-516's recoil detector and tight trigger parameters. They believed they could detect the as-yet-unseen charmed particles using solid-state detectors. Nash, who had invested years in developing the recoil detector and trigger parameters, was not ready to give them up. He pointed out that the physics of the new detectors was not yet completely understood. The dissenting members were in an awkward position, because they opposed the collaboration's most powerful member, one with the political power to overrule even group consensus.

The drafts of the proposal that the group circulated throughout January 1981 for their follow-up experiment reflected the team's fragmentation over the opposing experimental assumptions. By early February they had reached a compromise: their proposal would include a discussion of QCD and the photon-gluon production mechanisms and would state that the group planned to study both diffractive and nondiffractive photoproduction of charm.[30] The subsystems of the TPMS, including its recoil detector and trigger processor would remain almost unchanged. The directorate assigned number 691 to the proposal and requested the preparation of an impact statement to help determine how much new equipment and what new computing resources would be needed, as well as what effect the experiment would have on the resource economies of other proposed Tevatron experiments.[31] Thus began the experiment string generated by E-516.

Nash reported on the status of E-516 at the June 1981 meeting of the PAC, after the experiment's final run ended. When he admitted that the tracking system reconstruction programs were not ready, Lederman understood that the experiment had not even begun its in-depth data analysis, a prerequisite for publication. After the meeting, Lederman informed Nash that he was deferring P-691 and would not grant approval until he received a clear statement of E-516's physics results.

In the meantime, the collaboration weathered an internal storm relating to its tracking system particle reconstruction programs. While experiments in high-energy physics rarely find the resources for building multiple detector subsystems that perform identical functions, they often write more than one reconstruction program for a detector subsystem. Software packages reflect the style of those who write them, and multiple versions can check each other or offer different approaches to interpreting the portion of the data promising the most interesting results. For over two years, Uriel Nauenberg and John Bronstein had been writing independent reconstruction software packages for the tracking system, drawing on Monte Carlo simulations of charmed-particle decays (briefly defined in chapter 5). They did not know that a junior member of the collaboration, David Bintinger, had been quietly working on a third software package for the tracking system. Bintinger had not wanted to offend Nauenberg and Bronstein, who were both senior to him, but as he neared completion, he found himself in a delicate position. He showed his work to Nash, who appreciated its merit. As spokesperson, Nash had to decide how to present Bintinger's efforts to the rest of the collaboration. Before revealing the matter to the entire collaboration, Nash planned a way to measure the quality and efficiency of the three programs.

Unfortunately, the news of Bintinger's program leaked before Nash released it. Nauenberg and Bronstein were indeed disturbed that Bintinger had moved in on their territory and that Nash apparently supported Bintinger. Nash responded to Nauenberg and Bronstein in a formal letter: "David's right to pursue the direction of his interests is, of course, the same right you both had when you started your reconstruction efforts."[32] Nash then proposed settling their disagreements with a shoot-out. He asked Bracker, who had technical expertise in computing hardware and software but was not directly involved in this specific effort, to develop quantifiable criteria for comparing the three programs and to serve as a referee. Bracker agreed. He refused to examine any of the programs before he had developed the criteria. In an effort to eliminate infighting over the definition of the criteria, he recorded daily progress on his task into a computer-based file accessible to all members of the collaboration. Through numerous iterations of the criteria, Bracker eventually brought Bronstein, Nauenberg, and Bintinger to the point where all agreed that the final criteria constituted an acceptable test of the software packages. At the shoot-out all three programs were run with identical data sets. When the smoke settled, only Bintinger's program was left standing.[33]

The E-516/P-691 Interface and the Computing Revolution

The group remained divided over how E-516's experience should be reflected in the design of P-691. Morrison, Appel, and others in the group opposed Nash and his shrinking group of supporters on the issue of retaining the recoil detector and tight triggering assumptions. To everyone's considerable relief, during the last three months of E-516's final run the experiment began to show preliminary evidence for charmed-particle reactions. This evidence came in the midst of the struggle over the recoil detector and triggering assumptions, at about the time that the laboratory notified E-516 that it would have to submit its annual report. The new evidence resulted partly from Bintinger's reconstruction code and partly from the group's continued tuning-up of its hardware and software.

The collaboration could now claim that its reconstruction and analysis had yielded "encouragement in the charm channels that are our primary interest."[34] On May 6, 1982, the team requested the approval of P-691, but it was apparent that after a year of data analysis, E-516 had reconstructed but 15% of its total data sample of 20 million triggered events; two-thirds had been subjected only to preliminary analysis. Inadequate computing was in part to blame, and Lederman recognized the need to improve Fermilab's computing facilities. But even after he became director, securing the resources for the expensive computing needs of large experiments was difficult. Wilson's frugality policy and the group's inability to secure enough off-line computing resources at Fermilab, as well as the considerable pressure on the group to publish, had forced E-516 to put its completed software reconstruction packages into production on a large IBM computer at the University of Ottawa.

The negotiations over E-516's computing power illustrate how a single experiment can bring major change to the overall facilities of a large laboratory. Five days after submitting the E-516 progress report, Nash submitted another proposal titled "A Program for Advanced Electronics Projects at Fermilab" and explaining how experiments grow naturally until they reach a point at which their data is limited by "the amount of computing time they anticipate will be possible to squeeze out of the system." The more data that are processed, the more likely the experiment will succeed. They were, therefore, "up against the technological barrier of computing power limitations."[35] In almost a confession of E-516's near failure to detect the presence of charmed particles, Nash stated that isolating low cross-section effects is "limited by how many (correct) decisions physicists can make in live time." Without referring to E-516, he cast the collaboration's difficulties in obtaining results within the framework of computing problems confronting all Fermilab experiments and proposed that Fermilab address the data problems of large physics experiments with an R & D program focused on the development of multiple parallel-processing computing systems based on powerful, specially configured, Intel 32-bit microprocessors. Such systems, he claimed, would offer Fermilab experiments a thousandfold increase in cost-effectiveness. The combined reduction in cost and increase in computing power would allow experiments to use less biased trigger assumptions, record more data on tape, and simultaneously accelerate the data analysis leading to publication.

Had E-516 been able to shift the bulk of its computing to off-line data analysis, the pitfall of working with faulty triggering assumptions could have been avoided, for many of the charmed particles that had penetrated the spectrometer undetected would have been recorded on magnetic tape. With regard to the P-691 proposal, Morrison's and Appel's argument that they should run the spectrometer with a less biased, open trigger could only be implemented if this type of low-cost, high-power computing were available. Thus, out of the near failure of E-516 emerged a concept for computing on a far greater scale than Wilson in an earlier period might have wished, a concept that would affect most of the laboratory's subsequent experiments.

Nash's proposal offered a long-term solution to the laboratory's computing problems, but it did nothing to address the situation E-516 faced as its members continued reconstructing and analyzing its data with less than adequate computing resources. The pressure to publish from the accumulated data was complicated by the fact that plans for P-691's detector had to proceed rapidly. The preliminary design for P-691, funded by the University of California at Santa Barbara, was firmed up under the leadership of Morrison and his postdoctoral fellow Paul Karchin.

By the fall of 1982, the group within E-516 opposed to using the recoil detector in P-691 had grown in numbers. Various members had begun to explore the new technology of the SMD which measured the tracks of secondary vertices immediately downstream of the experimental target and could provide an off-line filter for selecting events with short-lived particles, such as the charmed particle D^0.[36] Yet despite E-516's excellent progress on the SMD, the Fermilab directorate viewed E-516's lack of published data as a risk and continued to press the group for final data. Early in March 1983, more than two years after the completion of E-516's final run, Lederman asked E-516 to publish its data or withdraw its proposal for P-691.[37] This ultimatum made the next three months a tense time, for the group's annual progress report, due in May 1983, would be the basis for evaluating P-691 at the PAC meeting that June in Aspen. Fortunately, after three years of mining their nearly barren data tapes, E-516 finally had some preliminary results. Nash asked for a brief extension of the May deadline so that earlier in June the collaboration could present its preliminary results to Fermilab's Friday afternoon joint experimental and theoretical physics seminar, the "Wine and Cheese Seminar" (chapter 6), reserved for the most important research announcements.

At this point, the E-516 collaboration had to change the way it worked. During the design, construction, and installation phases of E-516, the technical and financial demands of building the spectrometer and bringing it into operation had forced individual members of the collaboration to focus their attention on the particular regions of the spectrometer they were responsible for. Although most members gained some knowledge of systems for which they had no direct responsibility, they typically did not understand the entire detector.[38] Now the collaboration had to function as a whole and would have to agree on what results would be presented in publications.[39]

For some time, Morrison had been trying to convince Mike Witherell at Princeton to join the faculty of the University of California at Santa Barbara. After accepting, Witherell took on the task of analyzing E-516's data tapes. He worked with a graduate student, Don Summers, who had helped with the design, construction, and reconstruction program of E-516's SLIC. The first E-516 data submitted for formal publication appeared in their paper on the decay of the D^0. On that basis Kris Sliwa presented "First Results from the Tagged Photon Spectrometer" at the Wine and Cheese Seminar on Friday, June 10, 1983. Sliwa showed the group's preliminary data for the D^0 decay, explaining that the analysis was in progress and that there were more data to come.[40]

The P-691 proponents hoped this presentation would ease Lederman's concerns and bring him to grant first-stage approval for P-691. They submitted the overheads from Sliwa's talk along with the current design of the experiment to Lederman. Nash outlined the group's plan both to use the SMD behind a short beryllium target and, perhaps simultaneously, to run the liquid-hydrogen-filled recoil detector. Included in the design were the existing trigger processor and a number of hardware improvements in most systems. Nash also revealed that Bintinger's charged-track reconstruction program was being used to analyze E-516 data on the new Advanced Computer Program (ACP), which resulted from his earlier computing proposal.[41] Two weeks later, Lederman informed Nash that P-691 would remain on deferred status until the group had formally submitted E-516's results to the customary professional refereed journal. Because Nash insisted that such publication was imminent, Lederman agreed to let him make a presentation at the November 1983 meeting of the PAC.[42]

By the following month, July 1983, Witherell was circulating drafts of his and Summers's paper on the D^0.[43] But the collaboration members, now back at their home institutions, could not agree on the reconstruction programs or on the design of P-691. Multiple drafts were circulated and recirculated without consensus, while the November presentation to the PAC drew closer.[44] Fortunately, by late summer of 1983 the group's study of the small sample of charmed particles that had accumulated during the last three months of the final data run convinced everyone in the collaboration that charm was rarely produced diffractively and that the P-691 configuration of the TPMS ought not to include the recoil detector. The plan to implement the new experimental configuration was supported by Karchin's and Morrison's design of the SMD and of a beryllium target/SMD configuration.[45]

The group elected Witherell as spokesman for E-691, but it did not yet notify Fermilab's management of this decision, for fear that the change in leadership might adversely effect the approval of P-691. On October 5, 1983, Witherell set an October 10 deadline for submitting the paper to *Physical Review Letters*. "It has gone through six complete typings since it was put together in July," he wrote. He had tried to reach a consensus on all the issues, but "I think it has hit the equilibrium point where for every change I make, I offend as many people as I satisfy." He decreed that from then on, "I will consider only changes that are typos, wrong numbers, or outrageous breaches of grammar inserted when the last changes were made."[46] He submitted the paper to *Physical Review Letters*; it was received on October 14, 1983.[47]

Even before the paper was formally accepted, the collaboration informed Lederman that it would be ready to make its November presentation to the PAC requesting approval of P-691.[48] Nash was scheduled to present it on November 10, 1983; the collaboration reassigned the task to Witherell. One week after Witherell's presentation, the thrice-deferred P-691 was granted Stage I (conditional) approval. Lederman's letter stated that he would review the readiness of the SMD before granting Stage II approval.[49] Ten days later, Nash sent his formal letter to the E-691 collaboration announcing the "rotation of spokespersons."[50] Nash would continue as a member of the collaboration, but Witherell would negotiate the problems ahead.

E-691 went on to much subsequent success, obtaining 10,000 fully reconstructed photoproduced charmed-particle events and the most precise measurements yet obtained of charmed-particle lifetimes. E-769, the follow-up experiment to E-691, would accumulate 6,000 fully reconstructed hadronically produced charmed events. (i.e., using a beam of π

and K mesons rather than a photon beam to produce the charmed particles). At that time, this experiment had the largest sample of hadronically produced charmed particles in the world. And E-791, the follow-up to E-769, went on to accumulate 100,000 fully reconstructed hadronically produced charmed-particle events, surpassing experiments that used the photoproduction mechanism.

The ambitious physics programs that employed the E-516 spectrometer thus transcended the goals and studies proposed by P-516, but the configurations of the TPMS used for E-516 and E-691 were almost identical. The plan to employ the same spectrometer that E-516 had used in E-691 had been stated as early as 1976: "This spectrometer will be ideal—and unique—for studying photoproduction when the Doubler comes into operation. It is probably the only existing P-East facility that will be able to operate at 1000 GeV."[51] The crucial differences between E-516 and E-691 were that (1) the SMD replaced the recoil detector and trigger processor, (2) a simpler and less restrictive trigger was used, and (3) much more off-line computing was available to handle the greatly increased amount of recorded data. E-516 had triggered the spectrometer with its upstream (i.e., closer to the accelerator) recoil detector and projected the particle tracks back toward a vertex that could not be precisely defined by this recoil detector. The different hardware used in E-691, together with its associated reconstruction software packages, allowed E-691 to reconstruct the decay vertices of charmed particles in the upstream portions of the spectrometer and project the tracks of particle decays to the downstream subsystems of the detector.

P-516 had also planned further study in which "the electron beam can also be used to transport pions into the Tagged Photon Lab," extending the program beyond the boundaries of the original experiment to conduct hadronically produced charm studies, an effort to underscore "the flexibility and long range benefits to the Proton East program that the spectrometer we propose would bring."[52] The primary beam was eventually converted to transport pions, enabling E-769's study of hadroproduction of charmed particles and then E-791's study of decays of hadronically produced charmed particles.[53]

The most obvious continuity in the E-516-691-769-791 string was its spectrometer. Although each detector configuration in this string was more complex than the previous one, all the changes were but modifications of the original structure; most subsystems of this detector remained intact throughout the string. The configuration for experiment E-769 was so closely based on that of E-516 that the collaboration included a diagram of the E-516 configuration in the P-769 proposal and simply explained the modifications in the text. By the time of the E-791 proposal, the spectrometer design had become well enough known to the Fermilab community that the proposal did not even include an image of it. There were fewer continuities in the group of physicists involved in the E-516-691-769-791 string than in most other strings at Fermilab. Appel was the only physicist who worked on all four experiments.[54] There was continuity in the physics goals, although they did evolve considerably. The scientific practice of transforming one experiment into another before the first had been completed helped the collaboration maintain its possession of the TPL, reducing displacement by competitors.[55]

The Rationale for Strings

By the end of the Lederman era, experiment strings, rather than individual experiments, had become the relevant unit for research in Fermilab's fixed-target program. Inseparable from the laboratory's procedure for approving running time, the experimental traditions we call strings can be construed as a response to the laboratory's organizational infrastructure for dealing with larger experiments. In the case of the E-516-691-769-791 string, which lasted for about fifteen years, the size and complexity of the TPMS detector proved a good investment. The TPMS did not become a facility for general use, as promised, but it did yield a long-term program of results, for each new experiment in the string had a decreased risk of rejection by Fermilab's directorate. The string that began with E-516 thus established a continuous program of studying charmed particles that accumulated progressively larger samples having more precise statistics for finer-grained charmed-particle measurements.[56]

Wilson had celebrated risk, both to cut costs and for reasons of aesthetics. But as larger, more complex, and more costly facilities were needed in the experiments performed during Lederman's tenure, the laboratory's attitude toward risk shifted. Strings minimized risk by capitalizing on the most useful and successful investments in important commodities, for example, protons, detectors, work spaces, and scientific expertise. (As computers tended to be upgraded regularly, or replaced, they were never a very long term investment.) With each iteration of the spectrometer, the laboratory came to view the collaboration associated with the TPMS as more competent, a group with a record demonstrating its ability to make good on its promises. Experiments were more easily approved and assigned higher priority if they were part of a successful string. Because the tagged photon beamline was used for all four experiments in the string that E-516 began, the need for each experiment to design, construct, commission, and pay for a new beamline disappeared. By the time P-769 was submitted, it had been ten years since the design for converting the secondary beam to a pion beam had been established. By using the TPMS as the basis for all four experiments, a well-understood detector could be used in the follow-up experiments without expending the time and performing the many tasks involved in commissioning a new spectrometer. And because most of the detector's subsystems remained intact through all four experiments, much of the software for reconstructing particle events could be reused with only simple modifications of code. In the case of E-691, the collaboration's ability to produce results shortly after the end of their data run was a direct result of the decision not to modify most of the components of the E-516 spectrometer.

Experiment strings removed certain social uncertainties. The accumulation of group expertise became a resource for the follow-up experiments. The continuity of institutional affiliation helped professors who needed access to high-energy beams to train graduate students. The longevity of experimental strings could help university-based physicists publish results more regularly, and thus attract better graduate students. In the new era of megascience, a graduate student could, during a period of four or five years of work with a string, gain an understanding of the full process of experimenting by analyzing data from an experiment whose running time was over, taking data from an experiment then running, and helping to design and propose a future experiment.

Experiment strings put some pressure on Fermilab to revert to an earlier model of big science. In biasing the approval of experiments toward "homesteaders" entrenched in existing strings, Lederman was subtly discouraging new prospective "explorers." Out of the context that Wilson created for supporting the outside-user-dominated laboratory that Lederman described in his TNL paper emerged an economy that challenged the empowerment of users. That this subversion did not readily occur can perhaps be explained by the conscious pressure Fermilab felt to appoint younger, next-generation experiment spokespersons from outside the laboratory. This centrifugal shift of power toward outside groups competed with the growing dominance of strings in the laboratory's experimental program, subjecting it to pressure as Fermilab slowly shifted from fixed-target to colliding-beams experiments.

TWELVE

Megascience Realized: Colliding Beams, 1967–1989

Fermilab's collider brings the art of the accelerator builder to a new pinnacle of technological achievement, and in so doing, brings to the scientific community a magnificent new tool for the advancement of knowledge.

LEON M. LEDERMAN[1]

On the other track in the rise of megascience at Fermilab were the experiments performed with particle beams colliding inside supremely complex detectors. Such experiments differ dramatically from those of earlier decades, for they typically last at least twice as long as academic spans (such as the time to advance to a Ph.D. or a tenured teaching position). Like the large fixed-target experiments, colliding-beams studies lead to follow-up inquiries, but colliding-beams detectors are so large and costly that no one even thinks of dismantling them at the end of a run; the experiments are effectively unending. Composed of hundreds of researchers, collider experiments are substantial institutions existing within the larger laboratory.[2] While prior historical studies of large detector collaborations have explored the impermanent and fragile nature of such groups,[3] our study

Portions of this chapter draw on an unfinished manuscript by Kyoung Paik, "The Origin of CDF and the Fermilab Collider Project," prepared in November 1993, when she was a graduate student in the Department of History at the University of Illinois at Urbana-Champaign. We thank Ms. Paik for contributing this work to the body of materials on which this book stands.

of Fermilab's early experience with colliding-beams experiments highlights the robustness of these experiments, which are as large and long-lasting as many entire laboratories.

CDF and DZero are Fermilab's two colliding-beams experiments. CDF, which initially stood for the Colliding Detector Facility and later for the Colliding Detector at Fermilab, came to life in the early 1980s. At that time it was one of the world's largest physics experiments, both in its physical apparatus and in the size of its collaboration.[4] DZero, Fermilab's second colliding-beams experiment, and named after this experiment's Main Ring location, was designed to be comparable in energy to CDF, but it was built later, with fewer resources. As DZero was authorized to be more innovative than CDF, it was often referred to as the "modest but unique and imaginative" detector.[5] Unlike the prominent European colliding-beam experiments, CDF and DZero did their best to govern themselves democratically, achieving their scientific accolades in the distinctly American style of megascience that they pioneered within Fermilab's tradition of the truly national laboratory, the TNL. In the process the TNL became a "TIL," a "truly international laboratory." Wilson's concept of the "world laboratory for world peace" (chapter 13) would be realized at Fermilab as physicists, computer experts, and engineers from around the globe, including many from developing nations and the Southern Hemisphere, joined research teams cooperating at the highest-energy facilities in their regions. The codiscovery in 1995 of the top quark, by Fermilab's CDF and DZero, remains the outstanding achievement of this level of megascience in high-energy physics.[6]

Dreams of Colliding Beams, 1967–1977

Wilson was among the first to refer explicitly to reaching higher energy by colliding beams of particles. He indicated the idea in a note scribbled in 1948.[7] Rolf Wideroë patented the idea, in May 1953, noting in his memoirs that he came to the idea in 1943. But it would take decades before the idea of colliding beams could be applied in an actual accelerator.[8] The motivation came from many sources, among them accelerator builders such as Gersh Budker in the Soviet Union and Bruno Touschek in Italy and, in the United States, Donald Kerst and a group inspired by Gerald O'Neill that included Panofsky and Richter. An important step toward funding colliding beams was taken by the Ramsey Panel in 1963, when it recommended that "after a suitable study" storage rings should be constructed at Brookhaven.[9] The key technical problems of building colliders were identified at a conference in Novosibirsk that Budker organized in 1965, and at a subsequent meeting held at SLAC in 1966. At the SLAC meeting, Willibald Jentschke of DESY met with Sidney Drell, Panofsky, Richter, and others to develop the concept that would lead to the historic colliding-beams machine in Germany known as DORIS.[10] By then scientists at many institutions of the world, including MURA, Brookhaven, SLAC, the Cambridge Electron Accelerator (CEA), CERN's Intersecting Storage Ring (ISR), the world's first hadron collider, and the accelerators in Frascati in Italy, Orsay in France, and Novosibirsk in Russia, were assessing the possible benefits of colliding-beam machines.

In the early days at Fermilab, the possibilities of colliding beams of protons with beams of protons or antiprotons were discussed at both the Berkeley 200 GeV design study in 1965 and the Oak Brook summer study in 1967.[11] But despite fertile and provocative suggestions, most physicists were still skeptical about building a colliding-beams facility because of many unsolved technological problems. For this reason, Wilson and Goldwasser discouraged active work on colliding beams until the Main Ring had been completed.[12]

Wilson paved the way for future colliding-beams development at NAL when in 1968 he asked Lee Teng to organize a design study for proton-proton colliding-beam storage rings as part of an effort to determine realistic cost estimates for future expansion. Teng invited a group of experts to the study; they met in Aspen during the first two weeks of August in 1968. Wilson even asked the engineering firm of W. M. Brobeck and Associates to work with the physicists to prepare cost estimates for various alternatives. The discussions continued over the next months, considering both 100 on 100 GeV and 200 on 200 GeV collisions; the higher-energy scheme used superconducting magnets.[13] The principal conclusion was that a proton-proton colliding-beam storage ring would be both "practical" and "useful" for physics experiments, but the general opinion at this point, according to Wilson, was that "the art of superconductivity had not advanced to a stage where one could responsibly risk large sums of money on it,"[14] Wilson, with characteristic verve and optimism, estimated that a 100 GeV × 100 GeV colliding-beam ring could be built for about $75 million, with another $25 million to spend on the facility and experimental apparatus.[15] In 1971, Richard Carrigan was the first to respond, with a proposal to build two superconducting rings in the Main Ring tunnel intended for colliding-beams work, but this prescient suggestion was not seriously considered, because the technology for building superconducting magnets was not refined enough then to allow even making a cost estimate.[16] In any case, the serious problems encountered in commissioning the Main Ring during the last months of 1971 and early months of 1972 (see chapter 6) soon diverted the attention of NAL physicists from colliding beams.

Once the Main Ring began operating, the AEC and URA encouraged Wilson to explore the issue of colliding beams. By this period several storage rings were in operation or under construction at sites around the world.[17] CERN's 30 on 30 GeV *pp* Intersecting Storage Ring (ISR) was commissioned in 1971. CERN had been investigating the possibility of a proton-proton colliding-beam storage ring as early as 1957.[18] In 1974 Wilson responded to the URA board's request to design storage rings for proton-proton collision experiments with two possible schemes.[19] In one plan, the so-called Electron Target Project, directed by Tom Collins, the old CEA would be installed tangent to the Main Ring and used as a storage ring to supply 3–4 GeV electrons to collide with the 400–500 GeV protons emerging from the Main Ring.[20] NAL's Long Range Planning Committee rejected this idea because it felt the energy would not be high enough to make the effort worthwhile.[21]

The committee endorsed Wilson's other colliding-beams scheme, a joint NAL-Argonne plan known as POPAE (Protons on Protons and Electrons), in which protons or electrons would be stored in two storage rings inside a common tunnel. Employing the phenomenon of superconductivity, the protons in one of the rings, accelerated to 1000 GeV, would collide either with electrons in the other ring, or with protons at 1000 GeV. The plan included building an electron storage ring that would allow 20 GeV electrons to collide with the 1000 GeV protons. These designs were carefully explored by Collins, Donald Edwards, Robert Diebold, and others.[22] Unfortunately, POPAE became what Wilson termed "a political fiasco" by entering a heated competition with Brookhaven's proton collider ISABELLE, which in 1974 was planned for 200 on 200 GeV.[23] The fact that Wilson's cost estimate for POPAE was a factor of ten lower in cost per GeV than ISABELLE pressured Brookhaven to increase its intended energy to 400 on 400 GeV by July 1974; the move effectively killed POPAE. In June 1975 POPAE was officially dismissed by a HEPAP subpanel chaired by Francis Low. The blow hit both ways, for ISABELLE's ambitious attempt to raise its design energy so dramatically contributed to its eventual demise (see chapter 13).[24]

Despite POPAE's rejection, Wilson continued to encourage colliding-beam studies at Fermilab, with many supporters and numerous suggested schemes from inside and outside the laboratory. For example, a month after Carlo Rubbia's July 1975 letter to Wilson proposing colliding protons in the Main Ring, Richter of SLAC and Cline of the University of Wisconsin made similar suggestions.[25] Other colliding-beams schemes included (1) constructing a small ring whose protons would collide with the protons from the Main Ring and (2) colliding the antiprotons produced in an antiproton source yet to be constructed.[26] To help him decide among the many colliding-beams suggestions made at this time, Wilson asked Tollestrup to organize his two-day "Modest Colliding-Beams Meeting" to examine this issue.

Tollestrup's workshop in January 1976 (discussed in chapter 8), attended by physicists from Fermilab, Columbia, Harvard, the University of Wisconsin, and other institutions, was Fermilab's first definitive step in developing its colliding-beams program.[27] The discussions covered both proton-proton and proton-antiproton collisions, but no choice between them was made. Peter McIntyre presented ways to achieve proton-antiproton collisions, where the cross sections are higher, making use of the technology of electron cooling to cool and concentrate an antiproton beam. McIntyre and Rubbia also introduced the idea of producing antiprotons by a fixed target with protons, separating the antiprotons from the protons, and then concentrating them for collision by cooling using the stochastic method that CERN engineer Simon van der Meer had developed in 1968 for increasing beam density in the ISR. Nevertheless, at this point most of the physicists at the meeting felt that proton-proton collisions offered the more promising path for Fermilab because of the difficulties involved in creating antiprotons.[28]

A second major question pursued at Tollestrup's meeting was whether the collisions should take place in the Main Ring or in the superconducting Energy Doubler ring, then under development but not yet approved for construction. While not settling this issue, the discussions did bring a clear recognition that colliding-beams experiments would be the most important initial use for the future Doubler. For that reason, the Doubler was redesigned to be more effective for colliding-beams work. The advantage of having the two beams in proximity with each other implied that it was best to construct the Doubler below the Main Ring, rather than above it, as originally planned (see chapter 8).[29]

Goldwasser's call in May 1976 for experiment proposals employing collisions in the Main Ring–Energy Doubler system brought five colliding-beams proposals to the table of the Program Advisory Committee in June.[30] Three proposals concerned proton-proton collisions;[31] the other two, both involving Rubbia, proposed proton-antiproton collisions in the Main Ring.[32] The PAC rejected all five colliding-beams proposals, explaining that it was too early to decide on the type of collider to build, given that the Doubler's design was not yet finalized.[33] The committee did, however, endorse establishing a colliding-beams department and devoting research to the study of antiproton cooling. Wilson encouraged colliding-beams study as long as the work did not require large sums of money. By this time, roughly a dozen Fermilab physicists, including Tollestrup, were involved in the discussions as a side interest. All realized that any real progress would require substantial funding, not just for the colliding-beams accelerator but also for a colliding-beams detector.

In October 1976, Tollestrup wrote to Wilson to ask him to provide support for the development of detector designs, for "at present no one at Fermilab is committing enough time to imaginative instrumentation development. We would like to furnish a nucleus to stimulate fresh thinking of what is required for this new type of physics and perhaps carry out experimental tests of new hardware, if appropriate." He asked only that two researchers be authorized to concentrate on designing the detector, explaining that "at this stage, small, highly motivated groups are the most effective tools for getting initial designs." In considering the scale of the collaboration that would be required, he added that "the social problems of its ultimate use will undoubtedly be enormous (and humorous). We need not solve these problems yet."[34]

Wilson responded by establishing the Colliding Beams Department on November 17, 1976, within the Research Division. He appointed James Cronin head of the new department. A respected experimental physicist from Princeton, Cronin was on an extended visit to Fermilab, having joined the faculty of the University of Chicago in 1971. He and his Princeton colleague Val Fitch would win the 1980 Nobel Prize for the discovery of violations of symmetry principles in the decay of neutral K mesons. Cronin supported the PAC's earlier decision in June that year to reject the five submitted colliding-beams proposals. As his associate head of the new Colliding Beams Department, Cronin appointed Fermilab physicist James Walker, who had participated in the earlier collider discussions and was then spokesman for P-478, a colliding-beams experiment proposal that had been rejected.[35]

Unfortunately, the institutional structure that Wilson created for the Colliding Beams Department was "a real mess," as Tollestrup put it. For while the department, which began its work in January 1977, was in the Research Division, it shared some responsibilities with the Accelerator Division.[36] The Colliding Beams Department was responsible for construction of its detector at its BZero address, and for the design and specification of collider experiments in the Main Ring. The Accelerator Division and the Colliding Beams Department were jointly responsible for planning the simultaneous use of the Main Ring and Energy Doubler for collision work. There was thus plenty of room for confusion, because the responsibility for production of the protons and antiprotons as well as the management of construction of experimental enclosures fell to the Accelerator Division.[37] Wilson planned in time to move the Colliding Beams Department into the Accelerator Division, after the start of construction of the collider project, but the fact that this department began its life in the Research Division complicated progress.

Despite frustration over the cumbersome administrative structure of the new Colliding Beams Department, Cronin set out with enthusiasm to develop a proton-proton colliding-beams program. He firmly believed that the planned research was "the most natural first use of the doubler"[38] and that "the development of a colliding beam capability using the energy doubler" should be Fermilab's highest priority. Although the Doubler was faltering at this point, because of Wilson's inability to secure DOE funds to support it, Cronin felt that because the collider program had not yet requested large amounts of funding, the prospects of receiving small amounts would be good, at least during the years when DOE was funding the construction of PEP, the electron-positron storage ring at SLAC. Cronin projected that the entire collider physics program could cost roughly $65 million (about a fifth of its eventual cost), with $10 million for construction of the collision hall and detector.[39]

At the first meeting of Cronin's new department, on January 11, 1977, it was already clear that dividing the responsibilities for colliding beams between the Accelerator and Research divisions would handicap rather than help the group. To study the experimental components, Cronin's department needed a final version of the Doubler design from the Accelerator Division. To complete its design of the Doubler, the Accelerator Division needed to have the design of the colliding-beams detector.[40] The major issue of whether the collisions would be proton-proton or proton-antiproton further divided the Colliding Beams Department and the Accelerator Division. Unlike Rubbia, who had favored proton-antiproton collisions, Cronin continued to view proton-antiproton collisions as extremely risky because they would be based on experimental techniques that had not yet been developed. The plan of using Fermilab's Main Ring would probably not succeed, Cronin believed, because the quality of the field produced by the Main Ring magnets was not intended for storage ring use. He advised leaving proton-antiproton collision work to CERN, at least for the time being, and refining the known techniques of beam storage so they could work toward the time when proton-antiproton collisions could be achieved in the Doubler.[41]

Russ Huson, the head of the Accelerator Division, however, adamantly favored proton-antiproton collisions, as did McIntyre, Frederick Mills, and others. The decision-making process was not centralized enough to resolve this conflict, and the rift between Huson and Cronin widened.[42] Cronin felt he could not accomplish much without the facilities, manpower, equipment, and funding that Huson controlled.[43] To make matters worse, in the month before Cronin started his new position, Wilson had imposed a restriction on hiring for the new department. When Wilson stated in a labwide memorandum that work on the collider effort would be done on a voluntary basis, colliding-beams research dropped from most Fermilab physicists' plans.[44] It became clear to Cronin that organizing the colliding-beams effort would be an even greater challenge than anticipated. In fall 1977 he resigned, and his department proceeded to disband.[45]

Building CDF

In December 1977, while Cronin's department was phasing out its work, Wilson made new arrangements for Fermilab's colliding-beams effort. In a laboratory-wide memorandum, he announced that on January 1, 1978, he would create two new groups in Huson's Accelerator Division: the Colliding Beams Group under the direction of Stan Ecklund, and the Antiproton Cooling Group under Don Young and Fred Mills. Ecklund's group would be responsible for colliding-beams experiments and for the design and construction of the experimental area. Young's group would oversee construction and management of the antiproton ring and the antiproton target that would be added to the Main Ring. Both groups would work with Peter McIntyre's ongoing Internal Target Group, and related work would also be carried out by the laboratory's Research Division under John Peoples.[46]

To be sure that the administrative apparatus supported the entire colliding-beam effort this time, Wilson appointed Tollestrup head of the new Colliding Detector Facility Department within the Research Division. Tollestrup kept Jim Walker as deputy head and named McIntyre his assistant head.[47] This trio would focus on the research associated with the design, construction, and implementation of the detector for studying the collisions. McIntyre also continued his work on the antiproton source from his post in the Internal Target Group until 1981.

An attempted experiment to concentrate a 200 MeV proton beam by electron cooling offered useful information for the Armistice Day shoot-out in November 1978. At this point the electron-cooling scheme consisted of an electron gun, a collector, and a long solenoid for the electron beam to pass through, a design due largely to Mills. The plan called for beam to be extracted from the Main Ring and carried a distance to a small target station where 85 GeV protons produced 4.5 GeV antiprotons. These antiprotons would then be injected into the Booster, decelerated to 200 MeV, and extracted from the Booster into the cooling ring, where cooling and accumulating at 200 MeV would take place. Once a large number of antiprotons had been accumulated, they would be injected back into the Booster for acceleration in the opposite direction, and then into the same line that the protons had entered at 4.5 GeV at the start of the process. Finally, the antiprotons would be directed back to the Main Ring and accelerated to high energy in the backward direction. The work on this scheme continued until the fall of 1978.[48] It was done, according to Young, "in Wilson style with a minimum of funds and without interference with the high-energy physics program." Wilson had supported the early research on electron-cooling work using summer graduate students and visitors from the Soviet Union. This project ended with the Armistice Day shoot-out, when completing the Energy Doubler became the laboratory's top priority (chapter 10).[49]

The newly launched effort to design a colliding-beams detector developed slowly. Wilson, who in early 1978 was still director and head of the Energy Doubler/Saver, although he had officially resigned, was, as Tollestrup reflected, "very 'anti-group,'" preferring individual or very small team efforts. As a result, "there always was the underlying struggle to get an effective group going, as long as he was in charge."[50] The initial meetings were informal and small, with at most twenty people. Tollestrup hoped to create "a home for people who wanted to study what could be done with colliding beams."[51] Lee Holloway of the University of Illinois remembered the group as "just a bunch of guys sitting around talking." Many in the group had previously worked together. Diebold from Argonne had been head of the POPAE project. Collins had been involved in colliding beams since his time at the CEA. Walker, Cline, and McIntyre had participated in Tollestrup's January 1976 workshop and had submitted proposals for colliding-beam experiments. Most of those in the new CDF group had worked together in Cronin's Colliding Beams Department.

By mid-January 1978, the CDF group members had laid out their long-term physics goals and decided to explore two detector options: "a detector with magnet," which Cline would develop; and "a calorimeter type with no magnet," which Limon would explore. They planned to choose between the options over the next several months.[52] Walker and Collins formed a committee to finalize the collider parameters.[53]

The following April, one important decision had already been made: to build a magnetic detector. The appeal of the nonmagnetic detector had been that it appeared much less expensive and could be built faster. It might have been the right detector to build had Fermilab been in the race with CERN to find the W and Z particles. But once Fermilab withdrew from this race, the wish to proceed with the best detector for high-energy collisions outweighed the cost difference.[54] Now the group needed to settle on a specific magnet design. To help the group thrash out its ideas for detectors, CDF held a "Detectorfest" in mid-May of 1978. Tollestrup was optimistic that colliding beams might be available even earlier than planned. He felt that with the completion of the colliding-beams design, scheduled for August of 1978, it might be possible for proton-antiproton collisions to "be ready before the doubler."[55] But by July 1978, their progress had stalled. One reason was that Wilson had stepped down as Fermilab director. Tollestrup was convinced that lack of "the Director is the key to problems and that our future will be uncertain until a new Director assumes control."[56] Nevertheless, by September 1978 they had produced a preliminary "Conceptual Design of a Large, Thin Coil Superconducting Solenoid Magnet for Colliding Beam Experiments." The design did not represent the "unanimous opinion" of the group, but it was a basis for further discussion.[57]

The lull in the work of CDF continued until October 1978, when Lederman was named director designate and organized his Armistice Day shoot-out. It was still unresolved whether the collider should be a proton-proton or a proton-antiproton collider. This issue related to whether they needed to build another ring. For proton-proton colliders two rings would be required; a proton-antiproton collider required only one ring. A single-ring collider had the important additional advantage that it could quite easily be converted to run in fixed-target mode, unlike a double-ring proton-proton collider, although cooling the antiprotons was not trivial.

Soon the DOE also became interested in Fermilab's colliding-beams research. On one Friday afternoon after the shoot-out, as Fermilab theorist Chris Quigg recalled, DOE's Doug Pewitt phoned Lederman with an appealing proposition: if the laboratory would submit a proposal for colliding beams, DOE would very likely approve it. Fermilab's proposal was hurriedly prepared that weekend.[58] Meanwhile, Tollestrup encouraged individual subgroups of the collider effort to apply for additional funds for their own work, setting the stage for developing a political economy in the collider program in which authority was diffused among powerful outside groups with diverse sources of support. Before long, the University of Chicago and Rutgers groups won NSF grants. A Japanese group from Tsukuba and Italian collaborators from Pisa brought funds from their governments.[59]

As the new collaboration took shape, Tollestrup considered whether particular parts of the detector should be associated with individual institutions. To do otherwise, he decided, would cause the collaboration to suffer.[60] In February 1979, newly hired Fermilab physicist Hans Jensen distributed a preliminary "Construction Plan for the Colliding Beam Detector Facility at Fermilab," complete with goals, tasks, and construction needs. The individual institutional groups signed up for the components on which they wished to work on a first-come, first-served basis. Once these decisions were made, the building of prototypes and the testing of various materials began. The years 1979 and 1980 would be periods of great activity and substantial growth for CDF.[61] The planning of the experimental site, as well as of a prototype testing facility, a computer analysis department, and an engineering section of CDF all proceeded simultaneously. Lee Pondrom of the University of Wisconsin, who headed CDF's experimental area group, suggested that the experimental hall at BZero be designed so that, like PETRA (the electron-positron accelerator-storage ring) at DESY, the large detector could be rolled in and out of the beam area to facilitate maintenance and analysis in periods when the collider was not running.[62]

During the second half of 1979 plans for creating CDF evolved very rapidly. In July 1979 (as discussed in chapter 10), DOE authorized funding for construction of the Doubler, and in August, HEPAP endorsed R & D funding for colliding-beam physics. CDF's "vigorous recruiting effort" was in full swing by fall.[63] By the end of the year, the CDF collaboration was planning its request for DOE construction funding. Jensen had already started work on a draft agreement.[64] By then, CDF's organizational chart listed more than forty people, each assigned to one of twelve groups. In January 1980, CDF's design plans were presented to a review committee that would report to Fermilab's PAC. In June, Tollestrup reported that the review committee report "was generally favorable."[65]

Meanwhile, pressures were intensifying in Washington DC for a better project management plan. Edward Temple, Jr., the head of DOE construction efforts, exerted bureaucratic authority aimed to achieve stringent control over project budgets and schedules. The DOE had already begun pushing for tighter controls on the Tevatron (see chapter 10) and criticizing ISABELLE's management, which was struggling under its new goal to be a 400 on 400 GeV superconducting collider. Temple instituted new preconstruction procedures by which each new project could begin only after the approval of its formal conceptual design report, with a format that included detailed cost estimates and schedules. Strict oversight and exacting DOE reviews were crafted to ensure that all projects adhered to the plan during each stage of approval.[66]

CDF was becoming a substantial enterprise with its own independent voice within Fermilab. Lederman searched for the right team to address the collaboration's growing need for focus, as well as DOE's rigorous oversight regulations. He decided to shake things up by hiring an outsider. He recruited Roy Schwitters, a young member of the SLAC experiment that had codiscovered the charm particle in 1974 using the Mark I detector at SPEAR (the Stanford Positron-Electron Asymmetric Ring). Schwitters agreed to serve as CDF's associate head, alongside Tollestrup. He "will assume full responsibility for all aspects of the design of the Detector," Tollestrup explained in a memo to the group on October 20, 1980. "The authority to commit Laboratory funds or resources remains with me as does the administrative responsibility for CDF. Hans Jensen remains Assistant Department Head." Tollestrup also communicated that John Peoples would coordinate the design of the collision and assembly halls, that Fermilab physicist Dennis Theriot and engineer Wayne Nestander would assume responsibility for the preconstruction phase (the Title I design) to develop a conceptual design leading to preliminary approval, and that Tom Collins would serve as the liaison with the Accelerator Division. He added that "since both Roy and myself have other responsibilities, it will take patience and goodwill on everyone's part to make the system work."[67]

In the same memorandum, Tollestrup explained that the goal before them was to have a complete proposal by June of 1981, including both a conceptual design and a budget for the new detector. He listed the member institutions which were then part of the collaboration: the University of Chicago, Harvard, the University of Illinois, Purdue, Texas A & M, the University of Wisconsin, Argonne National Laboratory, Fermilab, Japan (KEK/Tsukuba), and Italy (Pisa/Frascati). "Roy and I feel that this set of institutions completes the CDF Collaboration. Should future needs make it desirable to add other institutions, the CDF Advisory Group will be consulted before any decisions are made."[68] Schwitters recalled that when he arrived in the fall of 1980 CDF was "a loose confederation of different factions with different ideas. The Italians and Japanese had just come on board." As the group lacked a "coherent picture," Schwitters felt his primary task was to "forge a coherent agenda and to force planning decisions." To help manage the day-to-day work of designing and building the detector, Schwitters appointed Theriot, who had a reputation for being extremely good at planning, as his deputy.[69]

Theriot and Schwitters proceeded to make an exacting plan and cost estimate for the CDF detector in an effort to gain the needed credibility for DOE to endorse the detector. The conceptual design report was due in the summer of 1981. Earlier that year, as the group prepared its detector plan, Schwitters looked elsewhere for cues that might help guide the budding effort. His attention focused on the Time Projection Chamber (TPC), a large detector then being built at Lawrence Berkeley Laboratory to be used at SLAC. "The TPC had grossly overrun its budget," Schwitters explained, and it had a serious technical failure, a short in the device's superconducting magnet which threatened the project's future. In sum, the whole project "was out of control." Schwitters could not afford to repeat TPC's mistakes, especially since ISABELLE was just then in crisis. Large high-energy physics projects were expensive enough that failures were remembered in Congress when it came time for approval of funding. For their own sakes, and for the sake of the field, managers needed to avoid both managerial and technical failures.[70] Schwitters sent Theriot to Berkeley to "find out what had gone wrong." Theriot recalled the TPC's managers telling him "that the most crucial thing you had to do was to carry through the design in stages."[71]

A draft of the conceptual design report was ready for distribution by mid-August, after several rounds of corrections.[72] Its centerpiece was a detailed description of the detector, which consisted of "electromagnetic and hadronic calorimetry over almost 4π solid angle around the interaction region. Fine-grain spatial segmentation has been matched to the large energies and high multiplicities characteristic of the high energy events expected. A large superconducting solenoid magnet containing drift chambers measures the momentum of charged particles and gives a visual reconstruction of the event. Muon chambers around the perimeter of the central detector and iron toroidal magnets at one end identify muons."[73]

The report also included a "codification of the effort" to build the detector, as Chicago experimenter Henry Frisch noted, a detailed explanation of the types of experiments the detector would allow, and a discussion of their importance to high-energy physics. In time, the collaboration built the detector (fig. 12.1) and, according to Frisch, "came close to realizing the physics goals listed."[74]

Like the experiment discussed in chapter 11, E-516, CDF's complex detector drove the social structure of the collaboration. The August 1981

12.1 The Colliding Detector at Fermilab, 1988. (Courtesy of FNAL Visual Media Services.)

report listed eighty-seven authors and twelve institutions, besides Fermilab. The groups that designed individual components took on the task of bringing their designs to life.[75] By this time, the collaboration had evolved into an effort as large as a small laboratory, with a budget that had grown to tens of millions of dollars and hundreds of employees arranged into subgroups led by its managers. Schwitters, Tollestrup, and Theriot coordinated all parts of the effort.

Schwitters also took on some of the jobs that had previously been performed by Lederman. Spending much of his time in Washington, he spearheaded negotiations with DOE on the collaboration's budget and schedule. When at first DOE balked at the CDF design, with its $40 million price tag, Schwitters, resorting to Theriot's approach of proceeding in stages, requested funds to build only a bare-bones version of the detector at reduced cost. The idea was that pieces would later be added to the detector and DOE could decide whether to increase its investment based on the performance of both the collider and the collaboration. DOE agreed to this plan. As Schwitters explained, the approach "aimed at being very realistic"; once DOE had made the commitment to fund CDF, "the collaboration would be in the position to perform well, and then be in line to get more funding."[76] Construction of CDF began on July 1, 1982, with ground broken at BZero and the crew "digging the deepest hole ever at Fermilab." The digging and construction continued until March 1983, when the collision hall was ready.[77]

The stage involved building the enormous 2,000-ton central detector, with its elaborate subdetectors and other components. They included a solenoidal magnet, tracking chambers, an electron magnetic shower counter, hadron calorimeters, muon chambers, forward and backward segmented time-of-flight counters, electromagnetic shower counters, hadron calorimeters, and a muon spectrometer. The advanced electronics for data acquisition and analysis included trigger counters. Coordination was an ordeal, because the effort proceeded not only at Fermilab but also at the home institutions of the collaboration members.[78] All had to comply with the new DOE accountability requirements.

As Schwitters was often away, Theriot coordinated the construction. Tollestrup solved the technical problems that arose. Theriot remembered learning how to break large tasks down into smaller ones, both in organizing the work and in managing time and money, "which was a whole new thing for physicists."[79] He tracked decisions, schedules, and budgets in a manner acceptable to DOE, review committee members, and the PAC. Frisch credits Theriot for being "list driven," "milestone driven," and "schedule driven." Theriot was sometimes considered "fierce," but CDF physicists endured his approach because they respected his knowledge and dedication. "He was somebody who understood what we were talking about and so we were willing to listen to him. We knew he was only doing what was necessary to get the job done," reflected Frisch.[80]

CDF hardly operated like a well-oiled, hierarchical machine, despite the designed structure that Theriot, Schwitters, and DOE imposed. Diffusion of power could not be avoided, because the university groups resisted the formation of any hierarchy above their own. and the key groups within the collaboration retained control of their own budgets. In Frisch's words, "The CDF rule is pretty much that people aren't told what to do, unless it's absolutely necessary for the sake of fiscal responsibility, the schedule, or safety."[81] Collaboration meetings became the primary means of solving problems and making decisions. They reminded CDF electronics engineer Tom Droege of a "Quaker prayer group on the prairie." As notes from the meetings confirm, "Anybody could come, anybody could talk, there was no agenda. Somebody would come up with a problem, there would be a discussion, and then somebody would volunteer to fix the problem." Schwitters, Tollestrup, and Theriot acted as coordinators and facilitators, rather than commanders.[82] Even though one of them might put pressure on a group, or even threaten to proceed without it if deadlines were missed, the making of decisions was usually consensual, at least among the assembled players.

From the beginning, there were doubts about whether the CDF collaboration could actually build its detector. Schwitters remembers being concerned when construction began on the central calorimeter, late in 1981. For this apparatus, forty-eight wedge-shaped modules had to be assembled to create four self-supporting arches resting on the base of the magnet yoke. Assembly of the first module took until the spring of 1982, when bids were going out for the construction of the collision hall. The building and integrating of the electronics in the remaining forty-seven modules would take three years.[83] Lundy managed the assembly-line work in industrial style. Plastics and phototubes for the calorimeter came from Italy and Japan; steel came from Purdue. Plastics from Japan arrived first at Argonne, where they were machined and then sent on to Fermilab. Schwitters described in Fermilab's 1983 annual report how "all of these pieces show up . . . at Fermilab, where they must be put together into an integrated module, outfitted with electronics and data acquisition capabilities and tested." In a later interview he confessed that he could not stop thinking: "And everything has to fit to better than a millimeter!"[84] Besides the central calorimeter, two other particularly large technical and organizational tasks were (1) the construction of the solenoid coil and (2) the enormous iron magnet yoke in the BZero collision hall (renamed the CDF Collision Hall in early 1985). Built in Japan during 1983 and 1984, the solenoid was shipped to Fermilab in mid-July 1984.[85] The high price of components forced the collaboration members to learn how to deal with financial issues within the formalized DOE guidelines.[86]

To grease the wheels of the emerging corporate-style megascience at CDF, standing "godfather" review committees were assigned to the detector components during the construction period. They would convey information between Schwitters and the construction groups, noting "any differences between the expected detector performance and the performance given in the CDF Design Report," to "express concern on technical aspects of the design," to "review the basic physics goals of the particular system in view of advances in the field," and to offer "recommendations on what aspects of the system could be delayed if the financial situation requires it."[87] The information gleaned from the godfather committees was often used in internal reviews of detector components, helping their managers provide detailed and accurate information to DOE and Fermilab.[88] With its network of committees, the godfather committee system improved communication and organization within the collaboration, offering not only information exchange but also a decision-making apparatus that would continue throughout the construction and use of the detector, laying the groundwork for CDF's semi-independent future.

The Antiproton Source

As the collider project gained momentum between 1980 and 1983, the need to build its antiproton source and produce a beam of antiprotons became urgent. What was required was a device to create and separate the needed antimatter particles, store and stack them, accelerate them, and ultimately steer them into the CDF detector for collision with protons from the Doubler. The 1977–1980 antiproton production attempts using a small electron-cooling ring, whose beam would be decelerated down to 200 MeV, had not panned out.

Lederman seized the opportunity to reorganize in late 1980 when John Peoples expressed interest in leaving the Research Division and work instead on developing the antiproton source. Shuffling his personnel cards in early 1981, Lederman appointed Peoples to succeed McIntyre (who had moved to Texas A & M) as leader of the Accelerator Division's Internal Target Group and also be the deputy head of Young's Antiproton Source Project, which was no longer in the Accelerator Division, answering directly to Lederman. In the spring of 1981 Lederman appointed a committee headed by Cornell's Maury Tigner and including Bjorn Wiik of CERN and DESY to give a technical review of the existing antiproton source plan.[89] The Tigner Report stressed the need for a new simpler design offering higher luminosity.[90] Further reshuffling continued that fall, as Peoples became project manager for Tevatron I, directing the construction of the antiproton source, Young became deputy project manager, and Fred Mills was made associate head.[91]

Alessandro Ruggiero and others pressured Peoples to change the cooling method. Together Ruggiero and Peoples developed an antiproton source design featuring stochastic rather than electron cooling, drawing on recommendations made by Tollestrup. By March 1982 the new design was ready. In the new scheme, antiprotons would be produced when high-energy beams hit a metallic production target. Then 8 GeV antiprotons would be collected in the so-called Accumulator and eventually, using stochastic cooling, prepared for injection into the Tevatron.[92] During five days of exacting investigation, DOE's Ed Temple, Mel Month,

12.2 Rich Orr, on left; Father Tim Toohig, on bended knee; and John Peoples, on right, celebrating construction of the antiproton source, 1984. (Courtesy of FNAL Visual Media Services.)

and Gordon Charlton examined every technical aspect of the new design. Robert "Obie" Oberholtzer, a specialist brought in by Peoples, gave credibility to the improved design. At the end of the long review, Temple agreed to support the new antiproton source (fig. 12.2). Approval came in May 1982.[93]

The design required the excavation of a new tunnel and the installation of magnets and advanced accelerator systems to run the stochastic cooling hardware. The systems were built and tested with the help of partners from Argonne, LBL, Wisconsin, CERN, and Novosibirsk. CERN physicists, including Bruno Autin, shared their knowledge of van der Meer's stochastic cooling technique. The Novosibirsk team contributed a prototype that allowed Fermilab to build a lithium lens for focusing and collecting large numbers of antiprotons.[94]

After its groundbreaking ceremony on August 16, 1983, harsh winter conditions, along with a range of technical and economic problems, impeded the construction of the antiproton source (fig. 12.3). In one crisis, a contractor went bankrupt. In Fermilab's 1983 annual report, Lederman optimistically remarked: "There is gold to be mined in the hills of $p\bar{p}$ physics at our higher energy,"[95] but in the philosophical finale of the next year's report he referred to the difficulties of building the antiproton source. He tried Wilson's style of coping with hardship by using a quote from the early seventeenth-century poet Francis Quarles:[96]

We gape, we grasp, we gripe, add store to store;
Enough requires too much; too much craves more . . .
Thus we, poor little worlds! with blood and sweat,
In vain attempt to comprehend the great.

Carefully synchronized and in harmony with all of the technical accelerator elements of the Main Ring and the Tevatron, the antiproton source was finally ready for use in August 1985.

A System Emerges

Fermilab's proton-antiproton collider was dedicated officially in a ceremony held on October 11, 1985. The dedication address by Secretary of Energy John Herrington affirmed the basic philosophy of the laboratory: "Searching for the hidden rules of nature's game is a very important thing to do." He described the new collider as a "truly remarkable accomplishment," and confirmed that "the support of basic research by our government is an essential element in our nation's long-term economic health." He stated that the achievement "represents a major step forward which restores U.S. leadership in facilities for high-energy physics research."[97] He also tied the work that would be performed with the new collider to broad themes of pursuing frontiers: "This project represents one of the best examples of how government support of basic research advances our nation's goals. It provides a frontier that

12.3 Aerial view of Fermilab showing the High-Rise (Wilson Hall), the Booster ring below, in the center of the photo, and the larger rounded triangle of the antiproton source, 1999. (Courtesy of FNAL Visual Media Services.)

challenges our young people to seek and apply new knowledge."[98] Lederman added: "We are privileged to look where no one has ever looked before. The new domain of energies our Collider will explore will yield very valuable data, and may well have great surprises in store." John Peoples expressed his exuberance: "I'm excited, exhilarated, and worn out."[99]

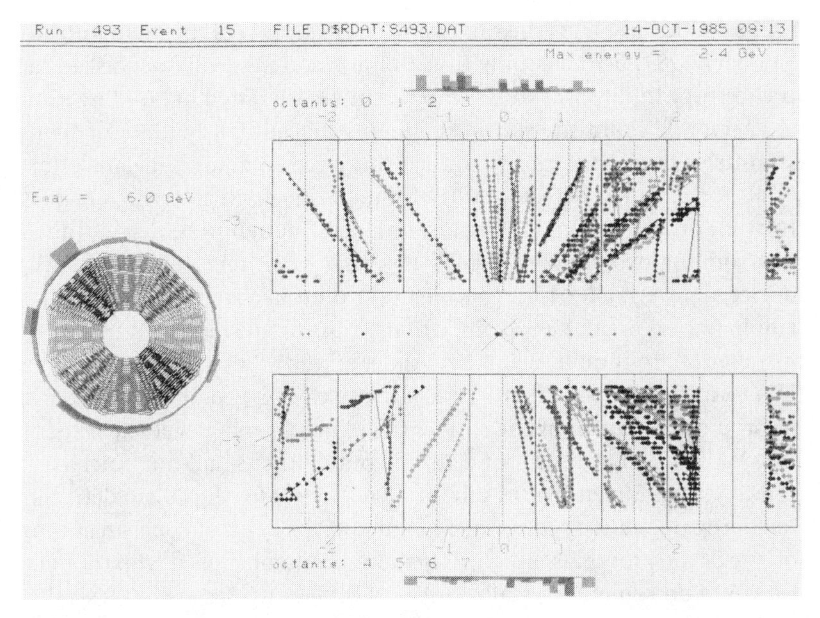

12.4 First collisions in the Tevatron, October 13, 1985. (Courtesy of FNAL Visual Media Services.)

On October 13, 1985, the corps of discovery working on the antiproton source, in collaboration with the Accelerator Division and the Colliding Detector team from the Research Division, achieved the first proton-antiproton collisions at 1.6 TeV in the center-of-mass system, the highest energy ever achieved in a laboratory. A new machine had been born, opening a new energy domain (fig. 12.4). "I remember being somewhat amazed," Frisch later confessed. "Everything worked, and it worked well."[100]

After its short test run in 1985, CDF was turned off while the detector was completed, and while the construction of components for subsequent upgrades proceeded. By 1988 the detector as described in the 1981 design report was finally working. Frisch remembered thinking that Rubbia's collider experiment at CERN, UA1, was turned off and "we had the energy regime to ourselves. It was glorious."[101] Two years after CERN's discoveries of the $W^{\pm}$ and Z^0 particles (see chapter 13) Fermilab regained the lead at the frontier.

An Alternative Megascience Springs to Life: DZero

While CDF approached completion, Lederman became interested in building a second colliding-beam detector which would be complementary to

CDF and located across the ring at the DZero section of the Main Ring. In February 1981, before issuing his call for proposals, Lederman asked for preliminary input from those who might be interested in building such a detector.[102] From the start, Lederman believed that because of "constraints of both time and funding," the second colliding-beams effort "calls for a modest detector built by a modestly sized group." In fact, as he went on to explain, it "would be modest enough to be discontinued after one or two runs (of 3 to 4 months) to be replaced by a newer device, or (eventually) by an e-p (electron-proton) adventure." To further emphasize his point, he explained that "stress should be placed on ease of installation, minimum debugging time, and maximum innovation."[103] In a form suitable for the new era of megascience, Lederman was trying to promote a version of the sort of small, quick, and clever experiment that Wilson had favored, but it turned out neither small nor quick.

More than fifteen groups submitted responses for the DZero detector, including preliminary ideas for experiments. As Bernard Pope, an author of one of the proposals, noted, "There were a number of us who thought we could do something really cute and different." Because CERN's discovery of the W and Z particles was still two years in the future, they "thought it would be possible to scoop CERN, even." Over the next two years, these proposals received scrutiny from the PAC and Fermilab's management. CDF's "godfathers" were asked to offer advice on the DZero proposals.[104]

Following a DZero workshop in June 1983, the PAC advised against approving any of these responses, on the grounds that "none of the groups making proposals was sufficiently strong to carry out the project in a timely manner."[105] This temporary roadblock challenged Lederman to organize a collaboration from among the contenders. The PAC members had requested, and Lederman had agreed, that Paul Grannis would lead the new DZero project and have "the authority to select the collaborating groups and make the design decisions which would yield a detector with the attributes suggested by the PAC."[106] Grannis was a professor at the State University of New York at Stony Brook who, with Mike Marx of Stony Brook, had submitted one of the early proposals. Whimsically named LAPDOG, this proposal describing a nonmagnetic detector with electron calorimetry and muon detectors had a cartoon dog, biting where he could not scratch, as its logo (fig. 12.5).

To form the DZero enterprise, the rejected LAPDOG proposal was joined in a forced marriage with two other rejected proposals, one put forth by Ernie Malamud and Dan Green of Fermilab stressing muon detection, and a third from Pope of Michigan State University. As Grannis

12.5 Lapdog logo of the DZero experiment, 1985. (Courtesy of FNAL Visual Media Services.)

told Pope in late June, a "DZero project" had been set up, which would combine all the proposals "in the best possible way." Grannis would work with Peter Koehler, head of the Research Division. That summer, the nascent collaboration began intensive discussions to formulate a general design for the new experiment and to advise the laboratory on the scope needed for its experimental hall. Thus, on September 1, 1983, yet another user-empowered institution within Fermilab had been launched. Grannis accepted Lederman's offer, thanking him for the laboratory's "early support" of the DZero venture.[107] Lederman noted in his job offer to Grannis that heading DZero was a "tremendous opportunity" as well as a "tremendous challenge."[108]

The transformation of DZero from Lederman's original notion of a small and clever experiment to one that in time rivaled CDF in its scale was a sign of the times. Because of CERN's experience with UA1 and UA2, everyone, including Lederman, realized from the start that, as Grannis remembers, "we weren't going to do the physics that we were trying to propose by being truly small and clever. We weren't going to be an ice-box sized experiment." Grannis's interaction with the PAC convinced him that they understood the basic problem with the experiments that were originally proposed for DZero: "They were too limited—they didn't fully embrace all of the bigness that they should." What was needed was a "full scale rival to CDF."[109]

Recruitment for DZero was particularly challenging because CDF had started first. "By the time we came along," according to Grannis, "they imagined that we were just going to be late, and CDF would do everything before we got there. In fact, the general sentiment was that we were a ridiculous exercise."[110] To make matters worse, by the time DZero was starting, other large collider collaborations, such as the SLAC Linear Detector (SLD) experiment and the ZEUS experiment using the new HERA I (the Hadron Elektron Ring Anlage) detector at DESY were forming. There were not enough people to build and use the envisioned detector. Because of the stiff competition for collaborators, Grannis started out with a group of roughly fifteen researchers, which included the original merged LAPDOG group, with Michigan State's group, now headed by Maris Abolins. Although Grannis and his early recruits did exclude a few potential members, they solicited "everybody suitable we could think of to come and join us." In the process "we couldn't really afford to be picky and say, 'well, you've got to come in with 5 senior people, and have experience with big projects, and have 5 post-docs if you're going to join.' We were happy with one senior person, a couple of post-docs and some students. And most people didn't have previous experience with the largest detectors."[111]

With its group of researchers unschooled in doing the largest physics projects, DZero struggled to find a niche within a community that already included CDF. The inherent difficulties in the task were already apparent in July 1983, when DZero was completing its design to submit to the PAC. By this time, satisfying DOE requirements called for a design report with specific cost estimates and staff assignments. Pope later explained: "we faced the problem that our original *raison d'être*—of mounting a small, clever experiment to discover the W and Z particles—was several years *passé*. We knew that the physics coming out of CERN made it clear that we needed a big detector that does more. But 'oops—there already is a big detector that's doing more, right here at Fermilab.'"[112]

An initial impulse had been to build a state-of-the-art electromagnetic calorimeter to enhance DZero's energy measurement capability. As a first step, the group thought of building a calorimeter with as-yet-untried scintillating lead glass, an idea proposed by the Michigan State group and favored by others as well. After a few months of discussion, the group began to worry that this option would not work, and the group looked for a new type of calorimeter.[113] Word then came from CERN that uranium–liquid argon calorimetry gave what Pope called "astonishingly good energy resolution." An appealing aspect of this technology was that it placed hadron and electromagnetic calorimetry on nearly the same footing. At the same time, he recalled, the "first buzz coming from CDF was that their scintillation calorimeters were giving them a lot of trouble."[114] The group discussed switching to uranium–liquid argon, but despite the advantages, as Grannis later explained, the idea of making such a switch was "daunting. This would mean putting together a liquid argon calorimeter with uranium that was bigger than any that had ever been built. The previous largest liquid argon detector was the MARK I calorimeter at SLAC, which had one central device and two ends. The central one worked, but the 2 ends never worked properly—they just didn't work!" The situation was worse because they "were sitting there, four years behind CDF, trying to catch up. Although I began to feel that it would be a great device, if one could make it work, trying to make it work might be a good way to delay DZero to the point that it might not happen, or it would happen so late that it would be irrelevant."[115]

DZero's November 1984 design report explained: "There are three major components to the detector: the central detector system which measures tracks and gives electron identification; the calorimeters, composed of five separate uranium–liquid argon (ULA) detectors spanning most of the solid angle; and the large solid angle muon identification and momentum measuring systems." In time the two calorimeters nearest the beamline on each side were subsumed into more extensive end calorimeters. The central detector had no magnetic field, unlike CDF. The calorimeters consisted of alternating plates made of depleted uranium and liquid argon. All components were "supported on a single rolling platform table which transports the experiments between the assembly hall and the collision hall." In addition, the detector included "one additional system of scintillators near the beam line, used for luminosity monitoring and fast vertex coordinate location."[116]

DZero's 1984 design report showed DZero to have about the same number of collaborators and universities as CDF (DZero had seventy-one collaborators from twelve institutions; CDF had eighty-seven collaborators from twelve institutions). DZero, however, had more institutions with only one or two representatives and lacked a large number of elite universities within its collaboration. As with CDF, the DZero effort was shaped by the contours of the detector, with institutions working alone or in groups to build the individual components.[117] According to Grannis, DZero had more collaboration than CDF among institutions working on particular detectors. For example, of the twelve groups specified in the DZero design report, six worked on the experiment's calorimeters.[118] DZero collaborators also had to provide complicated cost estimates. In a DOE review in November 1984, DZero's estimate for its new detector was a cost similar to CDF's (the entire cost was $45 million, as compared with CDF's cost of $40 million two years before). This considerable cost and scope caused DZero to be, like CDF, a mini-institution within then-seventeen-year-old Fermilab.[119]

DZero continued to be bedeviled by its struggle to survive alongside its older brother. In the 1984 DOE review, Grannis directly confronted the difficulties of this situation in a presentation titled "Why DØ if CDF?" He explained: (1) DZero, unlike CDF, could make full use of the CERN experience in shaping its detector design; (2) since DZero was nonmagnetic, it could "freely optimize calorimetry, electron-hadron separation and resolution" and provide "complete muon coverage"; (3) the two large Fermilab detectors were "sufficiently different to provide checks on new discoveries"; (4) DZero could provide a prototype for detectors for the Superconducting Super Collider (which was just then being put forth as the next, larger high-energy physics accelerator); and (5) CDF and DZero were "sufficiently complex that breakdowns are unthinkable." Collisions from Fermilab's collider "should not be wasted."[120] These reasons would continue to justify DZero's existence throughout its development.

The younger collaboration faced intense time pressure because of its fears that before DZero came to life CDF would have "skimmed off the cream" of scientific discovery in the Tevatron energy range. Recognizing this fact when recommending DZero in the fall of 1983, Fermilab's PAC had urged that "the design, the group, the funding, the laboratory commitments, and ancillary engineering manpower be marshaled with all deliberate speed."[121] Grannis quoted the statement in a letter to Fermilab's management. He had already stressed the necessity of proceeding with speed in a July 1983 letter: "My own strong conviction is that a DZero detector must be in place at the Collider turn-on."[122]

When DOE approved DZero in 1984, its collaborators were caught in double jeopardy. CDF's existence and activity challenged DZero to meet an almost impossible deadline, but at the same time it blocked DZero's development because, as Grannis explained, "CDF was in the process of building, so there weren't enough engineers and techs to go around. Our problem was the lack of Fermilab resources." Lederman was not in a position to obtain additional resources, for, as William Wallenmeyer, DOE's director of the Division of High Energy Physics, noted in a March 1984 letter, other large detectors were also being built at other laboratories, creating a serious funding crunch. SLAC's SLD was now in a race with the LEP collider at CERN to study the Z boson. In view of the SLAC-CERN competition, a February 1984 HEPAP meeting recommended pushing SLD and letting DZero fall behind. DOE, in fact, saw "little likelihood of meeting all of the stated needs for major new detectors" and gave Fermilab a restricted funding "envelope" into which funding for both CDF and DZero had to fit. "They weren't going to slow CDF down," so much of the essential support just "never materialized," said Grannis.[123]

The job of building the DZero detector was complicated because of its use of depleted uranium. Grannis remembered thinking: "How do we get the stuff? Well, the government owns it, so we've got to get it somehow out of the government." Grannis appealed for help to DOE's Division of High Energy Physics, and by May 1985, the process for procuring uranium had begun.[124] Constructing DZero in time for the collider turn-on proved impossible; groundbreaking for the DZero building did not even begin until the summer of 1985. In October 1985, as CDF was observing its first collisions, and as several detector components were being built at a variety of locations, the DZero collaborators were still straining to make arrangements to procure the necessary uranium.[125] Three years later, as Pope remembered, there was a period "of worrying. There wasn't much money, and CDF was getting what there was. I remember feeling depressed because they were already producing results and we weren't even finished yet." As DZero struggled to complete the fabrication of its components, collect and assemble pieces on-site, and begin component installation, Grannis complained to DOE reviewers that lack of funding had slowed progress.[126]

Power was less diffused in DZero than it was in CDF. There was no Quaker-prayer-group-style decision making in the DZero workplace. Sometimes decisions were simply made by Grannis, by his deputy spokesman Gene Fisk, or later by Hugh Montgomery, who became cospokesman in 1993. At other times, they appointed an advisory panel to help decide between alternatives or solve a problem. "The panel would make recommendations and the spokesman would decide." Grannis explained: "People had decided that having a decision maker was a good thing, so they stood behind me. And that's the tradition that has continued at DZero." Like CDF, DZero had a political organization that was semi-independent from that of Fermilab. Compared with CDF, DZero was composed of "more universities of the land grant type, without the prestigious names."[127]

Because of DZero's lower profile, the collaboration enjoyed a more congenial relationship with Fermilab managers than CDF did.[128]

Eventually, DZero (fig. 12.6) kept pace with CDF in its construction phase, albeit several steps behind. Fermilab's second collider detector started roughly four years after CDF and was completed about four years later, in 1990. The detector was placed in the Tevatron in February 1992, saw its first collision in April, and began taking its first postcommissioning data in late August 1992.[129] Although it came to life later than expected, its collaborators were upbeat. As Grannis noted in a letter to *Science*, "Although the funds have not come as rapidly as we would have liked, most of the physics menu envisioned for DZero at its inception still remains." He was particularly excited because "the top quark awaits discovery."[130] And when the time came to make that discovery in 1995 (see the epilogue), DZero was able to play its part.

DZero, "the little collider that could," was like CDF in many ways. The two experiments had similar price tags, and both ended up with several hundred collaborators. They experienced the same challenges that came with scale, and they operated under the same, exacting DOE accountability rules. In addition, they were both user-empowered efforts that operated as semi-independent institutions with outside users sharing management and decision-making duties within the institution of Fermilab.[131] DZero differed from CDF as well. An article in the September 2000 edition of *FermiNews* provided a humorous insider summary of the difference. "Sometimes it seems positively unfair," the article explained. "Right up front, there's the issue of photogenics. . . . CDF has always had those drop-dead good looks. . . . Not DZero. DZero looks like a big metal tank." CDF also benefited from a better location. "Time and again, the tour for the Senator, the Congressman, the Secretary includes a stop at close-to-the-High Rise CDF." In short, "It would seem as if CDF got it all: the looks, the great address, the old-line pedigree." And yet, from the beginning, DZero collaborators understood the value of making themselves heard. "It's as if they knew they had to do something to hold their own with CDF vamping over there across the ring. They learned to communicate."[132] DZero, the collaboration that had been launched deep in CDF's shadow, had its own distinct and understated style.

Reflections on Collider Megascience

One of the differences between the megascience of the experiments at CDF and DZero and those in the fixed-target program was the power

12.6 DZero detector, 1988. (Courtesy of FNAL Visual Media Services.)

wielded by the university groups in the collaborations. Fermilab managers often had to bow to the authority of the powerful outside user groups within the two collider organizations, not only when it came to planning, building, and running their enormous detectors but also in managing them. As the outside users had their own power base, the colliding-beams experiments could operate politically and economically independent from the overall laboratory, as semiautonomous mini-institutions within Fermilab.

The subgroups from different institutions (mostly universities) shared responsibilities for designing, building, and commissioning detector components. The character of these subgroups was driven not only by the nature of the components and their designs, but also by institutional arrangements, such as contracts and MOUs (memorandums of understanding). For reasons of speed, efficiency, and competition with other groups engaged in similar searches, the subgroups had to work together as a larger coordinated team, trusting and relying on each other's efforts for the common good. All performed construction, analysis of data, and other experimental tasks simultaneously, with the specific work of each subgroup dependent on the success of the others. The same was true of their work together on experiment proposals, funding reviews, and writing papers. Because so many subgroups were coordinating in a single project, meshing the schedules of individuals and assuring them of reliable funding became even more important than it had been in earlier days of high-energy physics.

The management of the collider collaborations thus became vastly more complicated than the management of fixed-target experiments, partly because of the greater expense and complexity of the individual components of their experiments. More opportunities arose for power struggles within the collaboration, and also between the collaboration and the laboratory within which the experiment functioned. Managers of the collaborations faced increasing pressures for accountability, resulting in greater formality in accounting procedures. Relationships with Washington grew more complicated as budgets grew larger and accountability increased. The laboratory's relationship with Washington changed as well, for the collaboration leaders had their own relationships with funding agencies, such as DOE and NSF, independent of the laboratory. Occasionally problems developed between the collider collaborations and Fermilab's management, especially as the laboratory's operating budget became increasingly tied to its collider program and as the power of the collaborations' managers grew. In the changed Washington environment that demanded carefully documented plans, the early days when spontaneity and taking risks were celebrated at Fermilab receded into the past.

There are as many conflicts, paradoxes, and ironies in the colliding-beams track to megascience as in the fixed-target program. The work of the large colliding-beam experiments had become the kind of bureaucracy that Wilson had fiercely opposed. As in experimental strings, there was a mismatch between the fifteen- to twenty-year time scales of colliding-beams research and the spans of academia. In colliding-beams work, too, the increased scale of equipment, collaboration, size, and duration of experiments altered what it meant to be a high-energy physicist while raising new questions about what constitutes professional accomplishment. Among the many new questions that needed to be addressed were: How should credit be distributed among several hundred scientists working together? Should an assistant professor be tenured if the only part of an experiment that he or she worked on was building equipment, or data analysis? Such questions prompted a professional identity crisis in high-energy physics that has not yet been resolved.[133]

Among the differences between the megascience of experiment strings and collider research was that there was no competition for proton beam in the Tevatron I program: the collider simply got almost all of it. The taking of data was still measured in "runs," but termination of runs was now temporary (e.g., to allow time for maintaining the accelerator or for detector upgrades). The apparatus was no longer disassembled; the experimental hall was not readied for the next users. In these ways the experiments no longer ended. The economy of experimental real estate became a monopoly, and physicists stayed with their groups for longer times, in some cases for life. The collider collaborations "became our career," reflected Frisch.[134]

Thus, in the western suburbs of Chicago, the city that Mark Twain had described in 1883 as one "where they are always rubbing the lamp, and fetching up the genii, and contriving and achieving new impossibilities,"[135] the genii imposed conditions. Some of the practices of megascience narrowed the explorable frontier.

THIRTEEN

The Super Collider Affair, 1982–1989

One takes up fundamental science out of a sense of pure excitement, out of joy at enhancing human culture, out of awe at the heritage handed down by generations of masters and out of a need to publish first and become famous. When the cost of pursuing this enterprise is high, it is fair to ask why society should support it.

LEON M. LEDERMAN[1]

At the same time that Lederman oversaw the completion of Fermilab's Doubler and colliding-beams program, he also looked beyond those futures and explored adding a much more powerful accelerator to Fermilab's facilities. His musings mingled with international plans dating back to the 1950s to build a worldwide collaborative accelerator too costly for any single country to afford. Wilson and Lederman were both involved with planning this worldwide accelerator, whose energy and design changed over the years, eventually settling as an American 20 TeV on 20 TeV proton synchrotron, the Superconducting Super Collider (SSC).

Cold War Origins, 1955–1975

After the icy first decade of the Cold War, when relations between Western and Communist powers began to thaw, sev-

This chapter is largely based on Kolb and Hoddeson 1993 and Hoddeson and Kolb 2000. For critical comments on earlier drafts of the material presented in this chapter, we are grateful to R. A. Carrigan, F. T. Cole, W. B. Fowler, E. L. Goldwasser, D. Jovanovic, L. M. Lederman, W. O. Lock, R. E. Marshak, J. R. Orr, W. Panofsky, J. Peoples, L. Pondrom, C. Quigg, and the late R. R. Wilson.

eral efforts kindled an internationalist spirit among physicists. From 1950 on, the series of Rochester International Conferences on High-Energy Physics brought together particle physicists from many countries. From 1950 to 1954 CERN was organizing as a prototype for international cooperation in physics. The Joint Institute for Nuclear Research (JINR) at Dubna became the Soviet Union's model for international collaboration in 1956. The Geneva "Atoms for Peace" conference, held in 1954, aimed at reopening scientific communication during the Cold War.[2] Similarly, the Pugwash Conferences on Science and World Affairs began in 1957 as an arena for scientists to discuss international problems of nuclear weapons and world security. That year the International Union of Pure and Applied Physics (IUPAP) established its Commission on High Energy Physics to "encourage international collaboration among the various high-energy laboratories to ensure the best use of these large and expensive installations."[3]

In this context Wilson, Lederman, and other internationalist physicists conceived an optimistic vision of a worldwide particle accelerator. Too expensive for any single nation, the cooperative project was to be a model for peaceful international collaborative ventures. The idea of a "world accelerator for world peace" was discussed at a meeting of the IUPAP Commission on Particles and Fields held in Kiev in July 1959 in conjunction with the Rochester conference.[4] A study committee formed by a number of American and Eastern and Western European scientists elaborated on the idea of the world accelerator when they met again in September 1959 at CERN. That month, USSR premier Nikita Khrushchev's visit to the United States, and his successful meetings with President Dwight D. Eisenhower (1953–1961), encouraged further dialogue about cooperative scientific projects, as did another meeting between V. S. Emelyanov, head of the USSR Administration of Atomic Energy, and John McCone, chairman of the U.S. Atomic Energy Commission (AEC) (see chapter 2). The agreement they signed on November 24, 1959, provided for simultaneous, reciprocal short-term "exchanges of information and visits of three to five scientists." They also considered "the design and construction of an accelerator of large and novel type."[5]

Six months later, as a consequence of the Emelyanov-McCone Agreement, a delegation of five physicists—Wilson, Robert Bacher, George Kolstad, Edward Lofgren, and Robert Marshak—set out for the Soviet Union "specifically to explore the joint construction of a large accelerator."[6] The meeting took place, but the opportunity for progress was lost when on May 1 the Soviets shot down an unauthorized American U-2 reconnaissance plane flying over Russia. Relations turned hostile when the

United States refused to apologize.[7] "People on the street would shake their fists at us," Wilson recalled. Even "the physicists would barely speak to us."[8] Trust further eroded when a summit meeting in Paris collapsed, shattering Eisenhower's hopes of resolving the Cold War.

Yet even in the fearsome climate of that period's nuclear arms race, a small group of internationalist physicists continued to dream that "somehow," as Wilson later wrote, "in building and operating a World Laboratory we would not only be exploring nature, but we also might be exploring some of the ingredients of peace."[9] The historic meeting that Wilson organized in August 1960 at the Tenth Rochester Conference to explore 1000 GeV "ultrahigh-energy accelerators in a world-wide context" (see chapter 2) became the historical root not only of Fermilab but of a series of planning discussions for the large international accelerator that in time would transform into the Superconducting Super Collider.[10]

Through the 1960s and 1970s, physicists from CERN, the United States, and the USSR continued to discuss the worldwide accelerator project, which they hoped, as Robert Marshak wrote, "might pave the way to future fruitful international collaborative efforts."[11] Unfortunately, domestic and political issues during the presidential administrations of John F. Kennedy (1960–1963), Lyndon B. Johnson (1963–1969), and Richard M. Nixon (1969–1974) held this goal in suspense. Another window for discussing international cooperation opened in June 1973 when President Nixon and Soviet party general secretary Leonid Brezhnev signed a historic accord designed to alleviate the economic impacts of that period's energy crisis. The prospects for international cooperation in high-energy physics improved when this agreement identified basic research on the fundamental properties of matter as one of three areas (behind thermonuclear fusion and breeder reactors) particularly useful for "expanded and strengthened cooperation for mutual benefit, equality and reciprocity between the U.S. and the U.S.S.R."[12] Over the next two years several other developments improved the conditions for successful international cooperation in physics. For example, in February 1974 and again in October 1974, the U.S.-USSR Joint Committee on Cooperation in Peaceful Uses of Atomic Energy, called for in the Nixon-Brezhnev accords, assembled to address the implementation of programs for cooperation in research on the fundamental properties of matter. Also, from September 1973 to July 1975, agreements were drafted for the Helsinki conference in August 1975 that resulted in the signing of the Helsinki Accords, an understanding that placed scientific cooperation and the free flow of information within the context of human rights. Thus, by 1975 many steps had been taken to build the world accelerator, but there was still no concrete plan for doing so.

Birth of the VBA, 1975–1980

By late 1974, some of the physicists involved in the talks about international cooperation had reached the limit of their patience. Like the stalemated Paris peace talks, the meetings about a world accelerator had produced only talk.[13] Addressing this frustration, the renowned theoretical physicist and former director general of CERN Victor Weisskopf, then serving as the organizing committee chairman, announced in his invitations that the next meeting of the international cooperation seminar would be different and be "an important contribution to international collaboration in our science."[14]

Sparks flew at this meeting held in New Orleans in March 1975. Wilson recalled the discussions being "spontaneously interrupted by a number of impassioned speeches to the effect that a world laboratory along the lines of a world-wide CERN would be necessary and desirable if we are to push into the multi-TeV region of proton energy."[15] Lederman, then the director of Nevis, endorsed the plan and proclaimed in his position paper that, "the world community of high-energy physics [should] bite the bullet and organize together to bring this 10 TeV machine to realization." He bestowed on it the name "Very Big Accelerator," or VBA.[16] The elusive and quixotic world accelerator was born.

Others at the meeting emphasized the potential contribution that the international accelerator would make to worldwide cooperation. Edwin Goldwasser, then serving as secretary of the IUPAP Commission on Particles and Fields, remarked: "If the world is to survive and flourish, its people, with their different cultures, politics and economics will have to work together much more closely than in the past. Many of us believe that high-energy physics may provide a small but useful prototype for the broadening and deepening of such cooperation activities."[17] The participants recommended the formation of a study group for the VBA, to be led by Weisskopf. They discussed the potential designs of the VBA at numerous meetings held over the next several years, continuing to project the philosophical underpinnings of the VBA to the scientific community, despite the many political uncertainties of the mid-1970s (e.g., the Watergate scandal, Nixon's resignation, the fall of Saigon, and energy shortages following intensified Arab-Israeli conflicts). In November

1975, Wilson wrote in *Physics Today*, "Such an undertaking might well provide some of the experience in international living so necessary for human survival—a candle in the darkness."[18] Yet by that year, which saw the reorganization of the AEC into the Nuclear Regulatory Commission (NRC) and ERDA, the spirit of cooperation between governments and physicists had almost vanished.

At the international collaboration seminar held in Serpukhov in May 1976 (fig. 13.1), participants pursued the general scale of the VBA. They conceived the machine either as a 10–20 TeV fixed-target proton accelerator or as a complex of 100 GeV *e+e-* (electron-positron) storage rings.[19] Two months later, at the Rochester conference held in Tbilisi, the IUPAP Commission on Particles and Fields agreed to sponsor an official subcommittee to organize future meetings and to study both the VBA and future regional facilities and collaborations. It was named the International Committee on Future Accelerators (ICFA).[20] Lederman began to discuss the VBA openly. When he spoke about it at the 1977 Particle Accelerator Conference in Chicago, he suggested selecting New York City, then near bankruptcy, as the potential site (fig. 13.2). Many necessary facilities were already in place, he joked, "high-rise international headquarters, educational resources, pre-tunneled terrain, and the usual degree of inaccessibility."[21]

But progress on the VBA continued to be plagued by political intricacies and the fact that while physicists from each region expressed support, actual funding had to come from national treasuries. As each region's plans for higher-energy machines evolved, the VBA dissolved into the distance.[22] American physics advisory panels worried that the leadership in high-energy physics was shifting to Europe. One panel, headed by Sam Treiman of Princeton, expressed the concern that without increased funding "the U.S. program will inevitably lose its eminence" and its ability "to compete with Western Europe."[23] Another, led by Maury Tigner of Cornell, concluded that "we must redouble our efforts to improve the cost effectiveness of our accelerators if the needs of U.S. particle physics are to be met in the resource limited situation."[24] This was the same period when Wilson's frustration about the lack of funding for Fermilab's Doubler resulted in his resignation.

The U.S. physics program faced serious competition from CERN, then led by Herwig Schopper, previously the director of DESY, in Hamburg. CERN's new plans for a Large Electron-Positron (LEP) collider challenged ICFA's concept of an electron VBA.[25] From a physics point of view, it was important to build a more powerful instrument to investigate the emerging physics agenda. The theory of grand unification

13.1 ICFA meeting in Serpukhov, Russia, 1976. (Courtesy of Lederman Collection.)

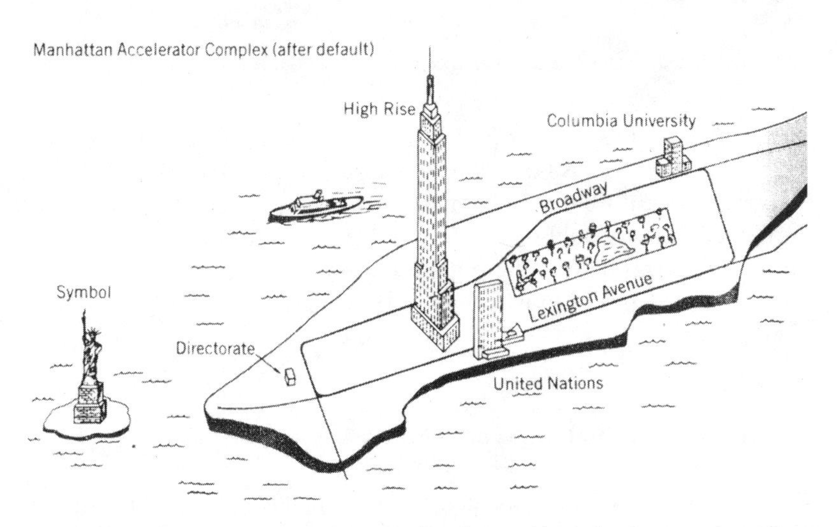

Lederman's proposed site for the VBA. The Broadway and Lexington Avenue subway lines would provide ready-made tunnels for the accelerator's straight sections. The high rise is the Empire State Building, and the stirring symbol is the Statue of Liberty. Inaccessibility to the public is assured by the traffic jams at toll bridges. Justification for the cession of Manhattan Island to the VBA complex by the US Government could be financial default by New York City.

13.2 Proposal for the VBA in New York, 1977. (Courtesy of Lederman Collection.)

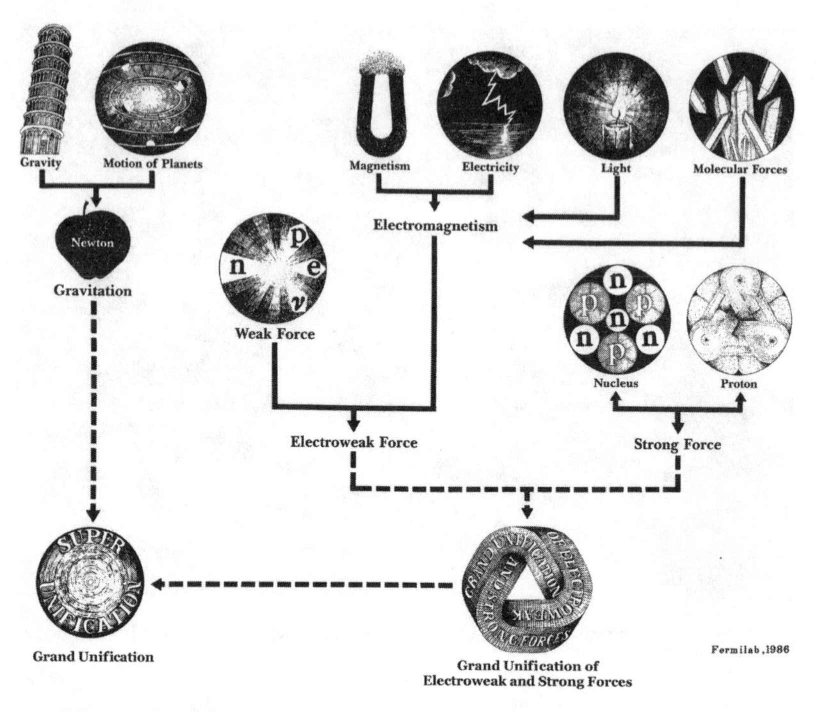

13.3 Representation of the Grand Unified Theory, 1987. (Courtesy of Angela Gonzales.)

(fig. 13.3) described a bleak desert of events between 300 GeV and 10^{15} GeV, one in which, as Carlo Rubbia expressed in March 1981, "*nothing* happens . . . until we reach the mass range of leptoquarks."[26] Physicists estimated that it would take at least 1 TeV in the center-of-mass frame to penetrate this desert. The highest-energy American accelerator then under construction, Brookhaven's ISABELLE, was aimed at a lower energy.

The American Initiative, 1980–1982

The science policy scene changed abruptly in November 1980 with the election of the former film and television actor and governor of California Ronald Reagan as president of the United States. Reagan appointed as his science adviser the bold and articulate forty-one-year-old physicist George A. Keyworth, who had been selected from the upper management of Los Alamos. Lederman, by this time Fermilab's director, reflected that Keyworth came "from out of the West in some sense," and announced that "he wanted exciting projects and quality."[27] Keyworth's passion for fundamental research resonated with that of Reagan, who, as Keyworth later explained, had a "huge faith that man could solve any problem with his own creativity." Under Keyworth, projects to keep America strong and competitive received higher priority.[28]

Keyworth feared that "mediocrity was beginning to creep into our profession." He wondered, "Where is the Wilson?" Early in his term, Keyworth concluded that ISABELLE did "not fill any unique bill." He felt that ISABELLE's energy was too low to address the "really interesting theoretical questions." It also bothered Keyworth that Brookhaven was justifying ISABELLE on regional grounds, as a project for the East Coast. "When you have facility-based research you are on the road to mediocrity," he asserted.[29]

ISABELLE had in fact become a concern to many high-energy physicists. The 1981 HEPAP subpanel on long-range planning, chaired by George Trilling of Berkeley, worried that ISABELLE's high budget might "bleed the rest of the program."[30] By February 1982, the subpanel concluded that ISABELLE must be discontinued if funding for it at the highest level was not forthcoming. HEPAP then recommended that to maintain a vigorous high-energy program "another major facility must be started in the mid-1980's so as to be available for research by 1990," a program capable "of exploring new frontiers." The completion of the Doubler was identified as the "highest immediate priority,"[31] while the absence of urgency and international funding stalled the VBA.[32]

The Snowmass Desertron, 1982

In July 1982, a large group of physicists gathered for a workshop in Snowmass, a ski resort high in the Rocky Mountains of Colorado. There, at the first major workshop of the American Physical Society's Division of Particles and Fields, Lederman rallied the physicists with a bold and romantic proposal. He described the "Desertron," a new machine resembling the VBA in many features of its design, as an oasis in the barren region between the weak energy scale (less than 1 TeV) and the unifying energy realm described by the grand unification theory (10^{15} GeV). One target for the Desertron to aim for was the Higgs particle, the key to understanding the mechanism of spontaneous symmetry breaking, concealed at then unreachable energies. Recalling Wilson's idealistic dream for international harmony, this "machine in the desert" held some of the VBA's international appeal. Lederman said he felt "the perceived political difficulties of establishing a new site would evaporate under the glow of wide community enthusiasm." But in every other way the Desertron would be an American machine built on American soil, a machine worthy of bold and adventurous explorers, which would be "the best possible move for U.S. HEP [high-energy physics] and indeed world HEP."[33] Lederman described a scenario in which the caravans of high-energy physicists from around the world joined in some isolated region of the United States to reveal fundamental truths of nature while demonstrating peaceful collaboration. He advised: "Let us not grow old, querulous and overcautious in facing our future."[34]

Lederman asked his audience: "Are we as a community growing old and conservative, and is there a danger of quenching the traditional dynamism we have surely enjoyed in the past three decades? How can we break out of the aging lab and inadequate lab site constraints—how can we creatively leapfrog the world and get to the multi-TeV domain soon?" His long-range plan, which he called "Slermihaven II," would combine the best features of SLAC, Fermilab, and Brookhaven and explore $\bar{p}p$ collisions at 20 TeV and ultimately 40 TeV. Utilizing the superconducting technology developed for the Doubler, the American Desertron held the dual promise of responding to CERN's expected capability to "pave the LEP tunnel with superconducting magnets" and also explore the desert beyond the frontier with a "great leap forward." For as Lederman argued, "any program we choose must be compared to LEP." A bold and dramatic response to CERN was essential to put the United States back on track.[35]

After Snowmass, many American physicists agreed that the United States needed a multi-TeV proton-proton collider. Lederman planned a Desertron workshop to be held early in 1983. Because of the rivalry between Fermilab and Brookhaven Tigner advised holding this meeting at Cornell, a neutral setting.[36] Meanwhile, the outlook for basic research improved under Keyworth and Reagan's second energy secretary, Donald Hodel, who came into office in November 1982, after dentist James Edwards resigned.

The evanescent mirage of ICFA's world accelerator faded for American physicists over the next six months as plans for an American 20–40 TeV collider came into focus. These plans responded to a series of developments at CERN. In January 1983, CERN announced the UA1 collaboration's discovery of the $W^{\pm}$ particles. That month, sooner than Lederman had predicted, Schopper spoke in Japan about his plan to install a 20 TeV–range hadron collider (LHC) in the LEP tunnel, Also that month, James Cronin wrote to William Wallenmeyer at the DOE from a conference in Rome, where he had heard a presentation of the results from the CERN collider, that the future for high-energy physics is "very bright indeed. At present we have the technology to reach 20 TeV in the center of mass for a *pp* or $\overline{p}p$ machine." But CERN's recent success, combined with the fact that Fermilab's Tevatron I program would reach 2 TeV in 1986, brought Cronin to caution Wallenmeyer: "Given these facts I believe we should look ahead. Time will have passed an ISABELLE-like machine by. We should proceed now with plans for a new hadron collider, even if a new laboratory is required."[37] This message seemed all the more urgent in June 1983, when CERN discovered the Z^0 particle.[38] "Our world leadership in high-energy physics has been dissipated," Keyworth noted.[39]

In the meantime, more support gathered for the Desertron. A new HEPAP advisory subpanel led by Stanford physicist Stanley Wojcicki formed in February 1983 and concentrated "on what is needed to reestablish American leadership and preeminence in the field during the next decade." The Wojcicki panel deliberated during its sessions from February to July 1983. Keyworth encouraged this panel to "think big."[40] Meanwhile, several physics workshops held at Berkeley and Cornell focused on the capabilities and possibilities of collider detectors. Extrapolating costs from the Fermilab Doubler, the conferees estimated the price of the Desertron, excluding detectors, equipment, contingency costs, and preoperating expenses, at about $2.7 billion (1983 dollars).[41] They concluded that the proposed collider was affordable, without the cumbersome international pooling of resources planned for the VBA. Moreover, construction could be underway by 1987. ISABELLE, then struggling as the CBA (Colliding Beams Accelerator), was at risk.

Among the continuing supporters of the Desertron was Wilson, now back at Cornell. He wrote Lederman on May 20, 1983: "We have both dreamed of, and worked for, a World Particle Physics Laboratory (WPPL). I suspect that now is a propitious time to realize both that dream and a natural patriotic desire to have the WPPL located in the United States." Wilson clung to the idealistic underpinnings of the VBA: "I would hope the government would have been convinced how desirable it would be to have this leading edge of culture and technology located here in our country for selfish reasons as well as for the idealistic reasons of mutual understanding and peace. Were that so, the President might be disposed to invite the participation of all nations to join with us."[42]

That spring relations between the United States and the USSR became as difficult as they had been during the 1960s. Manhattan Project and Livermore physicist Edward Teller, who did not see an optimistic future, advised Reagan to develop the Strategic Defense Initiative (SDI), which the media called "Star Wars," after the popular Hollywood film series (1977–1983) portraying a tale of good versus evil in the universe. In a famous speech edited by Keyworth and delivered in March 1983 to the National Association of Evangelicals, Reagan borrowed language from the Star Wars films and called the Soviet Union an "evil empire."

Reagan set the U.S. defense budget on a path to double. Funding for basic research would increase with it, although inflation, recession, and the growing costs of research and technology offset any benefits.[43] As Keyworth became increasingly involved with defending SDI, Lederman assumed the role of spokesman for the Desertron. He was joined by a new SSC supporter: Alvin Trivelpiece, the director of DOE's Office of Energy Research and the scientific adviser to DOE secretary Hodel.

Trivelpiece, who oversaw the SSC project from his base in Washington DC, was a plasma physicist who drew wide praise as a logical and cautious administrator. He had the ability to bridge diverse communities—politicians, industrialists, journalists, and scientists. Trivelpiece's deputy, James Leiss, the associate director for high-energy and nuclear physics, coordinated the funding for all the high-energy machines.

The SSC Proposal, 1983

The discussions of the Wojcicki subpanel held during June and July at Woods Hole and later at Nevis were shaped by a number of factors. First, the panel recognized that the Desertron offered a way for American physicists to respond with something innovative to CERN's demonstration of $\bar{p}p$ collisions, which resulted in the observation of jets in 1982, the $W^{\pm}$ in January 1983, and the Z^0 in June 1983.[44] Second, the panel realized that ISABELLE's energy of 0.4 TeV per beam was too low to address the questions then posed by theory, which were directing the community to explore the 1 TeV energy scale. For proton colliders achieving 1 TeV in the center of mass required at least 10 TeV in each colliding beam. It also seemed clear that the VBA had little prospect of materializing, although it had been extremely useful in formulating the 20 TeV concept. Fermilab, having recently demonstrated the successful operation of superconducting accelerator magnets, and with the new Doubler poised to begin work in mid-1983, was in a favorable position to mass-produce SSC magnets.[45] Superconducting technology promised dramatically to reduce costs and allow the Desertron to be created on a more habitable site than a desert. It might even be, as Drasko Jovanovic called it, a "Prairietron!"[46]

The Wojcicki subpanel's recommendations in July 1983 changed the course of American particle physics: (1) build the American accelerator called the Superconducting Super Collider, a machine with a diameter about 30 kilometers [18–20 miles], with 10–20 TeV protons in each of its colliding beams, at a cost estimated then at less than $2 billion (1983) spread over twelve years; (2) complete both the Tevatron at Fermilab and the Stanford Linear Collider (SLC) at SLAC and upgrade CESR (the Cornell Electron Storage Ring); (3) not approve Fermilab's Dedicated Collider (an alternative proton-antiproton/electron-proton collider that made use of the Tevatron as an injector for protons and antiprotons); (4) discontinue ISABELLE/CBA; and (5) support advanced accelerator technology.[47] The subpanel's executive summary stressed that the SSC "provides the promise of important and exciting advances in the field of elementary particle physics." HEPAP chairman Jack Sandweiss's letter transmitting the subpanel's report to DOE emphasized that the SSC "would be the forefront high-energy facility of the world and is essential for a strong and highly creative United States high-energy physics program into the next century."[48]

The concerns presented by the Wojcicki panel helped clinch Washington's decision to redirect part of the ISABELLE appropriation toward an American 20 TeV collider, free of the politics of high-level international collaboration. With DOE's subsequent acceptance of HEPAP'S recommendations in December, Congress planned to "divert" $18 million of ISABELLE/CBA's $23 million funding for the SSC.[49] American physicists were authorized to move forward in their new venture. And Fermilab, under Lederman, planned to serve as the flagship vessel for this frontier voyage.[50]

The supporters of ISABELLE were both disappointed and angry with the decision to discontinue ISABELLE, especially as the project's magnet problems had by this time been solved.[51] The Wojcicki subpanel had been divided on whether to discontinue the project, but the full HEPAP membership endorsed the recommendation to terminate the machine because its energy had become too low for work on the forefront of particle physics. The recommendation also provoked other members of ICFA, who felt that the United States had aggressively co-opted their VBA. Despite efforts by Americans to argue that the SSC would be an international center analogous to CERN, the European and Japanese members of ICFA considered the Woods Hole recommendation a "'nationalistic approach' . . . detrimental to . . . high energy physics as a whole"[52] They recalled that some of the same physicists who now championed the SSC—for instance, Lederman—had helped to conceive the VBA.[53]

The Outpost of a New Frontier: The Reference Designs Study and the Central Design Group, 1983–1988

The next phase of the SSC, and of Fermilab's efforts to continue its pursuit of the energy frontier, took the form of work at an outpost, a temporary staging ground erected at the edge of unknown territory.[54] Here the explorers organized their forays into the unknown and stored their necessary provisions, including knowledge, plans, and designs. Because the outpost is temporary, those posted there risked losing their position in the rush to the frontier. The inevitable fall of the outpost (real or figurative) marks a critical moment for historians to probe, for it reveals the forces involved.[55]

The SSC's outpost years, during the second half of Lederman's tenure as the director of Fermilab, divide into the period of the Reference Designs Study (RDS) from 1983 to 1984 (referred to by the DOE as phase zero) and the period of the Central Design Group (CDG) from 1984 to 1988 (referred to by the DOE as phase 1). Reminiscent of Fermilab's own preliminary outpost, both the RDS and the CDG were based in Berkeley—in fact, at Lawrence's historic laboratory. In this "fantastically exhilarating" atmosphere, in the words of Fermilab theorist Chris Quigg, who worked in the CDG as deputy director for operations, physicists, many from Fermilab, planned their exciting new adventure.[56] Even at this stage there were trends in the SSC's story that in hindsight can be recognized as undermining. First, much of the support was drawn from the budgets of existing high-energy laboratories (Brookhaven, Cornell, Fermilab, and SLAC). This support came after several prominent accelerator physicists wrote to the directors of these laboratories to ask that each contribute to the SSC by allowing a few members of their staffs to do early ad hoc design work. The DOE endorsed this imperfect plan, as no additional funds were required to support it. The department offered $19.5 million of additional support for the RDS, and this support was distributed among the existing laboratories as well as the new Texas Accelerator Center, a recent entrepreneurial creation of Houston industrialist George Mitchell and high-energy physicist Peter McIntyre, who by this time had moved to Texas A & M and gained the reputation of a rogue inventor.[57] The problem that gradually surfaced was that the commitments made by individual laboratories to the cooperative SSC venture were contingent on the SSC work not conflicting with the goals of their own laboratories.[58]

The RDS group, which included about fifty from Fermilab's staff—more than from any other laboratory—assembled in Berkeley from De-

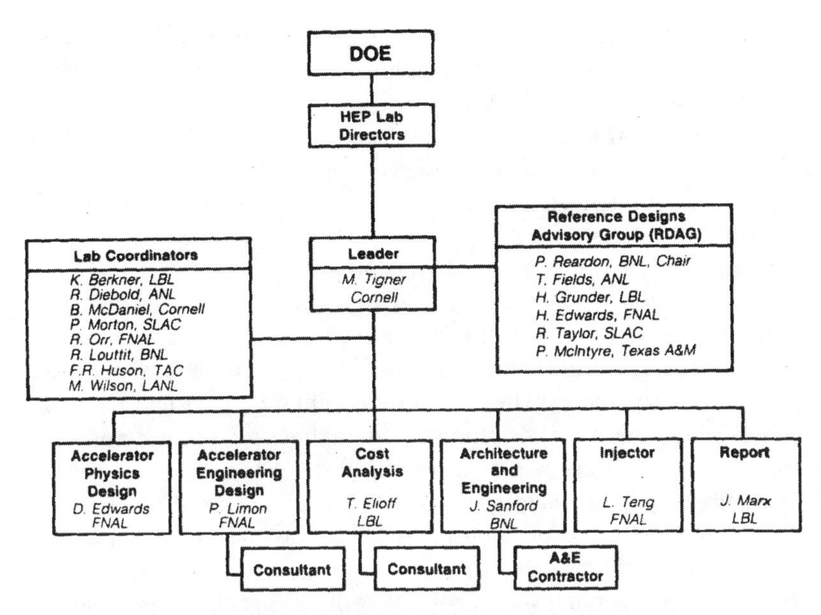

13.4 Organization chart of Reference Designs Study at LBL, 1985. (Courtesy of Lawrence Berkeley Laboratory.)

cember 1983 through May 1984 (fig. 13.4). The leader was forty-six-year-old Maury Tigner of Cornell, a product of the Cornell accelerator building tradition associated with Wilson and Boyce McDaniel. One of Wilson's former graduate students, Tigner had worked on all of Cornell's electron machines. He was also among the active participants of "Physics at the SSC" (PSSC), an open discussion group of researchers led by Bruce Winstein of the University of Chicago and Fermilab, which met to discuss physics questions affecting the new accelerator's design.[59]

The most crucial issue was how to design the superconducting magnets, for they determined the energy of the SSC. Three designs were explored: one, proposed by a joint Brookhaven and Berkeley team, featured superconducting wire windings and an iron yoke to reach the high value of 6.5 Tesla; another, submitted by a Fermilab group, was iron-free and would reach 5 Tesla; and a third, proposed by the Texas Accelerator Center, was a 3.5 Tesla superferric magnet that used significantly more iron and was to produce a more uniform, if lower, field in a larger ring. As the RDS judged all three designs feasible, work on all three proceeded with a plan to select the best design early in the next phase of the effort.[60]

On May 8, 1984, Tigner submitted the 441-page *Reference Designs Study* to DOE. Putting forth a six-year time table, the RDS raised its estimate of the cost of the SSC, which a year earlier had been set at less than $2 billion, to $2.7–$3 billion, pointing out that this budget excluded funding for "research equipment, preconstruction R & D, and possible site acquisition."[61] After approving the RDS, the DOE's review committee agreed that "all three basic SSC designs . . . appear to be technically feasible" and recommended increasing the cost estimate by $200 million.[62] Secretary of Energy Hodel approved the RDS in August 1984 and authorized an expenditure for fiscal year 1985.[63] To physicists grabbing the brass ring in support of the SSC, Hodel's approval of the Reference Designs Study was a sign that the vision of the new accelerator would be realized. The glowing rhetoric of Fermilab physicist Paul Mantsch expressed the excitement: "The moment was right for the great vision to become reality. The mood of optimism was on the Land. The technical progress that enriches the lives of the People has at its foundation a vital and dynamic program of scientific research. The People know little of quarks and leptons. But they know curiosity. They know asking questions and seeking answers. They share the spirit of pioneers at a new frontier. And they know it is good. For they were great People of a great Land where boldness and innovation is legendary. This, after all, is the Spirit of Snowmass."[64]

Washington determined that the SSC's design work required a formal management structure administered by a suitable organization. A natural choice for this organization was URA, by then a consortium of more than fifty universities throughout the country. In March 1984, DOE solicited URA for a management plan for the SSC's phase 1, the work of the CDG.[65] When URA's primary qualification, that it was the manager of Fermilab, seemed a possible conflict of interest, URA established an independent board of overseers (BOO), thus separating its "SSC BOO" from its "Fermilab BOO."[66] Hoping to create national support for the SSC, Trivelpiece decided to hold an open site competition.[67]

The second Snowmass workshop followed the announcement of the competition. Some 250 physicists convened there from June 23 to July 13, 1984, to address physics, magnet, accelerator, and detector design questions of the SSC.[68] By this time word had come from Guy Stever, URA's president, and from Boyce McDaniel, the chairman of URA's SSC BOO, that Tigner would stay on at Berkeley and lead the SSC's Central Design Group. Interest in the SSC heightened on July 3, when the UA1 collaboration at CERN led by Rubbia announced six candidate "top" quark events, a premature announcement that had a strong motivational impact, even though it was found wrong.[69] On the national scene, the site selection process began in November 1984, when Trivelpiece asked

Robert M. White, president of the National Academy of Engineering, to help coordinate the site selection.[70]

Shortly after that, personnel changes in Washington DC triggered a downward spiral, for when Reagan transferred Energy Secretary Hodel to the Department of the Interior in January 1985, the SSC lost a strong Washington ally. Tigner had to begin again to win DOE's top support, dealing now with California real estate developer John Herrington, who succeeded Hodel in February 1985. When Stever retired from his URA post, URA selected a president qualified to navigate the new, more politicized Washington scene. Edward Knapp, an experienced Washington statesman, fit this requirement. He had been a member of the nuclear physics staff at Los Alamos since 1958 and director of the National Science Foundation from 1982 to 1985. In August 1985 Knapp was appointed URA's sixth president.

While magnet development progressed at the CDG, organizational difficulties arose from the 1984 memorandum of understanding signed by the directors of the existing high-energy laboratories. This MOU specified that the CDG would coordinate the design work of participating laboratories. But as DOE distributed the funds directly to each laboratory, the arrangement reduced Tigner's authority to manage their work.[71] Another organizational difficulty concerned the magnet selection. During the summer of 1985, CDG had reduced its magnet choice to two candidates: the "high-field cos θ" (or "conductor-dominated") magnet, developed at Berkeley and Brookhaven with Fermilab's help, and the superferric design proposed by the Texas Accelerator Center. But the Texas camp grumbled in September when CDG announced that it favored the high-field cos θ magnet.[72] The DOE decided that all three of the laboratories backing the high-field magnet—Brookhaven, Fermilab, and Berkeley—should be jointly involved in the SSC magnet development in a complex partnership in which Berkeley would work on the superconducting cable and wire development as well as on the construction of short magnets, Fermilab would focus on cryogenics systems and testing, and Brookhaven would concentrate on building longer magnets.[73]

Uncertainty hung over the SSC project through 1986, as physicists and SSC supporters shared in the nation's anxiety about budgets and spending.[74] When political attention turned to reducing the national budget deficit, every recipient of federal funding worried that "the automatic GRH (Gramm-Rudman-Hollings) ax" might fall on their work.[75] Lederman's letter to the *New York Times* in March 1986 expressed the fading hope for new sources of support for the SSC. He was concerned that the GRH process would work in such a way that "we reduce the deficit even if it destroys the country."[76] Tigner and Lederman had expected the SSC to be funded with "new" money, offered in addition to the existing "base program" supporting high-energy physics, but there was no sign that such money would come soon.[77] Lederman suggested building a less expensive SSC, which would collide protons against antiprotons and require only a single ring of magnets. A review of this suggestion concluded that such a collider would be less reliable, have inadequate luminosity, and not save enough to make the risks cost effective.[78]

Trivelpiece was the strongest Washington advocate of the SSC from late 1985 through early 1987. His task was to translate the conceptual design report for President Reagan, whose approval was necessary for a project of the SSC's scope. At this time none of the expected increases in federal revenues had materialized, and there was still no money for new projects.[79] In response to Trivelpiece's briefing on the SSC on January 29, 1987, the president pulled an index card from his pocket and read a short poem, Jack London's "Credo":

I would rather be ashes than dust!
I would rather that my spark should burn out in a brilliant blaze than it should be stifled by dry-rot.
I would rather be a superb meteor, every atom of me in magnificent glow, than a sleepy and permanent planet.
The function of man is to live, not to exist.
I shall not waste my days trying to prolong them.
I shall use my time.

When asked to explain what the poem meant, Reagan invoked Kenny Stabler, the football quarterback of the Oakland Raiders who had gained fame for calling risky offensive plays. Reagan explained that the poem meant "Throw deep!" The next day, Secretary Herrington announced Reagan's official support for the SSC as a $4.5 billion federal construction project.[80]

Changing Winds: URA's Unsolicited Proposal

Several months before Reagan's authorization of the SSC, Goldwasser, the highly regarded physics statesman and Wilson's deputy, joined the venture. After leaving his post as deputy director of Fermilab in 1978, he had served for some years as vice-chancellor of the University of Illinois.

He then decided to go west and work as the CDG's associate director for development, arriving in September 1986. The managing organization for the SSC had not yet been chosen, so Goldwasser drafted an unsolicited proposal to DOE making the case that URA was the natural management organization.[81] His elegant and confident words reflected Goldwasser's judgment that everyone would save time, money, and effort by awarding a "sole source" contract to URA. Physicists had frequently used this approach successfully in negotiating their government contracts.

On March 2, 1987, Knapp transmitted Goldwasser's fourteen-page document to Trivelpiece. The proposal explained that URA was "uniquely qualified to provide the broad representation of the nationwide academic community together with the management experience that is essential to the success of an SSC laboratory and research program."[82] After establishing the legality of letting an unsolicited contract, Trivelpiece shepherded the proposal through DOE, where it arrived at a critical meeting with top DOE administrators. Herrington's undersecretary of energy, Joseph Salgado, "had pen in hand ready to sign the unsolicited proposal," but he wavered.[83] He asked Trivelpiece whether it was urgent to resolve the issue of the management organization just then. Trivelpiece could not say it was, and the matter was deferred. Had Trivelpiece told Salgado that resolving the issue was urgent, the course of the SSC might have followed the trail blazed by CDG.[84] Instead the matter sat in suspense, without action, while the SSC project entered a turbulent period.

The ground beneath the CDG had begun to shift. Once the site selection process reduced the number of qualified sites in January 1988 (from forty-three proposed sites in twenty-five states to seven—Arizona, Colorado, Illinois, Michigan, North Carolina, Tennessee, and Texas), Congress proceeded to criticize the SSC's increasing budget. Close congressional scrutiny of the SSC began to whittle away support for the SSC once the excluded states lost interest, while a cluster of economic and political events (including the Iran-Contra trials in the summer of 1987 and the stock market downturn that October) had lasting effects that threatened the government's support of the SSC. In an attempt to spur DOE into action on Goldwasser's proposal, URA submitted a slightly revised version on February 22, 1988. Still there was no progress.[85]

Technical problems threatened the outpost as well. The new magnets were not functioning as expected, and there were problems arranging the magnets into a unified system. Moreover, the collaboration between Brookhaven and Fermilab on magnet development was suffering. Brookhaven bore a lingering resentment toward Fermilab after the cancellation of ISABELLE. And as Fermilab was committed to developing its

Tevatron programs, it could not easily accept more responsibility.[86] In a brute force attempt to tackle this magnet collaboration, Tigner appointed a four-member Magnet Management Group, consisting of himself, Goldwasser, Limon from Fermilab, who was CDG's head of accelerator systems, and Victor Karpenko, a former Livermore engineer, then leading the CDG's magnet group. Every four weeks this "gang of four," as they called themselves, would visit Brookhaven for two days, and then go to Fermilab for two days.[87] Such an arrangement might have worked in wartime, when all groups shared the same defense goal, as had been the case during the Manhattan Project.[88] But the SSC's magnet goals were not sufficiently unifying. "It was a tricky business," Goldwasser recalled. "We would just be there hovering over what they were doing, meeting at their meetings, and impressing upon them that their responsibility was to do what we asked." The business was "even trickier because there was a different personality there every time—sometimes Tigner, sometimes Peter Limon, sometimes Karpenko, and sometimes Goldwasser," he explained.[89] This unwieldy attempt to coordinate magnet work presaged more serious difficulties on the road to the SSC.[90]

Within months of the appointment of the "gang of four," Karpenko announced his retirement, and Goldwasser agreed to serve as the temporary head of the CDG's magnet group for six months. Goldwasser felt that Peoples, then deputy head of Fermilab's Accelerator Division, would be better at the job, and he managed to persuade Peoples to accept responsibility for CDG's Magnet Division for one year. "I was a little bit like the plumber who'd been called in to fix the leaks and the toilets that are overflowing during a dinner party," recalled Peoples, who joined CDG at the end of December 1987. "I wasn't exactly invited to dinner."[91]

Peoples established an empirical program to study the SSC's magnet problems through extensive testing. Drawing directly on Fermilab's Doubler experience, he demanded that many magnets, each with slightly different parameters, be designed, built, and tested, so one could separately explore all the possibilities. He laid out a one-year testing program. By the early summer of 1988 good magnets were being produced.[92] Peoples came to view the magnet problem as "philosophical"—the result of a misalignment between the management practices of the national weapons laboratories (like Karpenko's laboratory at Livermore) and the national research laboratories (such as Fermilab).[93] He recognized what Tigner also understood, that while physicists typically can work with members of the industrial culture, they often clash with members of the military-industrial culture.[94] Karpenko was "used to one kind of management and really not used to trying to cajole people . . . how you have to do it in this [physics] business."[95] There was a mismatch, too, as Tigner later recalled, between those who understand engineering as having "all relevant parameters well within the range of current practice" and physicists aiming to "achieve some novel purpose or achieve some advance in performance or specific cost. It is the latter that 'frontiers' are all about," he reflected.[96]

Seeing the Elephant

The metaphor of "seeing the elephant," as employed by American frontier writers, stems from the nineteenth-century discourse of American farmers who spoke with awe about the sight of the elephant in a traveling circus. The expression grew synonymous with opening one's eyes, or "seeing the world." It was used during the California Gold Rush by those heading west, often as a way of indicating the possibilities of misfortune in their travels. The expression was also applied by Civil War soldiers, by voyagers to the western prairie, and, as early as the third century BC, when Alexander the Great's forces defeated the elephant-mounted army of King Porus, to denote reaching the breaking point in a venture, the point at which one has seen enough or all that one can endure. It was the moment when one sensed the monumental reality of an ordeal and, typically, decided to turn back.[97] By the end of 1988, CDG saw the elephant.

The year 1988 began auspiciously with a second strong endorsement of the SSC from President Reagan, who in his March budget requested $363 million for SSC construction in a ceremony held in the Rose Garden. "They had kids from every state," recalled Quigg. "We gave each kid an SSC T-shirt, and arranged for two of them to present one to the President." Reagan stressed adventurous themes of research. Quigg remembers being "very moved by the notion that our equations and ideas had, for fifteen minutes, gotten the attention of the most powerful leader on Earth . . . and caused him to say things that were noble and true."[98]

Many leaders of American politics, science, and industry spoke in favor of the SSC during the first half of 1988. Lederman and Quigg gathered fifty-eight supporting statements in a compendium published by URA in July 1988 and titled *Appraising the Ring: Statements in Support of the Superconducting Super Collider*.[99] Optimism was hard to maintain in the face of the budgetary crises then weighing heavily on the physicists. Quigg remembered frantic calls from DOE to CDG causing fears of layoffs, or even project termination, and creating great stress.[100]

In July 1988, Knapp learned that the DOE had made a decision on the SSC's management organization. Rather than accept Goldwasser's 1987 proposal, which had not been acted on for over a year, the DOE had decided to hear competing bids for the SSC's management and operations contract. Responding quickly to this bureaucratic move, Knapp established two tracks of business: one to prepare URA's response to DOE's official "request for proposal" (RFP); and the other to identify the best candidate for director of the SSC. Knapp decided that the URA should respond to DOE's RFP with a Department of Defense–style proposal typical for military and industrial contractors, perhaps because of the rumor that Martin-Marietta, a huge military-industrial contractor, was preparing a proposal to manage the SSC.[101] He also worked on a strategy for selecting the SSC's director.

To help URA write its proposal, Knapp hired N. Douglas Pewitt, a physicist who had served as an analyst in the Office of Management and Budget and who had been Keyworth's deputy in 1982, when Keyworth encouraged the Wojcicki panel to "think big." Pewitt was familiar with all the political and budgetary requirements for writing a successful military-industrial proposal of this kind. After an unsuccessful attempt to have the proposal written in Berkeley, Pewitt established the writing effort in rented office space in an industrial park located in St. Charles, Illinois, some fifteen minutes from Fermilab, in an effort to maintain distance from Fermilab and avoid any claim of impropriety.[102] Working through September 1988, the writing group of about twenty produced a heavy-duty, six-inch-thick, three-volume package, which URA submitted on November 4, 1988.[103] Notably different in its form from Goldwasser's fourteen-page unsolicited proposal, this URA proposal encompassed every aspect of daily site operations, including staffing, construction, magnet industrialization, corporate interface, wages and salaries, fringe benefits, detailed costing procedures, relations with other national labs, and property management. "Bureaucratically it was gorgeous," Knapp reflected. Building the SSC was looking more "like building an aircraft carrier than like building a particle accelerator."[104] Ultimately the DOE accepted the only bid submitted, the URA proposal. (See fig. 13.5.)

As part of its response to the DOE's RFP, URA named the upper management team of the SSC. They chose Roy Schwitters as director. This was a considerable blow to both the CDG and its director, Maury Tigner.[105] Although Schwitters had had important management experience in high-energy physics, while serving as a cospokesman for CDF in the early 1980s (see chapter 12) and as a member of URA's SSC BOO, he had been involved only peripherally in the SSC's design work at the Berkeley outpost. It was the responsibility of Panofsky, as chairman of URA's SSC BOO, to inform Tigner. He did so on the Sunday before Labor Day in

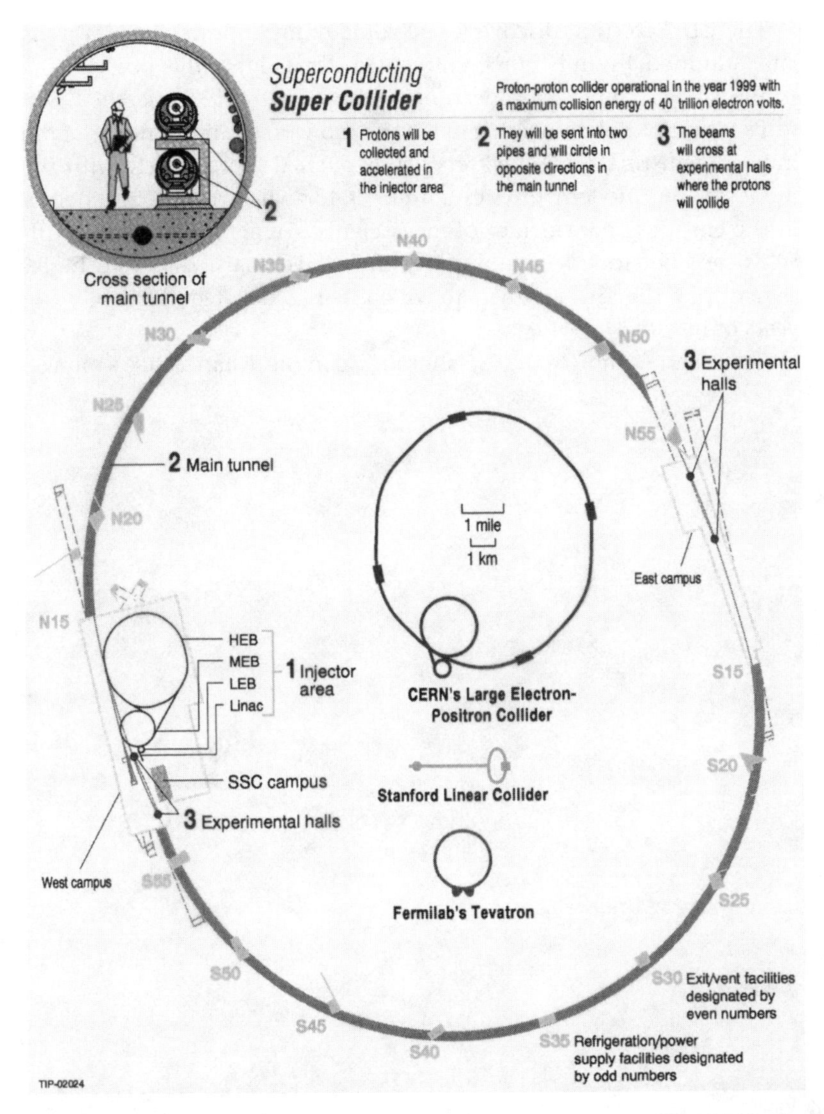

13.5 Drawing of proposed Superconducting Super Collider, 1988. (Courtesy of Fermilab Visual Media Services.)

1988.[106] Quigg recalled that Tigner had a habit of whistling hymns when things got bad. When they were extremely bad, "he would walk about absent-mindedly whistling the tune to 'Once to every man and nation comes the time to decide.'"[107]

The CDG corps of discovery had come to an impasse. As Schwitters later admitted, he and Tigner represented a "classic conflict of styles."[108] While Tigner's approach to leading a group of physicists matched Wilson's vision of the pioneering explorer who worked in the mode of the independent tinkerer, Schwitters's style aligned more closely with the model of the modern physics administrator who worked confidently in the emerging bureaucracy of megascience. Tigner tried to work with Schwitters but found it impossible, and on February 20, 1989, he resigned from the SSC, a project to which he had devoted more than five years of his life.[109]

Lederman's vision of the SSC shimmered in the distance, like a mirage.

Epilogue: Light on the Horizon, 1989–1995

It is a profound and necessary truth that the deep things in science are not found because they are useful; they are found because it was possible to find them.
J. ROBERT OPPENHEIMER[1]

Just before noon on November 10, 1988, nearly a thousand members of the Fermilab community from all its ranks poured into the Ramsey Auditorium of Wilson Hall. Lederman had sent an invitation to each member of his staff to a talk he planned to deliver at high noon. He expected to announce the success of the Illinois site proposal for the Superconducting Super Collider. But the news about the competition's outcome had leaked to the media, and many employees had already heard it on the radio as they drove to work. Waxahachie, Texas—not Batavia, Illinois—would be the home of the SSC. Word was spreading quickly, like wildfire on the prairie. The news had a numbing effect.

Many Fermilab employees had no heart for work that morning. Small groups gathered in the corridors of the High-Rise to ponder the implications and talk through their disappointment. The lab's users would conduct their research wherever the SSC would be built, but most of the in-house staff was tied to Illinois. They wondered about their job security and the uncertain future. Many could not believe the news and desperately hoped it was just another rumor.

Discussions continued in the auditorium while the assembled group waited to hear from their director. Most spoke in hushed tones as they found seats. Staff and users were joined by media representatives. Frank Cole, one of the few who had been in the first wave of NAL physicists from Berkeley, recalled the confusion, anger, and disbelief expressed twenty-two years earlier, when members of the Lawrence Berkeley Laboratory learned that Weston had been selected for the 200 GeV project.[2]

At noon, Lederman arrived at the entrance to the auditorium. He had the full length of the aisle (seventy steps—about 200 feet) to descend as he slowly progressed toward the stage. The audience waited silently as he moved through, climbed to the stage, and walked to the podium. A month earlier he had been celebrated in that same auditorium for winning the Nobel Prize in physics together with his Columbia colleagues Jack Steinberger and Mel Schwartz. Considering that joyous news a harbinger of good fortune, the Fermilab community tended to believe they would win the SSC.

Lederman had prepared a surprise for his people. He hoped it would relieve their distress. That morning he had asked that a large Stetson hat be placed under the podium, out of view of the audience. Mysteriously, this hat had been delivered to him the previous week. Now he surprised his audience by pulling it out and placing it on his head (fig. Epi. 1). As he faced them, they burst into laughter. He had calculated well. He waited patiently, allowing the laughter to subside. Then he removed the Stetson.

Lederman said with a grin, "I don't know if we have such a thing as a *prairie* hat." Again the audience laughed.[3] Only then did he begin to read his prepared statement, one of two he said he had written the night before. "As you know, the U.S. Department of Energy selected Texas as the site for the SSC. I think we're all disappointed at the decision for it not to go to Illinois. I personally believe that it would have been much easier to build a machine in Illinois and harder to build it in Texas."

He went on, making a strenuous effort to assure the staff that Fermilab would not abruptly end. "Now, I know some of the questions on your minds. What happens to Fermilab?" He spoke briefly about the success of the "new machine here, the Tevatron," and the need for its program of physics to continue as "the prime tool for keeping physics in the United States going over at least the next decade, if not the next fifteen years." He mentioned the importance of keeping a "lively physics program going until then," and he referred to upgrades and the fact that the SSC would not be ready to publish papers in physics until roughly 2000. "So we have a lot of things to do between now and when the SSC comes on the air."

He also spoke about the older laboratories that had not closed, mentioning SLAC, Brookhaven, and CERN. He noted that "Argonne is 40 years old and going strong. . . . I can't imagine that this Laboratory, with

Epi.1 Leon M. Lederman with gift Stetson Hat, 1988. (Courtesy of FNAL Visual Media Services.)

all of its facilities, and especially in this high-tech area, won't be devoted to some kind of scientific research."

Someone in the audience asked, "Is there any chance that Fermilab would be put in a box and shipped to Texas?" Lederman replied, "Move the apparatus? We won't let 'em." Someone else inquired, "Will you be going along with the SSC?" Lederman sidestepped. "I think we'll all be involved in it. Fermilab has a major role to play in the R & D program." Another asked, "Do you foresee any layoffs?" Lederman answered directly, "No I don't see any layoffs." Unclear himself whether he believed what he was saying, yet certain of his need to say it, Lederman forecast a likely build-up at Fermilab to provide necessary upgrades to facilities for training the next generation and to do the work that the SSC would require.[4]

He continued with many more comforting comments. He referred to new projects and increases in funds for science and education. He told the audience that he felt there was a good chance the SSC would be built, although it would probably be "funding limited." A member of the audience asked about Lederman's other prepared speech. "Can we hear the other one, the one announcing the success of the Illinois proposal?"

After more laughter and applause, Lederman joked, "I think I lost it. It was a great one."

After responding to a few more questions, he put the Stetson back on his head. "Call me Tex," he drawled in his hoarse voice when the laughter subsided. The audience roared again. He explained that the hat was a gift from friends at Rice University. "When I got this hat I suspected something was going wrong. I can't wear it, because my horse shies every time I put it on." After still more laughter he said, in closing, "One thing is certain, this laboratory will laugh no matter what happens, and that's why I love you all."

But there was little laughter at Fermilab in the months following the announcement that Waxahachie would be the SSC's home. Two days earlier, George H. W. Bush, from Texas, had been elected to succeed President Reagan. Many non-Texan physicists and politicians now looked for ways to withdraw from the SSC project. Much of the collective knowledge of the Central Design Group was lost, as few of them were asked to join Schwitters in Texas. Even fewer elected to do so.[5] Of the Fermilab staff who had worked at the SSC's outpost, only Father Tim Toohig moved from Berkeley to Waxahachie.[6] The announcement of Waxahachie as the SSC's site marked the end of seeing the SSC as Fermilab's future.

Fermilab also had to adjust to the loss of its charismatic leader. Just a month earlier, in the midst of the laboratory's celebration of his Nobel Prize, Lederman announced that he would be stepping down the next summer. He planned to take a teaching position at the University of Chicago. There had been wide speculation that Lederman's resignation was somehow connected with the SSC. Many colleagues thought he would be offered the directorship, given his active participation in the conception and nationwide selling of the project. He subsequently denied any connection between the SSC and his decision to step down. In any case, Schwitters had already been named director of the SSC by the time Lederman stepped down. On August 10, Harry Woolf, chairman of the URA board of trustees, had written to Lederman asking whether he would accept another term as director of Fermilab. "I should like also to reiterate," Woolf wrote, "that this appointment, if you accept, does not in any way preclude or prevent your possible selection as the Director of the SSC. This is, of course, especially pertinent should the SSC come to Fermilab."[7] There is no evidence that Lederman seriously considered Woolf's offer.

The somber mood at Fermilab continued into the directorship of John Peoples, Jr., Fermilab's third director. Peoples had come to Fermilab in 1971 from previous positions at Columbia and Cornell. His physics research and hands-on experience over twenty years with the Fermilab

Epi.2 John Peoples and Leon M. Lederman, at announcement of third Fermilab director, April 1989. (Courtesy of FNAL Visual Media Services.)

accelerators, coupled with his strong, intuitive, and confident manner, had brought him to leadership roles at the group and division levels of Fermilab's Research and Accelerator divisions. As recounted in earlier chapters, Peoples had managed the complex effort to build the antiproton source for Fermilab's collider program. In 1987–1988 he had lent his magnet expertise to the SSC's Central Design Group. When Peoples returned from Berkeley in September 1988, Lederman promoted him to deputy director. Peoples took the reins in July 1989. (See fig. Epi. 2.)

Plotting Fermilab's course, Peoples proceeded to realign priorities and open new directions, in addition to accepting all the usual tasks that Lederman handed off (e.g., the budget, the research program including the collider experiments and the fixed-target runs, securing computing strength for a new generation of needs, building international alliances, contributing to SSC research and development, improving the security of the site, protecting the environment, achieving the transfer of technology to industry, and supporting science education). By now his duties also included building the Main Injector, a substantial upgrade of the laboratory's accelerator. This project would require deft diplomacy on all levels. Since local real estate opposition had hurt Illinois's SSC proposal, Peoples worked with physicist-administrator Ken Stanfield, his deputy director, to strengthen Fermilab's position by building local, state, regional, and federal support. Soon, during his first month as director, Peoples received good news. In July 1989, CDF reported a breakthrough measurement of the mass of the Z^0 particle as 90.9 GeV. It was a major step in studying the postulated Higgs field that gives mass to particles. The result offered a strong rationale for emphasizing colliding-beams research at Fermilab.[8]

But the course for Fermilab under Peoples proved extremely rough. An oppressive funding crunch continued through the early years of his tenure. By this time the idea that fundamental research could serve national defense had lost the political hold it had enjoyed during the Cold War. With the collapse of the Soviet Union in 1991 came a diminished need for a conventional defense rationale in Congress. The increasingly bureaucratic Capitol Hill could barely remember Washington's earlier appreciation for the physicists' contributions to World War II, a view that had fueled the dramatic increase of the physicists' funding during the 1950s and 1960s. No longer the golden boys, physicists were just another interest group.[9] The result was a reduced investment in basic research at all American high-energy physics facilities. When President Bush's secretary of energy, Admiral James Watkins, observed in October 1991, "In a time of budget austerity, the Department has to give highest priority to the most cutting-edge science,"[10] he meant applied projects, not fundamental physics. Despite Keyworth's vigorous efforts during the early 1980s to justify the support of basic research, by the start of the 1990s, most Washington politicians did not understand the connection between high-energy physics and the economics of the American people and the world.

Fermilab, Brookhaven, and SLAC all managed to survive the 1990s, albeit with minimal support of their existing programs. The SSC's projected budget suffered dramatically as it became clear that the idea that "new money" would support this project, a view widely expressed when Reagan endorsed the SSC, would not be borne out.[11] And when Congress and the Office of Management and Budget (OMB) looked for areas to trim in their effort to manage the multibillion-dollar budget deficit that had built up over the Reagan-Bush years, construction appropriations for the SSC were held hostage year after year. Support for the SSC from the congressional delegations of many states (including Illinois, New York, and California) also waned, especially as Fermilab, Brookhaven, and SLAC struggled at the expense of the SSC. By May 1992 the SSC was threatened with congressional cancellation.

Peoples recognized that in the precarious funding context of the early 1990s, Fermilab's budget could not support both a colliding-beams and a fixed-target program. The 1988 close of the 15-foot bubble chamber had been an early sign of the end of the fixed-target program.[12] In October 1991, Peoples announced that "our highest priority will shift from the fixed target experiments to the collider experiments." Invoking frontier rhetoric like that used earlier by Fermilab's first two directors, Peoples reassured his community that the data from CDF and DZero "promises to be the most exciting taken at any accelerator in the world, as these two detectors will be searching for phenomena in completely unexplored territory."[13] The decade of Peoples's tenure as director saw Wilson's conflicted version of big science, planted partly in the smaller "quick and dirty" experiments that Wilson favored and partly in the much larger institutions of megascience, give way to ever-growing colliding-beams collaborations. And just as early big science had grown rapidly with the lavish funding physics had enjoyed in the post–World War II years, megascience, at least in the form it took at Fermilab, responded to limited funding with the paradoxical changes in practice described in chapters 11 and 12. That last phase in the transition from Wilson's frugal version of big science to megascience was not witnessed firsthand by most of the physicists who had stood beside Wilson, for by then many of them had retired or left the laboratory.[14] The demographics at Fermilab changed rapidly in the early years of Peoples's administration, because of the appeal the SSC had for the younger generation, the rise of the colliders and the reduced scope of the fixed-target program, and financial incentives that Fermilab offered for early retirement while the laboratory struggled to balance its budget.[15]

Wilson had discouraged major computing innovations; they were implemented during his administration only occasionally and on an individual basis. Under Lederman a more formal plan to institutionalize computing had begun around 1983 (see chapter 11), resulting in new labwide computing initiatives and the construction of the Feynman Computing Center (see chapter 9).[16] Under Peoples the revolution in computing spread to all aspects of the work of Fermilab, including desktop computing, data acquisition and analysis, publication of research, and communication throughout the laboratory and the world beyond it.[17] As high-energy physicists learned to keep up with their PCs, workstations, and browsers, the latest hardware and software replaced chalkboards, pencils, and paper logbooks. The library and stockroom were automated. Secretaries who had previously provided typing and word processing moved their work from typewriters onto mainframe computers, such as the CYBER or VAX, and then back to their desks when they became equipped with PCs.[18]

Fermilab's working environment became, in some ways, more bureaucratic in the years when James Watkins was secretary of energy because of a new government initiative aimed at bringing the management of the national laboratories into compliance with standards put in place by Washington.[19] Various environmental, safety, and health infractions by the weapons laboratories had troubled the DOE and Congress. Watkins tried to remedy the situation by insisting that a single set of standards, patterned after those at nuclear power plants, govern all the DOE laboratories. Creating what Watkins called "the new culture," with a greater emphasis on such aspects as industrial safety, hygiene, emergency preparedness, quality assurance, self-control, energy awareness, management of safety, recycling, and environmental protection, he applied a one-size-fits-all approach to the oversight of government laboratories, similar to the approach used almost four decades earlier in Admiral Hyman Rickover's nuclear navy. Regardless of whether the work of the laboratory was aimed at research or defense, the oversight system emphasized stringent controls and assessments, often made by "Tiger Teams" that visited the labs. Watkins's ten-point initiative aimed to implement "full accountability in the areas of environment, safety and health."[20] He stated in 1989 that his plan was intended "to strengthen the Department's environment protection and waste management activities." Under Watkins, formal ES&H operations "took precedence over production or research objectives."[21]

The new Washington culture threatened to end the university-like atmosphere that Wilson and Lederman established at Fermilab. One inside observer reflected, "The Bob Wilson way of running things . . . was *very*, very different from how everything else in government was."[22] In 1991, a Fermilab committee known as ESHPAC, the Environment, Safety and Health Policy Advisory Committee, was formed to offer the laboratory's response to the new issues, now typically referred to as ES&H. Associate Director Dennis Theriot, who chaired the committee, explained that "going right down to the individual performing the task" is "the essence of the cultural change Secretary of Energy Watkins wants—everyone taking responsibility for safety."[23]

In February 1992 William Happer, director of DOE's Office of Energy Research, called Fermilab the "the current jewel in the crown of high energy physics, and the key facility for exploring forefront particle physics in the 1990s."[24] Yet an intense debate continued in Congress over DOE's budget for high-energy physics. Fermilab was in jeopardy, and the SSC was at risk. SLAC and Cornell had to compete for funds for a "B" factory. BNL was on thin ice, shifting from its canceled CBA for high-energy physics to nuclear physics with the construction of RHIC, the Relativistic Heavy Ion Collider.[25] Peoples knew that something had to be done to assure the future relevance of high-energy physics. He took charge and sounded the alarm.

In Washington DC the Illinois congressional delegation, including Robert Michel and Dennis Hastert, and the Illinois Coalition, an active group of Illinois business proponents boosting Fermilab in the wake of the state's loss of the SSC, responded positively to the call from Peoples to find a way to secure funds for Fermilab's plan to upgrade the Tevatron in the fiscal year 1993 budget. It was hoped that both CDF and DZero would observe $\bar{p}p$ collisions and that once the substantial upgrades of the Linac, Booster, and antiproton source had been completed, Fermilab would observe the top quark. The upgrades were designed to double the luminosity of the accelerator. Peoples announced, "If the mass of the top quark is less than 160 GeV/c^2, we will discover it. If it is greater, we will clearly show that it lies beyond that value."[26] Peoples had good reason to be confident, for only the Tevatron had high enough energy to find the long-sought heavy particle.

Meanwhile, in mid-May 1992 thirty "Tigers," representing both DOE and private industry, arrived on the site just as Collider Run I began. They would review compliance with regulations and standards in every aspect of operations. The laboratory was, however, well prepared for their visit. In their detailed investigation (fig. Epi. 3), the Tigers found "many islands of excellence" at Fermilab and nothing that "presented a significant risk to either public health or the environment."[27] Peoples said the laboratory would "expand these islands and plan our research so that it never compromises health, safety, or the environment." For the time being, this storm had passed, but a great deal of precious research time had been spent preparing for the Tiger Team's visit.[28] The heightened sense of oversight of the national labs continued for several more years as the Tevatron continued its search for fundamental physics.

During this intense period new storms gathered on the frontier. Fermilab continued its research and development for the SSC, expecting its scheduled "turn-on" in 1996. The lab was shaken that June 1992 by the news that the House of Representatives had voted to discontinue funding for the SSC project. Although the Senate restored the SSC's funding, the uncertainty of the SSC project was deeply troubling to the high-energy physics community. Many had hoped for increased government support of the SSC when Lederman was awarded DOE's 1992 Enrico

Epi.3 Cover of DOE report on Tiger Team visit to Fermilab, 1992. (Courtesy of Angela Gonzales.)

Fermi Award for a lifetime of achievement in the field of nuclear energy. Previous recipients of the Fermi prize had included Ernest Lawrence in 1957, Glenn Seaborg in 1959, Hans Bethe in 1961, J. R. Oppenheimer in 1963, Wolfgang Panofsky in 1978, Wilson in 1984, and Stanley Livingston in 1986. Neither Lederman's recognition nor the election in November 1992 of Arkansas governor Bill Clinton as president of the United States improved the SSC's funding prospects.

The initial design of the Main Injector followed in the tradition set by the early Doubler and colliding-beams programs in that its early work was performed by a small group studying how to enhance or replace the existing Main Ring with a better-performing accelerator.[29] With the Main Injector as part of the system, the Tevatron could ideally operate in both fixed-target and colliding modes at the same time, instead of alternating every few years. This was a strong selling point. In addition, by significantly improving the reliability of the Tevatron, the Main Injector dramatically increased the number of proton-antiproton collisions. The laboratory's physics program could now extend its reach to the search for rarer and higher-mass particles. Yet few believed the DOE's budget would provide for this possibility. To almost everyone's surprise, the Main Injector was approved. In early 1993, as soon as Peoples received word that this was imminent, he pushed forward to schedule the groundbreaking ceremony (fig. Epi. 4). In an event reminiscent of the official start of construction for the Linac in December 1968, the Illinois congressional delegation stood on March 22, 1993, in two inches of snow with a cold wind blowing, together with officials from URA and DOE, this time to break ground for the Main Injector.[30] "This isn't the first time that Fermilab began building for the future on a cold, snowy day," Peoples reminded the audience.[31]

Many physicists were disheartened, and some were insulted, when Clinton's new secretary of energy, Hazel O'Leary, a utilities company manager, continued Watson's pursuit of better laboratory management. Corporate-sector jargon found its way into the DOE discourse about "total quality management (TQM) awareness" and "DOE core values," with such buzz words and phrases as "customer orientation," "people are our most important resource," "teamwork," "leadership and quality," "highest standards of ethical behavior," and "accountability." Although many of these ideas had already entered the practice of physics, earlier officials had not preached them so blatantly. DOE's new TQM philosophy entered the language of the SSC's contract in August 1993, conveying O'Leary's intentions to provide "the best possible project management and financial systems expertise."[32] To meet the objective, Peoples and all DOE laboratory directors were required to attend training courses during the summer of 1993 at the Motorola/Milliken Quality Institute in Schaumburg, Illinois.[33]

Vigorous arguments about the SSC continued in Congress. O'Leary vacillated in her support of the SSC while lines of congressional representatives on both sides of the SSC faced off in debate. A second vote went against the SSC in the House of Representatives in June 1993. That same month

Epi.4 Groundbreaking for Main Injector, March 1993. *Left to right,* John Peoples, Representative Dennis Hastert, Senator Carol Moseley Braun, Senator Paul Simon, and Wilmot Hess of DOE. (Courtesy of FNAL Visual Media Services.)

Congressman John Dingell, chairman of the House Subcommittee on Oversight and Investigation, attacked not only the SSC but URA, which he accused of mismanagement in a hearing before his subcommittee, and for a "high level of arrogance . . . and intolerance toward government oversight." He also accused it of "lavish spending of taxpayers' money on luxuries and entertainment." Schwitters, and even O'Leary, challenged Dingell's claim, but opinions in Washington nevertheless turned more solidly against the SSC.[34] In September, a distinguished group of SSC supporters, including Nobelists Lederman, Richter, and Weinberg, gathered at George Washington University with the executive board of the APS to promote funding for and completion of the new collider. British physicist Steven Hawking sent his support via videotape. The meeting was planned for broad media coverage, but instead, the major news story that day was of progress toward peace in the Middle East. President Clinton had brought Israeli prime minister Yitzhak Rabin and PLO chairman Yasser Arafat together to sign the Oslo Peace Accords.[35]

In late September, as in 1992, the Senate restored the SSC's funds, and the issue was sent to the joint energy and water conference committee to determine the fate of the SSC, then 20% completed.[36] Fall of 1993 was an especially hard time for expensive science projects. A perennial favorite in Congress, NASA's International Space Station was roundly criticized, and its funding bill passed by only one vote. Following a second, larger vote against the SSC on October 19, 1993, in the House of Representatives, instructions were sent to the conference committee to acknowledge their decision to terminate the SSC. The only hope for rescue was by the conference committee.[37] It was the final showdown.

On October 21, 1993, one day after the NASA vote, the American high-energy physics community was stunned by the confirmation that the SSC would be terminated. They felt devastated and abandoned with the collapse of their frontier fortress. "SSC is dead!" wrote Drasko Jovanovic to the editor of the in-house newletter *FermiNews*. Like most Fermilab physicists, he experienced the cancellation as a painful loss. He lamented, "It feels like the loss of a child, full of promise and potential, conceived and nurtured by the 3,000 of us, and now—dead." From Washington, O'Leary commented: "I deeply regret the House decision, but we see no prospect of reversing it."[38] Part of the money requested to build what was to have been the new flagship for high-energy physics was now directed toward shutting it down. Schwitters resigned.

The DOE, with URA's approval, assigned Peoples to double duty; he would direct the shutdown of the SSC laboratory while still serving as Fermilab's director. With Peoples in Texas, Deputy Director Ken Stanfield was left in charge in Batavia. Peoples's immediate SSC responsibilities included closing out the contract, with its property and employment issues, and writing summary reports for the DOE. His effective performance in these painful tasks helped restore DOE's confidence in URA as a contractor. Peoples labored especially hard to find jobs for approximately two thousand four hundred SSC employees whose positions had suddenly vanished. Roughly a hundred employees who had formerly worked at Fermilab returned to the laboratory.

With the SSC's termination the VBA/SSC dissolved once again into the distance. Over the next several years the United States retreated from hosting the outpost for the distant high-energy physics frontier, although a 1994 HEPAP subpanel led by Sid Drell of SLAC found a way for

U.S. physicists to participate in the future Large Hadron Collider (LHC), then planned to be built internationally at CERN over the next decade.[39]

Peoples returned to Fermilab from Texas in the fall of 1994. He sought to reestablish the lab's position on the frontier despite unpromising budget projections. He expressed confidence "that Congress is prepared to support high-energy physics at a level of funding that is not too different from the level of funding that high-energy physics received prior to the SSC."[40] A set of provocative and diverse research initiatives were underway at Fermilab.[41] Besides the Main Injector and colliding-beams program, Fermilab's new prospects included two ambitious fixed-target experiments (KTeV and HyperCP) designed to explore both matter and antimatter in the universe, the NuMI program of neutrino research with the Main Injector (including MINOS, the Main Injector Neutrino Oscillation Search for neutrinos in a Minnesota mine), the Pierre Auger Observatory, an international cosmic-ray experiment, and Fermilab's first astrophysics experiment, the Sloan Digital Sky Survey, which aimed to map the universe.

Peoples was fairly confident that a new light would soon shine on the frontier. Both CDF and DZero had gathered an immense amount of data by the time the collider run ended in June 1993. The collaboration members hoped their results contained traces of the top quark. During Peoples's time in Waxahachie, the experimenters working on CDF and DZero had told their director that they believed they had seen the top quark. Its mass was at least 130 GeV/c^2. To firm the discovery both collaborations needed more time. It would take many months to sift through and analyze the vast amounts of data and "discover" the particle.[42] Almost a thousand physicists in the two collaborations had to be convinced; some had doubts, and many were reluctant to declare victory without certainty. The teams worked steadily, examining their data from all angles, questioning and proving in the process of understanding what the detectors and computers revealed.

Fermilab's Web site had been swiftly and deliberately constructed to send the news of the breakthrough across the world. When CDF and DZero cautiously announced in April 1994 that their preliminary data showed evidence of the top quark Washington responded immediately. Martha Krebs, the director of energy research for DOE, proclaimed the achievement as one that "represents the high vitality of our high-energy physics program in the United States." She added, "We lead the world at this energy frontier and the capabilities here at Fermilab are crucial to maintaining that leadership."[43] That August, a White House report titled *Science in the National Interest*, the first presidential statement on science

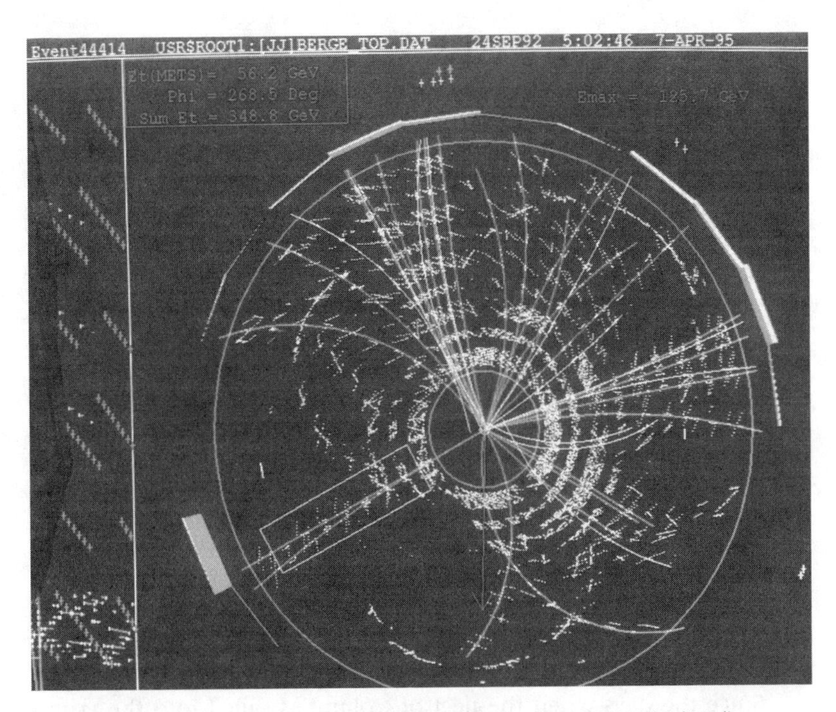

Epi.5 Data showing the discovery of the top quark, 1995. (Courtesy of FNAL Visual Media Services.)

policy in almost fifteen years, outlined the Clinton administration's strong support for fundamental research and emphasized the importance of basic science to the nation, ranging from science literacy to leadership at the "frontiers of scientific knowledge."[44]

The discovery of the top quark was proudly and officially announced on March 2, 1995, after months of planning. Both collaborations accepted that what they were seeing was really the top quark (fig. Epi. 5). Papers by the CDF and DZero collaborations had already been submitted to *Physical Review Letters* on February 24, 1995. The Fermilab press release explained that the top quark was the last of the six quarks predicted by the current theory and provided "strong support for the quark theory of the structure of matter." Secretary of Energy O'Leary stated, "This discovery serves as a powerful validation of federal support for science." She praised Fermilab: "Using one of the world's most powerful research tools, scientists at Fermilab have made yet another major contribution to human understanding of the fundamental nature of the universe."[45]

In the fall of 1998, Peoples announced that he would step down and return to research the following summer. He had served as Fermilab's director for two five-year terms and wished to accept the directorship of the Sloan Digital Sky Survey. Having followed Wilson's trail as a relentless builder and, like Lederman, doggedly determined to reach the highest energy, Peoples left a legacy to Fermilab that included both the next-generation Main Injector and the discovery of the top quark. His final contribution was to shepherd the turn-on of the Main Injector in June 1999. DOE secretary Bill Richardson and Speaker of the House Dennis Hastert turned the key empowering the new instrument for exploring yet another frontier (fig. Epi. 6). Under Peoples, and against the odds, Fermilab had succeeded in building a major new accelerator (fig. Epi. 7).

As Lederman had predicted in his Stetson at that tense November 1988 meeting on the day the SSC's site in Waxahachie was announced, life continued at Fermilab—indeed for longer than the decade and a half that Lederman promised. As of this writing in the spring of 2007, Fermilab remains the highest-energy physics laboratory in the world. Its research still pursues new frontiers of physics—revelations of neutrino mass, the Higgs boson, supersymmetry, and superstrings—that will only manifest themselves at higher energies with optimal vision and careful comprehension. But it is clear that the time will come soon, within a few years, when CERN's LHC will take Fermilab's place at the energy frontier.

"Since the days when the fleet of Columbus sailed into the waters of the New World, America has been another name for opportunity," reflected Frederick Jackson Turner at the end of his 1893 address at the Columbian Exposition.[46] Like Chicago's legendary symbol of rebirth at the exposition, a phoenix rising from the ashes of the 1871 fire, the American frontier was to be reborn in a continuing stream of frontier studies of the atom, the nucleus, and the cosmos. As a flagship vessel for exploring scientific frontiers in understanding nature's building blocks, Fermilab had responded well to the advice that Chicago's renowned architect and designer Daniel Burnham offered to the city in 1910: "Make no little plans; they have no magic to stir men's blood and probably themselves will not be realised. Make big plans; aim high in hope and work, remembering that a noble, logical diagram once recorded will never die, but long after we are gone will be a living thing, asserting itself with ever-growing insistency. Remember that our sons and grandsons are going to do things that would stagger us. Let your watchword be order and your beacon beauty."[47]

* * *

Epi.6 Turning on the Main Injector, 1999. Secretary of Energy Bill Richardson is on the left, accelerator head Steve Holmes in the center, and Congressman Dennis Hastert on the right. (Courtesy of FNAL Visual Media Services.)

Epi.7 The Tevatron and Main Injector rings, 2003. (Courtesy of Fermilab Visual Media Services.)

Epi.8 Geese on frozen Swan Lake, 1999. (Courtesy of FNAL Visual Media Services.)

Many, Many Years Later

In the spring of the year 9007, toward the end of the latest ice age, the receding North American glacier exposed what was once a suburban area west of metropolitan Chicago. The retreating glacier revealed evidence of a civilization that existed at the beginning of the twenty-first century. Archaeologists, looking for clues about the people who had inhabited the area, uncovered the remains of suburban dwellings, commercial malls, schools, and parking lots. They were perplexed when they discovered a 4.26-mile underground circular tunnel that had been buried under the ice for almost seven thousand years. They wondered what its purpose could have been.

Surrounding this ring was fossil evidence of almost 7,000 acres of agricultural fields, sections of primeval forests, restored native North American prairie plants, freshwater lakes, wildlife (including the extinct blue heron and Canada goose; fig. Epi. 8), curious winding paths and roads, unusual architecture, and even a herd of still-endangered *Bison bison*. Buried beneath the rubble were traces of other, smaller tunnels—some round, some elongated, triangular, and oval-shaped—and a maze of buildings containing remnants of an advanced technical civilization. It seemed out of place amid the easily understood suburban sprawl of the early twenty-first century.

As they excavated and mapped their findings, the archaeologists considered theories for the use of the "ring," including ideas that it was used for some sort of sporting event or was involved in some religious ceremony. Slowly the archaeologists recognized the artifactual remains of what had once been an early center for scientific research, a kind of base for frontier exploration, not of the unexplored land of nineteenth-century North America, but of the unexplored invisible realm of the twentieth-century atom. Historical documents found in the archaeologists' search explained that at the end of that century scientists traveled from around the globe to this ring to use a machine called the Tevatron to understand the nature of matter and energy.

The archaeologists had discovered the fabled Fermi National Accelerator Laboratory (Fermilab), once home to the earth's most powerful "accelerator," then popularly known as an "atom smasher." They were astonished to learn that people called "physicists," like ancient astronomers, alchemists, medicine men, philosophers, and wise shamans from other cultures, had once dedicated their lives to the pursuit of questions like What are the ultimate building blocks of nature? What is the stuff of which the universe is made? And what is the nature of this stuff? Perhaps their opinions changed when they realized that these scientists were exploring frontiers inside the atom, and that the people of this era had supported this expensive pursuit through a crude government mechanism called "taxes." Perhaps they wondered what could possibly have motivated this strange quest for knowledge with no obvious survival benefits. Why did this earlier culture take from its own wants to support such curiosity about the universe?

Authors' Statements and Other Acknowledgements

Lillian Hoddeson (Department of History, University of Illinois at Urbana-Champaign)

Several months after moving to Urbana, Illinois, in 1977, I received a phone call from Dick Carrigan, a Fermilab physicist. He asked whether I would be interested in helping Fermilab create its archives. Robert R. Wilson, the laboratory's director, wanted to document the laboratory's history, in fact, particle accelerators more generally. I was interested. At that time I was in the process of changing fields, from physics to history of science, and I had not yet found a position in my new field.

I had discovered my passion for history of science more than a decade earlier at Columbia University, where I took my Ph.D. in physics in 1966. After graduate school, I worked at Columbia with Samuel Devons to create a "history of physics laboratory" at Barnard College. I found it exciting to learn what was known of science in the past. A few years later, I studied the history of science with Thomas S. Kuhn, Charles Gillispie, and others in Princeton's Program in the History and Philosophy of Science. There was no turning back for me now.

Accepting Carrigan's offer, I was soon spending a week each month at Fermilab, gathering historical materials around the laboratory, sometimes in odd places, like old barns on the site. The collection would in time grow into the Fermilab Archives, which hold most of the material on which this book is based. The archives are presently cared for, lovingly, by Adrienne Kolb, Fermilab's archivist.

My early work at Fermilab was guided by a committee that Wilson himself chaired. It included Wilson's deputy director Ned Goldwasser, Fermilab's librarian Roger Thompson, and a number of staff physicists who were interested in history, including Carrigan, Tom Collins, Frank Cole, Don Young, and Fred Mills. Initially the discussions focused on building an archive, but they soon turned into an ongoing seminar about the history of accelerators, including the Fermilab machine. The committee also sponsored a program that brought eminent accelerator physicists to the laboratory for talks and oral history interviews. These visits always included a well-prepared lunch with the speaker, the committee, and other interested colleagues. The recordings made on those occasions are a valuable, as yet untapped, source for historians of science.

Believing that I had been hired primarily to create an archive, I visited a number of other collections around the country. I was especially eager to learn about proper procedures for archival storage and retrieval, something I had not learned about at Princeton or Columbia. On being advised by Joan Warnow of the AIP's Niels Bohr Library, I ordered a number of shiny flame-retardant boxes.

One day, shortly after the boxes arrived, Wilson came by the new History Room, located adjacent to the Fermilab library. I proudly showed him the new boxes and explained that they were fireproof. I also told him about my plans to computerize the collection. He was not pleased. I did not yet know about his mistrust of computers. Nor did I understand what Wilson really had in mind when he hired me. I soon found out.

"I'd been envisioning," he explained to me, "a comfortable study room where physicists and other staff can quietly learn about the great events in the history of accelerators." He furnished the room with his own oval table (one that Adrienne and I have used ever since). He joked about his vision including "a large dog sleeping peacefully in front of an open fire" in the History Room. It was to be a space for study and contemplation about the past and about the past's relation to the present and future of Fermilab. As for the new boxes, he decided to test one on the spot. He struck a match, and when he touched its flame to a corner of one box, it burst into flame! Fortunately, the flame soon died out. "Good," he muttered. "They seem to work."[1]

Afterward, I pondered Wilson's reaction. Was I approaching things too "professionally?" Was I focusing too much on the archives, too little on the history and the people who made it? I decided to interview Wilson, picking up the story where the AIP's Spencer Weart had left off in his earlier interview of Wilson on his life in physics before becoming Fermilab's director.[2] My interview with Wilson would become a series extending over many sessions and years, eventually spanning most of his career.

The first session, in early 1978, included Wilson's oft-told account of how and why he built the laboratory.[3] It was a colorful tale, full of adventure and passion, but it revealed little beyond what I had already read and heard. He had told the story many times. In fact, the narrative had improved over many tellings. I planned to probe deeper in the second session, conducted a few months later. I brought along a pile of the documents I had been collecting, and I drew on them as Wilson spoke, adding dates, names, numbers, and other detail. The interview went smoothly until I produced a document that conflicted with one of Wilson's favorite stories—as it happened, just as he was about to tell it. Wilson appeared upset. We were both embarrassed. But not for long, for an urgent phone call interrupted the session, and he left abruptly, as recounted in chapter 8.

He was even more upset when he returned. The rumor I heard later was that the phone call had confirmed URA's acceptance of Wilson's resignation as director of Fermilab. Wilson had not wanted this response from URA, for his resignation had been a gamble in his battle to gain support for the laboratory. I found, however, that the distress of the phone call loosened the grip of Wilson's narrative about Fermilab's early history. He found it easier now to adjust the account to fit information from the documents. His revised story was richer in detail. From that point on, Wilson and I became partners in reconstructing Fermilab's history. It was the start of the thirty years of work that culminated in this book. Sadly, Wilson passed away in 2000, before he could see the final product.

Some months later, I learned that my colleague from Columbia, Leon Lederman, would become Fermilab's second director. That made more work for me, because Lederman wanted to extend the mission of the laboratory's history program to include all of particle physics, not only particle accelerators. I accepted this exciting, admittedly impossible, goal at about the same time that I became a mother. Michael Baym, my first child, was born in July 1981; Carol Baym, my second, in January 1985. Within a few years my professional responsibilities expanded, too, when I began teaching history of science at the University of Illinois. I needed to cut back in my work on Fermilab's history.

Lederman came to the rescue and appointed Kolb as the laboratory's part-time archivist. A short time later Catherine Westfall began to help, too, by setting down an important part of Fermilab's history in her Ph.D.

dissertation. The three of us have worked together on Fermilab's history now for more than twenty years. For part of this time, Mark Bodnarczuk worked with us as well, while he studied philosophy of science at the University of Chicago and served as Fermilab's quality assurance officer. In addition, many University of Illinois graduate students in history helped for periods of time. They included Everett Carter, James Nelligan, Kyoung Paik, Nicole Ryovic Ranganath, Melissa Rohde, and Derek Shouba. Most importantly, May West, Fermilab's assistant librarian, was constantly available to help us find references, documents, and individuals with important stories to tell. We were shocked and saddened by the news of her sudden illness and death in April 2003. We greatly miss her insight, cheerful friendship, and generous help.

Adrienne Kolb (Fermi National Accelerator Laboratory)

I arrived at Fermilab in the fall of 1983 with my husband Rocky and our three children, Karen, Jeffrey, and Christine (then aged eleven years, three years, and eight months, respectively). Lillian and I had met earlier in Los Alamos, when she and a small team (which included Catherine Westfall) were writing about the atomic bomb and I was working as a part-time volunteer at the Los Alamos Historical Museum. Previously I had worked at Caltech's Development Office. As our family grew, my hopes of some day working at the Smithsonian with my degree in history from the University of New Orleans faded.

In 1982 Leon Lederman had visited Los Alamos. He asked Rocky whether he would like to start the first theoretical astrophysics group at a national laboratory. He suggested that Rocky spend a year at Fermilab to launch the effort. We agreed to explore this prospect: if we liked it, we would stay. We arrived just as the Twelfth Rochester Conference, held that year at Fermilab, ended. As we settled into 17 Sauk Circle in the Village, we wondered about the festive tents and the nineteenth-century look of the site, so different from Caltech and Los Alamos. It was the summer when the Wojcicki subpanel recommended the dissolution of Brookhaven's ISABELLE project and the start of the Superconducting Super Collider (SSC; see chapters 8, 10, and 13), and Fermilab was full of activity. An open house for the public the next month offered us more of an introduction. As Lederman rushed off to host one of the activities, he stopped on the steps of the High-Rise to welcome us into our new world.

About a month later, the Accommodations Office called and asked if I could babysit the child of a researcher at the lab, who turned out to be

Lillian. She was continuing her part-time work as Fermilab's archivist and historian. We renewed our acquaintance and she explained that she needed someone to care for her son while she worked in the History Room. When she returned at the end of the day, we talked about her work. I asked if she needed any help while we were at Fermilab—I could help for the year. To my surprise, she called the next day to say she had pitched the idea to Lederman and he had agreed to hire me as her assistant, the Fermilab archivist! Later I learned that putting spouses to work at the laboratory was not so unusual. It was a policy begun by Robert R. Wilson, following in the tracks of J. Robert Oppenheimer at Los Alamos.

I began work in November 1983, while Karen went off on the school bus to West Chicago and our two toddlers, Jeffrey and Christine, attended the Fermilab Children's Center, one of the laboratory's important family resources. My biggest source of information was May West, who answered all my questions and introduced me to people on the third floor of Wilson Hall, where both the library and the History Room were located. As fate would have it, Rocky's Astrophysics Group was also on the third floor. The laboratory was a welcoming environment, and I learned a lot from everyone I met. I couldn't believe our good fortune.

In those days, before Lillian started teaching at the University of Illinois, she came to Fermilab fairly often. The collections in the archives were growing rapidly, and library patrons and visitors often dropped by the History Room. It was important that a regular Fermilab staff member be available. We set out to catalog the History Room and the Catacombs, for which Wilson had set aside space in the basement of the High-Rise to house the laboratory archives. Within the year the catalog was finished and the History Room was arranged for historical research. Back then I read everything in order to understand and organize the collections for best use by researchers. Often boxes of files were anonymously dropped off at the door of the History Room, and I felt like a detective tracking down what the materials were, where they best belonged, and often from whom they had come.

By the end of our trial year at Fermilab, Rocky had hired postdoctoral fellows and organized visitors' programs as well as the lab's first Inner Space/Outer Space conference. My work in the archives had grown to ten hours a week. We decided to stay at Fermilab. More than twenty years later we were still there, but our children had grown up and moved away, thinking of Fermilab as their childhood home.

I met Catherine Westfall in May 1985, when Lillian, along with Laurie Brown and Max Dresden, hosted the Second International Symposium on the History of Particle Physics at Fermilab. Lillian now had Carol, her four-month-old daughter, along with her. I was working on a chronology of the history of the SSC; Catherine, based at Michigan State, was writing her graduate dissertation about Fermilab, which Lillian was helping to direct. The three of us have helped each other in our research and writing ever since.

During that 1985 conference I also met Joan Warnow from the AIP, who took time to explain the program she had set up back in 1975, appealing to DOE lab directors to preserve the history of their labs. It was Joan's proactive intervention that led to Wilson's setting up the early History Committee and hiring Lillian. Joan wanted me to learn from the best, so she encouraged Lederman to send me to Berkeley to visit Vicki Davis. Vicki, Joan, and Lillian opened my eyes to the world of the archives.

In 1992, May and I assisted Lillian with organization from Fermilab to help plan the Third International Symposium on the History of Particle Physics. We worked along with Michael Riordan, Laurie Brown, Max Dresden, Nina Stolar, Rene Donaldson, and others to hold the symposium for the first time at SLAC. Again the giants of particle physics came together with leading historians of science to recount the history of their scientific contributions.

Through all these projects, May West was our constant supporter, never too busy to find facts and resource material as only a librarian could. After May retired in 1995, she was a volunteer in the History and Archives House, the new on-site storage space developed in the Village that year. While Rocky and I worked at CERN for his sabbatical in 2001–2002, May anchored the History and Archives Project in my absence. She continued to work with me on the collections and to assist patrons of the archives until her untimely death in April 2003. She is heartachingly missed.

Sue Grommes has also given much support to the History and Archives Project, and me personally, during her career at Fermilab. Whether transcribing oral histories, editing conference proceedings, or assisting me with word processing, databases, and her many technical and practical skills, Sue has been a great relief.

My greatest thanks go to my husband, Rocky Kolb, the Master of the Universe.

Working with Fermilab's first four directors, Wilson, Lederman, John Peoples, and Michael S. Witherell, has been a privilege. Preserving their historical collections is an awesome and constantly inspiring responsibility. I believe there is gold in these boxes—treasure to be discovered and revealed.

Catherine Westfall (Lyman Briggs College, Michigan State University)

In 1983, as a Michigan State University graduate student, I was given an opportunity to write a history paper. I chose the most interesting subject I could think of, the founding of Fermilab. It was the mid-1980s and I did not yet own a personal computer, so I wrote the paper at Michigan State's National Superconducting Cyclotron Laboratory, where my nuclear physicist husband, Gary Westfall, still works. One of his colleagues brought me a *FermiNews* article on Fermilab's early history written by Lillian Hoddeson, the Fermilab historian. I was intrigued by the idea that someone could actually do such work for a living. Gary offered to fly me to meet Lillian, and also to see the laboratory that I had only heard and read about.

Lillian and I met in Fermilab's History Room in the summer of 1984. She was pregnant and had a toddler named Michael by the hand. They gave me a tour of the laboratory. I was impressed by the architecture of the High-Rise, by the art gallery on its second floor, and by the buzz of excitement in the cafeteria atrium. We ran into famous physicists left and right, first Leon Lederman, then James "Bj" Bjorken. Lillian asked if I'd like to work with her to expand my paper into a dissertation. I appreciated the opportunity to work on a fascinating topic with someone I was already comfortable with at so interesting a place. I accepted immediately and began commuting to the laboratory. The next year I agreed to join a team that Lillian had organized to write a history of the development of the first atomic bombs at Los Alamos. In the spring of 1988 I finished my dissertation about the founding of Fermilab. That summer I also completed my part of the Los Alamos history.

In 1989 Lillian and I received an NSF grant to write a history of Fermilab. The project was possible because by then Lillian and Adrienne had Fermilab's archives in place. But the research and writing would confront numerous interruptions, including the birth of my son, Forrest, in 1991 and my work at three other laboratories (Thomas Jefferson National Accelerator Facility, Lawrence Berkeley National Laboratory, and Argonne National Laboratory, where I worked as the laboratory historian).

Adrienne faithfully assisted us, not only with continuing archival duties, but with research and writing. Over the years the three of us have enjoyed an unusually warm and productive collaboration. We also benefited from the help of a number of Fermilab scientists, who took an interest in our work (and sometimes argued quite vigorously with our interpretations). I got to know Fermilab's first three directors, Robert R.

Wilson (when he came to visit), Lederman, and John Peoples. I remember Wilson sitting in the atrium, patiently and humbly outlining what he considered to be his many mistakes as director. I remember Lederman dropping by out of the blue to thank us for helping to preserve Fermilab history. I remember Peoples assuring me that the top quark "was in the bag" but insisting that he wasn't going to make an announcement until they had absolute proof in hand. It would be impossible to list everybody who was helpful (that would be a list as long as the author list in a high-energy physics paper), but I have to mention the late Frank Cole, who spent hundreds of hours explaining synchrotrons to me. Working on this book was often challenging, sometimes stressful, occasionally painful, but mostly it was pure joy.

Final Thanks

We are especially indebted to Angela Gonzales, Fermilab's resident artist from 1967 to 1997, who helped us express the spirit of Fermilab in this book with her unique artwork.

Many others besides those already mentioned helped us write this book. We are especially grateful to Mark Bodnarczuk, president of the Breckinridge Institute, for contributing his work about experiment strings, on which chapter 11 and part of chapter 5 are based. Most of this work can be found in his articles about the sociological consequences of experimentation in high-energy physics.[4] We are also very grateful for having had the use of Kyoung Paik's 1993 unpublished manuscript, "The Origins of CDF and the Fermilab Collider Project," written while Paik was a graduate student in the Department of History at the University of Illinois; some of Kyoung's work contributed to our present chapter 12.

Besides the archival and library assistance from May West, who offered us countless other kinds of nurturing, we have had a great deal of help from many physicists, historians, engineers, secretaries, administrators, librarians, and archivists, who offered their time, service, recollections, knowledge, and advice. Our thanks go to them and to the many friends, colleagues, and reviewers (most of them anonymous) who read countless early chapter drafts of this book or who spent long hours with us in interviews, or in other ways added to this book or corrected our many mistakes. It is impossible to thank everyone who helped us in these ways but we are nevertheless very grateful. For their extensive help we particularly wish to thank Nadezhda Shemyakina Anikeev, Jeffrey Appel, Cindy Arnold, Gordon Baym, Bruce Brown, Diana Canzone,

Richard Carrigan, Bruce Chrisman, the late Francis Cole, Jackie Coleman, James Cronin, Victoria Davis, Donald and Helen Edwards, William B. Fowler, Henry Frisch, Peter Garrett, Michelle Gleason, Edwin Goldwasser, Paul Grannis, Sue Grommes, J. L. Heilbron, J. D. Jackson, Drasko Jovanovic, Rocky Kolb, Barb Kristin, Donald Lamb, Leon M. Lederman, Jean Plese Lemke, Phillip V. Livdahl, Richard Lundy, Ernest Malamud, Fred Mills, the late Andrew Mravca, Luann O'Boyle, J. Richie Orr, Marilyn Paul, Jud Parker, John Peoples, Norman Ramsey, Lincoln Read, Jean Reising, Michael Riordan, David Ritchie, James Sanford, Cynthia Sazama, Robert Seidel, Frederick Seitz, Marilyn Smith, Lee Teng, Alvin V. Tollestrup, the late Father Timothy Toohig, Fred Ullrich, the late Robert R. Wilson, Ryuji Yamada, Taiji Yamanouchi, John Yoh, and Donald Young. We are most indebted to one anonymous reviewer for the University of Chicago Press for his or her most helpful critical comments, and to the editorial and production staff of the press, especially Jennifer Howard, Catherine Rice, and Erik Carlson, for their patience and encouragement in the years leading to this book's publication.

For providing a home for the research and writing of this book, and for many, many other kinds of support, we thank both the Universities Research Association and Fermilab, especially Fermilab's directorate and the Library and Visual Media Services in the Business Services Section. Lillian thanks the Department of History at the University of Illinois for granting her several leaves from teaching that allowed her to work on writing this book. Adrienne thanks Tullio Basaglia, Anita Hollier, and Corrado Pettenati from the CERN library and archives for their generous hospitality and assistance in 2001–2002. For a major grant to Hoddeson and Westfall to work on this book during 1989–1993 we thank the Program in History and Philosophy of Science of the National Science Foundation in the period when its program director was Ronald Overman, for its generous support (NSF grant no. DIR—90 15473).[5] For funds to support a graduate student assistant to work on this project during 1990–1991, we thank the University of Illinois Campus Research Board. For major support during 2004 that allowed us to complete the first full draft and bring our years of research and writing about the early years of Fermilab to completion, we thank the Richard Lounsbery Foundation.

Appendix: Fermilab Approved Experiments, 1970–1992

	Experiment	Spokesperson	Completion Date
Meson Area:			
4	Neutron cross section	Michael J. Longo	1974
7	Elastic scattering	Donald I. Meyer	1975
8	Neutral hyperon	Lee G. Pondrom	1976
12	Neutron backward scattering	Neville W. Reay	1974
22	Multigamma	George B. Collins	1974
27A	Neutron dissociation	Jerome L. Rosen	1974
51A	Missing mass	Eberhard Von Goeler	1974
61	Polarized scattering	Owen Chamberlain	1977
69A	Elastic scattering	Joseph Lach	1976
72	Quark	Lawrence B. Leipuner	1973
75	Quark	Taiji Yamanouchi	1973
81A	Nuclear chemistry	Sheldon Kaufman	1978
82	K^0 regeneration	Valentine L. Telegdi	1975
86A	Pion dissociation	Henry J. Lubatti	1976
90	Emulsion/protons @ 200 GeV	Wladyslaw Wolter	1972
96	Elastic scattering	David Ritson	1975
99	Associated production	Robert E. Diebold	1978
103	Emulsion/protons @ 200 GeV	David T. King	1972
104	Total cross section	Thaddeus F. Kycia	1977
105	Emulsion/protons @ 200 GeV	Prince K. Malhotra	1972
108	Beam dump	Miguel Awschalom	1975
110A	Multiparticle	Alexander R. Dzierba	1978
111	Pion charge exchange	Alvin V. Tollestrup	1974
114	Emulsion/protons @ 200 GeV	Piyare L. Jain	1972
116	Emulsion/protons @ 200 GeV	Jacques D. Hebert	1972
117A	Emulsion/protons @ 200 GeV	Osamu Kusumoto	1972
118A	Inclusive scattering	George W. Brandenburg	1977
147	Superheavy elements	Monique DeBeauvais	1975
156	Emulsion/protons @ 200 GeV	Kiyoshi Niu	1972
171	Emulsion/protons @ 200 GeV	Jere J. Lord	1972
178	Multiplicities	Wit Busza	1975
183	Emulsion/protons @ 200 GeV	M. I. Tretjakova	1972
189	Emulsion/protons @ 200 GeV	David Ritson	1972
216	Form factor	Donald H. Stork	1975
226	K^0 charge radius	Valentine L. Telegdi	1977
229	Detector development	Luke C. L. Yuan	1974
230	Multigamma	Michael J. Longo	1974
236A	Hadron jets	Paul M. Mockett	1977
248	Neutron elastic scattering	Michael J. Longo	1976
260	Hadron jets	Donald W. McLeod	1976
261	Detector development	Ching Lin Wang	1974
268	Inclusive photon	Joel Mellema	1976
272	Hadron dissociation	Thomas Ferbel	1979
290	Backward scattering	Winslow F. Baker	1978
305	Neutron dissociation	Bruno Gobbi	1975
324	Inclusive scattering	Howard L. Weisberg	1977
330	Particle search	H. Richard Gustafson	1975
335	Muon search	Orrin D. Fackler	1975
337	Di-muon	David P. Eartly	1975
350	Inclusive neutral meson	Robert w. Kenney	1977
357	Particle search	Donald I. Meyer	1976

	Experiment	Spokesperson	Completion Date
361	Lambda beta-decay	Lee G. Pondrom	1979
365	Particle search	David A. Garelick	1975
366	Particle search	Maris A. Abolins	1976
371	Superheavy elements	Mira Juric	1975
383	Inclusive K_S	Hans G. E. Kobrak	1978
395	Hadron jets	Walter Selove	1977
396	Hadron dissociation	Konstantin Goulianos	1977
397	Particle search	Jerome L. Rosen	1976
404	Inclusive neutron	H. Richard Gustafson	1977
415	Particle production	Lee G. Pondrom	1976
416	Particle search	Henry J. Lubatti	1975
425	K^0 regeneration	Valentine L. Telegdi	1976
426	Fragmentation particles	Katsura Fukui	1976
427	Detector development	Luke C. L. Yuan	1978
438	Neutron-nucleus inelastic	Lawrence W. Jones	1977
439	Multi-muon	David A. Garelick	1978
440	Lambda magnetic moment	Gerry M. Bruce	1977
441	Lambda polarization	Lee G. Pondrom	1977
451	Inclusive scattering	Donald S. Barton	1978
456	Form factor	Donald H. Stork	1977
468	Particle search	Phillip H. Steinberg	1977
469	Particle search	David Cutts	1978
472	Particle search	Kenneth C. Stanfield	1976
486	K^0 cross section	Bruce D. Winstein	1977
490	Particle search	Jack Sandweiss	1980
495	Ξ^0 production	Kenneth J. Heller	1978
505	Proton polarization	Samuel Peter Yamin	1978
507	High energy channeling	Edouard N. Tsyganov	1977
508	Emulsion/protons @ 500 GeV	Wladyslaw Wolter	1985
515	Particle search	Jerome L. Rosen	1982
524	Emulsion/protons $>$ 500 GeV	Richard J. Wilkes	1985
533	Pi-mu atoms	Gordon B. Thomson	1979
540	Particle search	Michael J. Longo	1978
555	Neutral hyperon	Thomas J. Devlin	1982
557	Hadron jets	Ernest I. Malamud	1984
577	Elastic scattering	Roy Rubinstein	1981
580	Particle search	Daniel R. Green	1981
584	Particle search	Bruce D. Winstein	1980
585	Kaon charge exchange	William R. Francis	1981
605	High mass pairs	John P. Rutherfoord	1985
609	Hadron jets	Walter Selove	1984
613	Beam dump	Brown P. Roe	1982
617	CP violation	Bruce D. Winstein	1982
620	Charged hyperon mag moment	Lee G. Pondrom	1980
622	Quark	H. Richard Gustafson	1980
623	Particle search	Daniel R. Green	1982
629	Direct photon production	Charles A. Nelson, Jr.	1981
660	Channeling	Walter M. Gibson	1982
663	Lambda polarization	Hans G. E. Kobrak	1981
668	Emulsion/π^- @ 800	Wladyslaw Wolter	1985
672	Hadron jets	Andrzej Zieminski	1992
704	Polarized beam	Akihiko Yokosawa	1990

	Experiment	Spokesperson	Completion Date
706	Direct photon production	Paul F. Slattery	1992
729	Emulsion/protons @ 1 TeV	Atul Gurtu	1985
731	CP violation	Bruce D. Winstein	1988
743	Charm production	Stephen Reucroft	1985
751	Emulsion exposure @ 1 TeV	Piyare L. Jain	1985
753	Channeling studies	James S. Forster	1985
755	Beauty & charm study	Richard D. Majka and Anna Jean Slaughter	1988
758	Emulsion exposure	Mitsuko Kazuno and Hiroshi Shibuya	1985
759	Emulsion exposure	Yoshihiro Tsuzuki	1985
762	Emulsion/protons @ 800 GeV	Shoji Dake	1985
763	Emulsion/protons @ 800 GeV	Takeshi Ogata	1985
764	Emulsion exposure	Hirotada Nanjo	1985
765	Emulsion/protons @ 800 GeV	K. Imaeda	1985
772	Dimuons	Joel M. Moss	1988
773	η_{00} $\eta_{\pm}$ phase difference	George D. Gollin	1991
789	Bottom-quark mesons & baryons	Daniel M. Kaplan and Jen-Chieh Peng	1992
792	Nuclear fragments	Kjell Aleklett and Lembit Sihver	1988
795	Warm liquid calorimetry test	Morris Pripstein	1991
Neutrino Area:			
1A	Neutrino	David B. Cline	1975
2B	30-inch hybrid	Gerald A. Smith	1974
3	Monopole	Philippe Eberhard	1974
14A	Proton-proton inelastic	Paolo Franzini	1973
21A	Neutrino	Barry C. Barish	1975
26	Muon	Louis N. Hand	1974
28A	15-foot neutrino/H_2 & Ne	William F. Fry	1975
31A	15-foot antineutrino/H_2	Malcolm Derrick	1977
34	Detector development	Richard W. Huggett	1974
37A	30-inch *pp* @ 300 GeV	Ernest I. Malamud	1973
45A	15-foot neutrino/H_2	Frank A. Nezrick	1976
53A	15-foot neutrino/H_2 & Ne	Charles Baltay	1981
76	Monopole	Richard A. Carrigan	1974
98	Muon	Herbert L. Anderson	1975
115	Long-lived particles	M. Lynn Stevenson	1974
121A	30-inch π^+ & *pp* @ 100 GeV	Richard L. Lander	1974
125	30-inch π^-p @ 100 GeV	Douglas R O. Morrison	1973
137	30-inch π^-p @ 200 GeV	Fred Russ Huson	1973
138	30-inch *pp* @ 400 GeV	Jack C. Vander Velde	1975
141A	30-inch *pp* @ 200 GeV	Thomas H. Fields	1972
142	Superheavy elements	Raymond W. Stoughton	1975
143A	30-inch π^-p @ 300 GeV	George R Kalbfleisch	1974
154	30-inch hybrid	Irwin A. Pless	1974
155	15-foot EMI test	Vincent Z. Peterson	1974
161	30-inch *pp* & Ne @ 300 GeV	James Mapp	1974
163A	30-inch π^-p & Ne @ 200 GeV	William D. Walker	1974
172	15-foot antineutrino/H_2 & Ne #172	Henry J. Lubatti	1976
181	Emulsion/protons @ 300 GeV	Arthur S. Cary	1973

	Experiment	Spokesperson	Completion Date
194	30-inch *pd* @ 100 GeV	C. Thornton Murphy	1976
195	Emulsion/protons @ 300 GeV	Yu K. Lim	1975
196	30-inch *pd* @ 400 GeV	Roderich J. Engelmann	1975
199	Massive particle search	Sherman Frankel	1973
202	Tachyon monopole	David F. Bartlett	1976
203A	Muon	Leroy T. Kerth	1978
205A	Emulsion/muons @ 150 GeV	Osamu Kusumoto	1973
209	30-inch *pd* @ 300 GeV	Fu Tak Dao	1975
211	Beam dump	Klaus Goebel	1973
217	30-inch π^+ & *pp* @ 200 GeV	Richard L. Lander	1974
218	30-inch $\pi^- d$ @ 200 GeV	Philip M. Yager	1974
228	30-inch π^+ & *pp* @ 60 GeV	Thomas Ferbel	1974
232	Emulsion/protons @ 300 GeV	David T. King	1973
233	Emulsion/protons @ 300 GeV	Jacques D. Hebert	1973
234	15-foot engineering run	Fred Russ Huson	1974
237	Emulsion/protons @ 300 GeV	Jere J. Lord	1975
238	Emulsion/protons @ 400 GeV	Jere J. Lord	1975
239	Long-lived particles	William Frati	1974
242	Emulsion/protons @ 300 GeV	Kiyoshi Niu	1973
243	Emulsion/protons @ 400 GeV	Kiyoshi Niu	1975
244	Emulsion/protons @ 300 GeV	Piyare L. Jain	1973
245	Emulsion/protons @ 400 GeV	Piyare L. Jain	1975
247	Particle search	Eric H. S. Burhop	1976
249	Emulsion/protons @ 400 GeV	Wladyslaw Wolter	1975
250	Emulsion/protons @ 300 GeV	Osamu Kusumoto	1973
251	Emulsion/protons @ 400 GeV	Osamu Kusumoto	1975
252	30-inch *pp* @ 100 GeV	Thomas Ferbel	1972
253	Neutrino	Luke W. Mo	1979
254	Neutrino	George R. Kalbfleisch	1975
255	Emulsion/muons @ 150 GeV	Piyare L. Jain	1973
262	Neutrino	Barry C. Barish	1974
264	Emulsion/π^- @ 200 GeV	Poh Shien Young	1974
265	Emulsion/protons @ 400 GeV	Poh Shien Young	1975
271	Emulsion/protons @ 200 GeV	Kurt Gottfried	1975
275	Plastic detectors	Wolfgang Enge	1973
276	Quark	Andreas Van Ginneken	1975
279	Emulsion/protons @ 400 GeV	David T. King	1975
280	30-inch *pd* @ 200 GeV	Thomas H. Fields	1975
281	30-inch hybrid	Gerald A. Smith	1975
285	Superheavy elements	Leon M. Lederman	1976
292	Emulsion/protons @ 400 GeV	Kurt Gottfried	1975
295	30-inch π^+ & *pd* @ 200 GeV	Gideon Yekutieli	1975
297	Quark	Lawrence B. Leipuner	1974
299	30-inch hybrid	Irwin A. Pless	1976
310	Neutrino	David B. Cline	1978
311	30-inch $\overline{p}p$ @ 100 GeV	William W. Neale	1975
319	Muon	K. Wendell Chen	1976
320	Neutrino	Frank Sciulli	1974
327	Detector development	Wade W. M. Allison	1975
328	Emulsion/π^- @ 200 GeV	M. I. Tretjakova	1974
329	Emulsion/protons @ 300 GeV	M. I. Tretjakova	1975
331	Di-muon	James E. Pilcher	1976

	Experiment	Spokesperson	Completion Date
336	Emulsion/protons @ 400 GeV	Takeshi Ogata	1975
338	30-inch $\pi^- d$ @ 360 GeV	Keihachiro Moriyasu	1976
339	Emulsion/π^- @ 200 GeV	Wladyslaw Wolter	1975
341	15-foot *pp* @ 400 GeV	Winston Ko	1975
343	15-foot *pp* @ 300 GeV	Roderich J. Engelmann	1976
344	30-inch $\overline{p}p$ @ 50 GeV	Laszlo J. Gutay	1976
345	30-inch $\overline{p}d$ @ 100 GeV	Gosta Ekspong	1976
346	Emulsion/protons @ 400 GeV	Gosta Ekspong	1975
356	Neutrino	Frank Sciulli	1979
362	Emulsion/π^- @ 200 GeV	Piyare L. Jain	1975
369	Particle search	Thomas B. W. Kirk	1977
370	Neutrino	David B. Cline	1975
373	Emulsion/muons @ 200 GeV	Piyare L. Jain	1976
374	Emulsion/protons @ 300 GeV	D. H. Davis	1975
379	Particle search	Stanley G. Wojcicki	1977
380	15-foot neutrino/H_2 & Ne	Charles Baltay	1979
382	Particle search	Louis N. Hand	1975
385	Emulsion/protons @ 400 GeV	Yog Prakash	1975
386	Emulsion/new particles	Jere J. Lord	1976
387	Emulsion/π^- @ 200 GeV	Richard J. Wilkes	1975
388	15-foot antineutrino/H_2 & Ne	Vincent Z. Peterson	1979
391	Muon	Leroy T. Kerth	1978
398	Muon	Richard Wilson	1976
419	Emulsion/protons @ 300 GeV	Giorgio Giacomelli	1975
421	Emulsion/protons @ 300 GeV	Venedict P. Dzhelepov	1975
423	Emulsion/protons @ 400 GeV	Hisahiko Sugimoto	1975
424	Emulsion/muons @ 200 GeV	Tomonori Wada	1976
428	Emulsion/protons @ 400 GeV	Jacques D. Hebert	1975
434	Emulsion /protons @ 400 GeV	Shoji Dake	1975
444	Di-muon	A. J. Stewart Smith	1978
448	Muon	William A. Loomis	1978
461	Emulsion/protons @ 400 GeV	Jere J. Lord	1975
462	Emulsion/protons @ 400 GeV	Giorgio Giacomelli	1975
463	Emulsion/protons @ 400 GeV	M. I. Tretjakova	1975
467	Test muon irradiation	Melvin Freedman	1976
481	Emulsion/π^- @ 300 GeV	Yoshiyuki Takahashi	1978
482	Neutrino	Barry C. Barish	1978
499	Emulsion/protons @ 400 GeV	Junsuke Iwai	1978
501	Test muon irradiation	Kenneth Lande	1976
502	Monopole	David F. Bartlett	1980
503	Emulsion/π^- @ 300 GeV	Takeshi Ogata	1978
506	Emulsion/π^- @ 300 GeV	Shoji Dake	1978
509	Emulsion/muons @ 200 GeV	T. Shirai	1976
525	Emulsion/π^- @ 300 GeV	Richard J. Wilkes	1978
531	Neutrino	Neville W. Reay	1981
536	Emulsion/neutrino	Kiyoshi Niu	1977
545	15-foot neutrino/D_2 & high Z	George A. Snow	1979
546	15-foot neutrino/H_2 & Ne	Fred Russ Huron	1978
547	Emulsion/protons @ 400 GeV	C. J. Jacquot	1978
553	Neutrino	Paul F. Shepard	1980
564	15-foot & emulsion/neutrino	Louis Voyvodic	1981
565	30-inch hybrid	Irwin A. Pless	1982

	Experiment	Spokesperson	Completion Date
568	Emulsion/π^- @ 300 GeV	Jacques D. Hebert	1978
570	30-inch hybrid	Irwin A. Pless	1982
573	Emulsion/π^- @ 300 GeV	Noriyuki Ushida	1978
574	Emulsion/π^- @ 300 GeV	Wladyslaw Wolter	1978
575	Emulsion/protons @ 400 GeV	Jere J. Lord	1978
576	Emulsion/protons @ 500 GeV	Jacques D. Hebert	1985
594	Neutrino	James K. Walker	1982
595	Particle search	Arie Bodek	1980
596	Particle search	Leon M. Lederman	1978
597	30-inch hybrid	J. James Whitmore	1982
610	Particle search	Thomas B. W. Kirk	1980
616	Neutrino	Frank Sciulli	1980
631	Nuclear calibration cross section	Samuel I. Baker	1981
632	15-foot neutrino/H_2 & Ne	Douglas R. O. Morrison and Michael W. Peters	1988
653	Particle search	Neville W. Reay	1988
665	Tevatron muon	Heidi M. Schellman	1992
673	Chi meson	John W. Cooper	1982
690	Particle search	Bruce Knapp	1992
701	Neutrino oscillation	Michael H. Shaevitz	1982
711	Constituent scattering	David A. Levinthal	1988
733	Neutrino interactions	Raymond I. Brock	1988
744	Charged interactions	Frank S. Merritt	1985
745	Muon neutrino	Toshio Kitagaki	1988
750	Multiparticle production	Ram K. Shivpuri	1985
770	Quad triplet neutrino	Wesley H. Smith	1988
782	Muons in 1-mm bubble chamber	Toshio Kitagaki	1990
790	Calorimeter for Zeus	Frank Sciulli	1990
802	Muons in emulsion	Lali Chatterjee and Dipak Ghosh	1991
Proton Area:			
25A	Photon total cross section	David O. Caldwell	1976
48	Muon search	Robert K. Adair	1975
70	Lepton	Leon M. Lederman	1974
87A	Photoproduction	Thomas O'Halloran	1978
95A	Photon search	Bradley B. Cox	1977
100A	Particle search	Pierre A. Piroue	1974
152B	Photoproduction	Clemens A. Heusch	1978
177A	Proton-proton elastic	Jay Orear	1977
187	Particle search	Leon M. Lederman	1973
258	Pion inclusive	Melvyn Jay Shochet	1979
284	Particle production	James K. Walker	1976
288	Di-lepton	Leon M. Lederman	1978
300	Particle search	Pierre A. Piroue	1976
325	Particle search	Pierre A. Piroue	1977
326	Di-muon	Melvyn Jay Shochet	1982
340	Emulsion/electrons @ high energy	Shoji Dake	1976
358	Di-muon	Wonyong Lee	1975
399	Emulsion/electrons @ $>$ 100 GeV	Robert L. Golden	1976
400	Particle search	James E. Wiss	1984
401	Photoproduction	Michael F. Gormley	1979
435	Muon search	Robert K. Adair	1976

	Experiment	Spokesperson	Completion Date
436	Di-muon	Robert K. Adair	1975
466	Nuclear fragments	Norbert T. Porile	1988
494	Di-hadron	Myron L. Good	1977
497	Charged hyperon	Joseph Lach	1981
498	Detector development	Charles R. Gruhn	1976
510	Emulsion/electrons @ high energy	Kiyoshi Niu	1976
516	Photoproduction	E. Thomas Nash	1981
537	Di-muon	Bradley B. Cox	1982
567	Particle search	Michael S. Witherell	1979
592	Nuclear scaling	Sherman Frankel	1978
608	Particle search	Charles N. Brown	1979
612	Photon dissociation	Konstantin Goulianos	1982
615	Forward search	Kirk T. McDonald	1984
619	Transition magnetic moment	Thomas J. Devlin	1982
621	CP violation	Gordon B. Thomson	1985
630	Charm particle	Jack Sandweiss	1982
650	Particle search	Robert C. Webb	1980
666	Emulsion exposure	Richard J. Wilkes	1981
667	Emulsion/π^- @ 500 GeV	Wladyslaw Wolter	1990
683	Photoproduction of jets	Marjorie D. Corcoran	1992
687	Photoproduction of charm & beauty	Joel N. Butler and John P. Cumalot	1992
691	Tagged photon	Michael S. Witherell	1985
705	Chi meson	Bradley B. Cox	1988
715	Sigma beta decay	Peter S. Cooper	1984
730	Emulsion/ϵ^- @ 250 GeV	Richard J. Wilkes	1984
747	Charged particles	Alan A. Hahn	1985
756	Magnetic moment	Kam-Biu Luk	1988
761	Hyperon radiative decay	Alexei A. Vorobyov	1990
769	Pion & kaon charm production	Jeffrey A. Appel	1988
771	Beauty production by protons	Bradley B. Cox	1992
774	Electron beam dump	Michael B. Crisler	1990
791	Hadroproduction heavy flavors	Jeffrey A. Appel and Milind Vasant Purohit	1992
797	Fine-grained electromagnetic calorimetry	H. Richard Gustafson and Rudolf P. Thun	1990
798	SSC detector test	Priscilla Cushman and Roger W. Rusack	1990
800	Magnetic moment	Kenneth A. Johns and Regina A. Rameika	1992
Internal targets:			
36A	Proton-proton scattering	Rodney L. Cool	1973
63A	Photon search	James K. Walker	1975
67A	Proton-proton missing mass	Felix Sannes	1973
120	Photon search	David B. Cline	1973
184	Particle search	Peter J. Wanderer	1974
186	Proton-deuteron scattering	Adrian Melissinos	1974
188	Proton-nucleon inclusive	Felix Sannes	1973
198A	Proton-nucleon scattering	Stephen L. Olsen	1977
221	Proton-proton inelastic	Paolo Franzini	1974
289	Proton-helium scattering	Ernest I. Malamud	1977
313	Proton-proton polarization	Homer A. Neal	1977

	Experiment	Spokesperson	Completion Date
317	Proton-nucleon inelastic	Rodney L. Cool	1975
321	Proton-proton inelastic	Juliet Lee-Franzini	1976
363	Particle search	Stephen L. Olsen	1975
381	Proton-nucleon scattering	Ernest I. Malamud	1977
418	Particle production	Felix Sannes	1975
442	Nuclear fragments	Frank Turkot	1977
522	Proton polarization	Harold O. Ogren	1978
552	Proton-nucleon scattering	Felix Sannes	1978
591	Particle search	Laszlo J. Gutay	1981
735	Particle search	Laszlo J. Gutay	1989
760	Charmonium states	Rosanna Cester	1992
Miscellaneous experiments:			
720	Free quark search	John P. Schiffer	1982
723	Gravitational detector	Adrian Melissinos	1985
766	MR tunnel neutrons	Joseph B. McCaslin	1985
776	Nuclear calibration cross sections	Samuel I. Baker	1988
777	MR tunnel neutrons	Joseph B. McCaslin	1987
778	Magnet aperture studies	Rodney E Gerig and Richard Talman	1991
784	Bottom at the collider	Nigel Lockyer	1992

	Experiment	Spokesperson	Completion Date	Location
Colliding-beam experiments:				
710	Total cross section	Jay Orear and Roy Rubinstein	1989	EZero
713	Highly ionizing particles	P. Buford Price	1989	DZero
740	DZero detector	Paul D. Grannis	1996	DZero
741	Collider detector	Melvyn Jay Shochet and Alvin V. Tollestrup	1989	BZero

Note: There were 386 approved experiments and 819 proposals during this period.
Source: Fermilab Research Program Workbooks, Program Planning Office.

Notes

INTRODUCTION

1. See e.g., J. Hughes 2002, pp. 105–121.
2. Much of the scholarship on "big science" is referenced and discussed in Westfall 2003; Galison and Hevly 1992; and Capshew and Rader 1992.
3. Galison 1987.
4. For a summary of these breakthroughs, see for example, Brown, Dresden, and Hoddeson 1989; and Hoddeson et al. 1997).
5. Norman Ramsey explained that in the United States, the change from BeV to GeV occurred over a period of years. "BeV had the serious disadvantage of being ambiguous since billion in the US is a thousand million and billion in England is a million million." Ramsey to Kolb and Hoddeson, February 20, 2004.
6. R. Wilson 1965.
7. Kaiser 2004.
8. R. Wilson 1968. By the early years of this century, cowboys, an important element in Buffalo Bill's Wild West shows, had assumed their place among the icons of frontier mythology. White, Limerick, and Grossman 1994; also R. White 1994, p. 10.
9. Goldwasser 1969; R. Wilson 1968.
10. Cronon 1987.

CHAPTER ONE

1. Turner 1994, pp. 58–59.
2. This is understandable given that the National Science Foundation is a direct outgrowth of the frontier rhetoric of Vannevar Bush. See, e.g., Kevles 1995.

3. Seidel 1983.
4. Turner 1956.
5. Turner 1956, p. 2. Turner's lack of precision in his definitions of the frontier provided his critics with "a happy hunting ground." Other definitions besides the hither edge of free land include the edge of settled territory, the line of settlement, the "West" itself, and a "form of society." See Billington 1966, p. 16.
6. Turner 1956, 14.
7. Cronon 1987; and R. White 1994, p. 2.
8. Cronon 1991b, 1987; Billington 1973, esp. chap. 18, pp. 444–471; 1966; R. White 1988, p. 672; Shoemaker 1993), p. A48; Margolis 1992. A dramatic testament to the robustness of the frontier image in popular culture is the continued popularity of Western movies, revised to fit present values; see Wilmington 1994.
9. See e.g., "Lewis and Clark . . . " 2002.
10. Billington 1973.
11. F. D. Roosevelt to V. Bush, November 17, 1944, reprinted in Bush 1960, pp. 3–4. See especially Alan T. Waterman's "Introduction," pp. vii–xxvi of Bush 1960.
12. Bush 1960.
13. There is now a large literature on American science policy. The crucial first source is Kevles 1978. See also Dickson 1984; and Greenberg 1967. For more on Vannevar Bush, see Reingold 1987; and for a different perspective, Hart 1998. Today's science statesmen often ask whether the scientific frontier of Vannevar Bush is in the process of closing. E.g., see Lederman 1991.
14. Westfall 2003.
15. Kennedy's speech of acceptance as presidential nominee to the Democratic National Convention on July 15, 1960, quoted from "Kennedy and Johnson Open the Campaign . . . " 1960, pp. 100–102, quote on p. 102.
16. Compton added, "The earth was not as large as he had estimated, and he arrived at the new world sooner than he had expected." Compton 1956, p. 144; Wattenberg 1988; Rhodes 1986, pp. 39, 4–442; Hewlett and Anderson 1962, p. 112.
17. Goldwasser 2000.
18. U.S. Congress, Joint Committee on Atomic Energy 1969, pp. 112–118, quote on p. 113.
19. The crucial issue of promoting big science is treated by Pinch 1986 and R. Smith 1989. For an excellent discussion of the trend for physics to become suburbanized, see Kaiser 2004.
20. Wilson brought Livingston from MIT, where he was director of the Cambridge Electron Accelerator, the first alternating-gradient synchrotron, to NAL in 1967 as a historic link between earlier accelerators and his new laboratory. Livingston wrote the classic textbook *Particle Accelerators* with John P. Blewett

(Livingston and Blewett 1962), and an early history of NAL in 1968 (Livingston 1968a). He was elected to the National Academy of Sciences in 1970 and retired to Santa Fe later in 1970. He died in 1986, but his famous 1954 graph, plotting the rise of energy in accelerators over time, endures.

21. Livingston 1968a, pp. 2–8.
22. Lawrence 1941; also see Heilbron and Seidel 1989, pp. 1–8. The frontiers of technology have been as compelling for Americans as frontiers of science. See e.g., Corn 1983; and McDougall 1985.
23. Findlay 1995, p. 34.
24. See Findlay 1995, pp. 37–38, 40. Battelle took over Hanford in 1965, and its mission shifted. See Westwick 2003, p. 271. Today the laboratory is called the Pacific Northwest National Laboratory.
25. Taylor 1956, p. 2. For a selected bibliography of frontier history, see Ward 1996; and especially Billington 1966; Cronon 1987, 1991a; R. White 1988, 1991a, 1991b; Slotkin 1992; Turner 1956; White, Limerick, and Grossman 1994.
26. The few exceptions include Coleman 1966; Dupree 1986; Findlay 1995; Kay 1993, p. 58–76; and M. Smith 1993. This point is discussed in more detail in Kolb and Hoddeson 1995.

CHAPTER TWO

1. Ramsey Panel 1963, p. 3.
2. Courant, Livingston, and Snyder 1952.
3. Seidel 1983.
4. In "fixed-target'" experiments a beam of particles hits a fixed, stationary target. In "colliding-beam" experiments a beam of particles collides with another beam of particles. For a given beam energy, colliding beams have a much larger energy reach. The energy available for a fixed-target experiment involving a beam of energy E and a target particle of mass m is the square root of the quantity $(2m^2c^4 + 2Emc^2)$. The energy available for two beams of energy E colliding with equal and opposite momenta is $2E$. Notice that the available energy for colliding beams is linear in beam energy, while for fixed-target experiments the available energy increases only as the square root of the beam energy. MURA Report 506, August 31, 1959.
5. Midwestern Universities Research Association, "Introduction," in "1959 MURA Summer Study on Design and Utilization of High-Energy Accelerator," MURA Report 465, pp. ii–iii; Tollestrup 1978; Hoddeson 1983, p. 14; and Tollestrup and Walker 1985. For more detail on the historical and physics origins of the 200 GeV accelerator see Westfall 1988.
6. Midwestern Universities Research Association, "Introduction," pp. ii–iii.
7. Teng had made the first suggestion in 1955 of a tandem-FFAG machine with a resonant beam transfer scheme. Lee Teng to L. Hoddeson, E-mail, December 15, 2003.

8. Wilson's earlier suggestion of a cascade had not yet been developed into a machine because it appeared to require more beam control than was believed feasible. Sands also credits Marcus Oliphant and T. A. Welton for conceiving cascade schemes. Sands 1959, p. 1; 1960, p. 4.
9. Sands 1959, p. 1. Also Sands 1960, p. 4; Midwestern Universities Research Association, "Introduction," p. iii. In 1959–1960, Wilson, Maury Tigner, and Ernest Malamud attempted to build such a system. If it had succeeded, it would have been the first ring-to-ring system. Unfortunately, this project, called the "Injectotron," failed. E. Malamud, private communication to Hoddeson and Kolb, 2005.
10. Sands 1959, pp. 1, 2.
11. Sands 1960, pp. 1, 16; Hoddeson 1983, p. 15.
12. "The Future of High Energy Physics at Caltech," May 17, 1960, unpublished document, collection of Catherine Westfall. We would like to thank Robert Walker for providing this document.
13. Lawrence died in 1958 and was not involved. By the late 1950s he had become more concerned with the practical applications of nuclear physics than with building accelerators more powerful than his 6.2 GeV Bevatron. Lofgren 1984b; Wenzel 1984; Judd 1984. Also see Edward J. Lofgren, "High Energy Accelerator Study," November 23, 1955, unpublished document, Lofgren Papers, Bancroft Library, University of California, Berkeley.
14. David L. Judd and Lloyd Smith, "Summary of Accelerator Studies since June 1960 at LRL," p. 1, McMillan Papers, Box 1, Bancroft Library, University of California, Berkeley. Also see David L. Judd, "The Development from 1952 to 1960 of Planning for a Larger Accelerator," September 29, 1960, unpublished manuscript, p. 7, Judd Papers, Bancroft Library, University of California, Berkeley. For more information on the development of the AGS, see Crease 1999a, pp. 219–228, 251–256; and for information on the PS see Hermann et al. 1990.
15. In World War II, McMillan contributed to both the radar and atomic bomb projects. In 1951 he shared the Nobel Prize in chemistry with Glenn Seaborg for their prewar discovery of the first two transuranic elements. Heilbron, Seidel, and Wheaton 1981, pp. 40, 42, 53–60.
16. Quotes, respectively, from Matthew Sands letter to Lloyd Smith, August 29, 1960; and Lloyd Smith, "Notes on a Visit to Cal Tech (Oct. 10–11, 1960)," unpublished document, Lofgren Papers.
17. Matthew Sands letter to John Blewett, March 27, 1961. Also Blewett letter to Sands, February 15, 1961; form letter from Robert Bacher and Matthew Sands, January 12, 1961, collection of Catherine Westfall; Sands to Bacher, January 30, 1961; and "The Future of High Energy Physics at Caltech," collection of Catherine Westfall.
18. Vasily S. Emelyanov to John McCone, July 18, 1960, in "Relevant Extracts from Exchange of Correspondence between Mr. McCone and Professor Emelyanov, High Energy Physics." Also see John McCone and Vasily S.

Emelyanov, "Memorandum: Cooperation between the United States of America and the Union of Soviet Socialist Republics in the Field of the Utilization of Atomic Energy for Peaceful Purposes," November 24, 1959, pp. 3, 30; John McCone letter to Vasily S. Emelyanov, August 31, 1960; and Buck 1983, p. 4.

19. R. Wilson 1960, p. 4.
20. Quotes, respectively, from notes taken by participants R. F. Mozley and W. D. Walker, "Conference on Super Energy Accelerators," unpublished document, p. 4; and R. Wilson 1960, pp. 3, 4–5, 6.
21. K. R. Symon, "Summary of Remarks Regarding Accelerators," enclosed with Symon to R. Wilson, September 7, 1960. For more information on the development of the bubble chamber see Galison 1989, pp. 213–251.
22. R. Wilson 1960, p. 6.
23. "Piore Panel Report—1960" 1965, pp. 124–127, 131, 135. See appendixes 4, 5, and 6 of this document for 1954, 1956, and 1958 National Science Foundation panels. The Piore Panel was first convened in 1958 to provide a special advisory board on high-energy physics for the executive branch. Advice on high-energy physics was also sought by the National Science Foundation, which convened panels in 1954, 1956, and 1958.
24. "Piore Panel Report—1960" 1965, pp. 125, 127, 128, 130; and Ramsey Panel 1963, p. 95.
25. A 70 GeV Soviet synchrotron was under construction at Serpukhov. "Minutes of September 16, 1960," unpublished document, p. 1; and "Desirability and Feasibility of New Accelerators of Large and Novel Type," report to U.S. and Soviet atomic energy authorities, n.d. Also see Lloyd Smith letter to Sands, August 24, 1960; "Resume of September 16, 1960 Meeting of US and USSR Scientists on Desirability and Feasibility of a New Accelerator of Large and Novel Type."
26. California Institute of Technology, "A Proposal to the Atomic Energy Commission for the Support of the Accelerator Design-Study Program of the Western Accelerator Group," April 1961, unpublished document, p. 2, 5, 12. Sands and Bacher had worked together at Los Alamos in the early years of the war, when Bacher led the Experimental Physics Division, which also included Walker and Wilson. Edward Lofgren, who would come to head the Berkeley design effort, also worked at Los Alamos. There were other connections among the physicists of this group. Before the war, Lofgren and Wilson had worked as Lawrence's graduate students along with McMillan and Seaborg. Hoddeson et al. 1993.
27. R. R. Wilson to M. Sands, April 25, 1961.
28. Berkeley had previously requested $200,000 for such studies for that year. The laboratory asked that the original $200,000 be increased to $500,000, and $2.5 million be granted for the following year, a total budget of $3 million dollars. Lawrence Radiation Laboratory, "Extract from LRL FY1963 Budget Submission, 4-21-61," unpublished document, Lofgren Papers.

29. California Institute of Technology, "A Proposal to the Atomic Energy Commission," pp. 5, 24; Lawrence Radiation Laboratory, "Extract from LRL FY1963 Budget Submission"; Tollestrup and Walker 1985.
30. California Institute of Technology, "A Proposal to the Atomic Energy Commission," pp. 5, 24; Lawrence Radiation Laboratory, "Extract from LRL FY1963 Budget Submission."
31. Quotes from Edward J. Lofgren memorandum to Edwin M. McMillan, April 6, 1961, Lofgren Papers. Also see Lofgren 1984a, 1987.
32. Edward J. Lofgren memorandum to Edwin M. McMillan, April 6, 1961.
33. California Institute of Technology, "A Proposal to the Atomic Energy Commission," p. 7.
34. Paul McDaniel letter to J. Robert Oppenheimer, May 26, 1961.
35. Robert F. Bacher and Edwin M. McMillan letter to Paul McDaniel, May 12, 1961, McMillan Papers, Box 5.
36. Quotes from AEC representative George Kolstad, untitled report, June 22–23, 1961, unpublished document, Lofgren Papers. Also see Paul McDaniel memorandum to Ellison Shute, July 3, 1961, McMillan Papers, Box 5.
37. Lloyd Smith letter to Matthew Sands, January 11, 1961; Sands to Blewett, March 27, 1961; and Lloyd Smith letter to Paul McDaniel and enclosures, May 10, 1961. "Accelerator Planning Group, Minutes of Feb. 9., 1961 Meeting," February 16, 1961, unpublished document, Lofgren Papers.
38. Yuan and Blewett 1961; Cole 1978; Schlesinger 1983, pp. 379–405; Peoples 2004.
39. Yuan and Blewett 1961.
40. Heilbron and Seidel 1989; McMillan 1984.
41. Edward J. Lofgren, "History of Accelerator Studies at Berkeley," February 22, 1964, p. 2, and "Minutes of Accelerator Steering Committee Meetings," November 10, 1961, Lofgren Papers; "Study Group on Super High Energy Accelerator, Committee on Experimental Matters," November 22, 1961, Green Papers, Brookhaven National Laboratory.
42. "Minutes of Accelerator Steering Committee Meetings," September 6, October 6, and November 10, 1961, Lofgren Papers.
43. Hayden Gordon, "Minutes of Meeting Held to Discuss the Organization of the Study of a Super-energy Accelerator," Lofgren Papers; and Brookhaven National Laboratory, "National Design Study for a Super-high-energy Accelerator," December 21, 1961.
44. Tollestrup and Walker 1985.
45. Edward J. Lofgren, "Notes on a Meeting to Discuss the Organization of a Study of a Super High Energy Accelerator," January 2, 1962, Lofgren Papers.
46. Lawrence Radiation Laboratory, "Proposal to the AEC for a National Design Study of an Ultrahigh Energy Accelerator," unpublished document, Lofgren Papers, pp. 5–6. Also see this source, pp. 9, 10, 13–15, 16.
47. Quotes from Brookhaven National Laboratory 1961; Atomic Energy Commission, "Appendix A," in "Initiation of Design Studies for a Proton-

Synchrotron of Several Hundred BeV," August 22, 1962, AEC 1096/1, p. 15, DOE Archives, Secretariat Collection, Box 1474, Washington, DC. The contrasting approaches to machine design were not new. Blewett points out that at Berkeley, in the period when the Bevatron and Cosmotron were being built, "machine design was guided mainly by experience and by model studies; not much detailed theoretical analysis had previously been found profitable." Blewett 1989, p. 10.

48. Haworth, a leader of the AGS design team, had left his post as Brookhaven director several months earlier to become an AEC commissioner. Quote from Glenn Seaborg, record of conversation, February 8, 1962, Seaborg Papers, Bancroft Library, University of California, Berkeley; see Westfall 1988, pp. 126–127. As Kennedy's science adviser, Wiesner was also chairman of the President's Science Advisory Committee and director of the Office of Science and Technology (OST) and of the Federal Council for Science and Technology (FCST). As part of the executive branch the commission worked closely with Wiesner, since all executive office budgets were approved by the president before being submitted for congressional approval. For information on the formation and function of the OST and FCST, see Katz 1978; Trenn 1983.
49. Glenn Seaborg, record of conversation, February 8, 1962, Seaborg Papers.
50. Letter from Paul McDaniel, April 19, 1962, in "Appendix B," by Atomic Energy Commission, in "Initiation of Design Studies for a Proton-Synchrotron of Several Hundred BeV," August 22, 1962, AEC 1096/1, pp. 17–18, DOE Archives, Secretariat Collection, Box 1474; McDaniel to McMillan, April 19, 1962, Lofgren Papers; and McDaniel to Maurice Goldhaber, April 19, 1962, McMillan Papers.
51. Edwin McMillan to Paul McDaniel, May 8, 1962, McMillan Papers, Box 1.
52. Maurice Goldhaber to Paul McDaniel, May 10, 1962, Green Papers.
53. Crease 1999a, pp. 263–267.
54. Atomic Energy Commission, "Appendix A," pp. 8, 9, DOE Archives, Secretariat Collection, Box 1474.
55. Atomic Energy Commission, "Appendix A," pp. 8, 9. Also see pp. 12–14.
56. Like the OMB, the BOB ruled on all executive-branch budgets, including that of the AEC. For more information on the evolution of the AEC budget, see Physics Survey Committee, National Research Council 1972, pp. 653–659. See also E. J. Lofgren, "Conference of McMillan, Lofgren, Laslett, Kolstad at AEC Germantown, Sept. 26, 1962," unpublished document, Lofgren Papers. Also see Tape 1986.
57. Quote from Lofgren, "Conference of McMillan, Lofgren, Laslett, Kolstad." Also Tape 1986.
58. Lofgren, "Conference of McMillan, Lofgren, Laslett, Kolstad"; and Edward J. Lofgren, "Conference with Haworth in Washington, September 25, 1962," unpublished documents, Lofgren Papers. For more information on the deliberations for the Bevatron and Cosmotron, see Seidel 1983.

59. As quoted in Lofgren, "Conference with Haworth in Washington, September 25, 1962."
60. Others on the Ramsey Panel were Philip H. Abelson, Owen Chamberlain, Murray Gell-Mann, Edwin L. Goldwasser, Tsung D. Lee, Wolfgang K. H. Panofsky, Edward M. Purcell, Frederick Seitz, and John H. Williams. The ex officio members were Randal M. Robertson from the NSF, who represented the Technical Committee on High Energy Physics of the Federal Council for Science and Technology, and David Z. Robinson, from the Office of Science and Technology. Ramsey Panel 1963, pp. 85, 104. Also see Ramsey 1978, p. 4097; Kevles 1978, p. 368.
61. Lawrence Radiation Laboratory, "Proposed Program for an Alternating Gradient Synchrotron in the 100 BeV Range," December 8, 1962, unpublished document, pp. I-2, IV-3, V-8, Lofgren Papers. Also see McMillan telex to Paul McDaniel, and Edwin McMillan telex to Paul McDaniel, October 2, 1962, both in Lofgren Papers.
62. Quotes from Lawrence Radiation Laboratory, "Proposed Program for an Alternating Gradient Synchrotron," p. III-3.
63. Ramsey Panel 1963, p. 106; Ramsey 1978. According to John Peoples, Wilson originally intended to build a half-scale copy of the 6 GeV CEA with the help of an outside vendor and the NSF approved the project, but he was persuaded to change course and use the money to instead build the 10 GeV machine.
64. Ramsey Panel 1963, pp. 94, 89, 90, 95. Also see Danby et al. 1962, p. 36.
65. Goldwasser 1985.
66. Lofgren, "Conference of McMillan, Lofgren, Laslett, Kolstad." Paul McDaniel letter to E. M. McMillan, October 25, 1962, McMillan Papers, Box 1.
67. Robert E. Bacher letter to Norman Ramsey, January 8, 1963, Lofgren Papers. Also Norman Ramsey letter to E. M. McMillan, December 19, 1962, Lofgren Papers; Goldwasser 1985.
68. Ramsey Panel 1963, p. 34.
69. Goldwasser 1985.
70. Crease 2005.
71. In a later "supplemental report," dated December 11, 1963, a majority of the Ramsey Panel recommended the return to 10 GeV for the FFAG. Ramsey Panel 1963, p. 34; "Supplemental Report of the GAC/PSAC Panel on High Energy Accelerator Physics." See also Edwin Goldwasser to Westfall, July 23, 1985; Ramsey 1978, p. 79.
72. Atomic Energy Commission, "Summary Notes of Briefing on the Report of the GAC-PSAC Panel of High Energy Accelerator Physics," May 9, 1963, p. 4, Seaborg Papers; and Ramsey 1980a.
73. Ramsey Panel 1963. According to Peter Westwick this gave a new meaning to the notion of a national laboratory. Westwick 2003, pp. 8–10.
74. Schweber 1997, pp. 64–57, quotes on pp. 649 and 647.
75. Schweber 1997, p. 658.

76. Goldwasser 1985, p. 3. Also see Ramsey 1978, 1980a.
77. Ramsey letter to Kolb and Hoddeson, February 20, 2004.

CHAPTER THREE

1. Robert Oppenheimer, foreword to Yuan 1965, p. 5.
2. Ramsey to Kolb and Hoddeson, February 20, 2004.
3. Besides Good, this committee included Frank S. Crawford, James Cronin, Leon M. Lederman, George E. Masek, Louis S. Osborne, Melvin Schwartz, Alvin V. Tollestrup, and William A. Wenzel; M. L. Good, "Report of the PSAC's User's Panel of High Energy Accelerators," unpublished and undated document, in Box A.3.i in Fermilab History Collection. Quotes from "Attachment C: Summary of Comments Solicited by the Office of Science and Technology from a Group of Users of High Energy Accelerators," in U.S. Congress, Joint Committee on Atomic Energy 1965, p. 121. Also see Wenzel 1984.
4. Edwin Goldwasser, private communication to Kolb, 2004.
5. The AUI member institutions were Columbia, Cornell, Harvard, Johns Hopkins, MIT, Pennsylvania, Princeton, Rochester, and Yale. Ramsey 1966.
6. Holl 1997, pp. 202, 218–219; Goldwasser 2004a; Crease 1999a, p. 201; Hildebrand, E-mail to Kolb, July 16, 2004.
7. MURA's defeat and its effect on the political background for the 200 BeV has been noted in many works, including Greenberg 1967; Jachim 1975; and Lowi and Ginsberg 1976. These three, however, ignore the major contribution of outside user tensions and obscure the distinction between the concerns of physicists and those of Midwestern politicians. Lofgren 1984b; Lederman 1984a; Wenzel 1984; unsigned draft to Edwin McMillan, October7, 1964, Edwin McMillan letter to Wolfgang Panofsky, November 2, 1964, and McMillan letter to Paul McDaniel, July 24, 1963, McMillan Papers, Box 5, Bancroft Library, University of California, Berkeley; William Fry letter to G. Kenneth Green, November 3, 1964, Green Papers, Box I, Brookhaven National Laboratory; William Fry letter and enclosures to McMillan, October 30, 1964, Lofgren Papers, Bancroft Library, University of California, Berkeley.
8. Lederman 1963. Crease quotes Brookhaven researcher Ken Green as saying that by the 1960s the New York laboratory was running the AGS so that "it makes relatively little difference whether the experimental team comes from BNL, from an outside institution, or is a mixture of the two." Clearly, some, including Lederman, did not take that view; Crease 1999a, p. 282, also p. 210.
9. Lawrence Radiation Laboratory, "200 BeV Accelerator Design Study," vol. 2, June 1965, pp. B1–B4, C1–C6; Brookhaven National Laboratory, "LRL-CERN-BNL Coordination Session on Super-energy Accelerators, January 13–15, 1964," unpublished document, Green Papers, Box II; Livingston 1968a, p. 7.

10. Keefe, Lambertson, and Laslett 1986.
11. Other regions, such as the South, did not have large numbers of high-energy physicists, and therefore there was no strong push for representatives from them. Edwin McMillan letter to Paul McDaniel, May 10, 1963, McMillan Papers, Box 5; Paul McDaniel letter to Edwin McMillan, May 14, 1963, DOE Archives, Secretariat, Box 1425.
12. All the candidates suggested by McMillan—Richard H. Dalitz from the University of Chicago; William B. Fretter from the University of California, Berkeley; Murray Gell-Mann from Caltech; Maurice Goldhaber from BNL; G. Kenneth Green from BNL; Vernon W. Hughes from Yale; Leon M. Lederman from Columbia; Wolfgang K. H. Panofsky from Stanford; Herbert York from the University of California, San Diego; and William F. Fry from the University of Wisconsin (designated as chairman)—eventually served on the committee, with the exception of Dalitz, who planned to be out of the country. He was replaced by Lee Teng of Argonne. Quotes from Edwin McMillan letter to Paul McDaniel, July 24, 1963, McMillan Papers.
13. Quotes from Edward Lofgren, "Notes of a Conversation with Paul McDaniel, Wallenmeyer, Severiens, Phil McGee, and Fitcher, a New Member Regarding the Accelerator Study, Germantown, January 16, 1963," January 17, 1963, unpublished document, Lofgren Papers. Also see State of California, "Camp Parks, Sierra Foothills," unpublished document, p. 1, Salsig Papers, LBL Archives.
14. Edward Lofgren, "Conference with Haworth in Washington," October 1, 1962, unpublished document, Lofgren Papers.
15. Waldman 1983. For more information on the development of the ZGS, see Paris 2003; and Holl 1997, pp. 196–211.
16. Greenberg 1967, p. 250. For more information on the formation of the Atomic Energy Commission and the National Science Foundation, see Hewlett and Anderson 1962, pp. 428–530.
17. Quote from Greenberg 1967, p. 250. Also see Waldman 1983; and Tape 1986.
18. Manson Benedict letter to Glenn Seaborg, July 24, 1963, p. 7, Seaborg Papers, Bancroft Library, University of California, Berkeley.
19. Goldwasser and Frederick Mills, who were MURA supporters, and Bernard Waldman, MURA's staff director until 1964, all admit that while they supported the FFAG, by 1963 they felt that exploring high energies was more important than exploring high intensities. Goldwasser 1985; Mills 1984; and Waldman 1983; Ramsey 1978; Judd 1984; personal communication from Lawrence W. Jones to Catherine Westfall, July 9, 1987.
20. Glenn Seaborg, record of conversation, July 17, 1963, and July 22, 1963, LBL Archives; Greenbaum 1971, pp. 155–156; Seaborg 1984; and Tape 1986. See Greenbaum for more details of the interaction among MURA, Argonne, the University of Chicago, and the AEC.

21. "Supplemental Report of the GAC/PSAC Panel on High Energy Accelerator Physics," December 11, 1963, one page, unpublished.
22. Kermit Gordon, memorandum to President Lyndon Baines Johnson, December 7, 1963, Seaborg Papers.
23. Glenn Seaborg, memorandum to Lyndon Johnson, December 14, 1963, Seaborg Papers.
24. Waldman 1983. Also Cole 1994; and Glenn Seaborg, diary, December 20, 1963, LBL Archives.
25. Quotes, respectively, from Glenn Seaborg, diary, December 20, 1963, Seaborg Papers; and Elvis Stahr letter to Hubert Humphrey, December 21, 1963, Atomic Energy, Box Ex AT/AT 2, White House Central Files, LBJ Library. Also see Lyndon Baines Johnson letter to Hubert Humphrey, January 16, 1964, Secretariat, DOE Archives, Box 1424.
26. Glenn Seaborg, record of meeting, January 14, 1963, and January 17, 1964, and record of conversation, January 17, 1964, and January 23, 1964, Seaborg Papers; Johnson letter to Humphrey, January 16, 1964, Secretariat, DOE Archives, Box 1424; Waldman 1983; Greenberg 1971, pp. 263–264; and Jachim 1975, pp. 82–99.
27. Glenn Seaborg, record of conversations, January 23, 1964, and January 17, 1964, Seaborg Papers.
28. Galison, Hevly, and Lowen 1992, pp. 72–73; Traweek 1988, p. 127–128.
29. Quotes, respectively, from Glenn Seaborg, record of conversation, January 23, 1964, and March 2, 1964, Seaborg Papers. For information on the founding of SLAC, see Leslie 1993, pp. 181–187; and Galison, Hevly, and Lowen 1992.
30. According to BNL records, Edward Reynolds, the president of American Universities, Inc., sent a copy of "Report of the High Energy Study Group" to 180 members of the scientific community. The report discussed a number of topics, including BNL plans for converting the AGS to higher intensity, the 600–1000 GeV machine, and a meson factory. Edward Reynolds, form letter, February 21, 1964, Brookhaven National Laboratory, Director's Office Files, Box 9. Text quotes from Milton G. White, letter "To Friends of Fundamental Particle Physics," April 13, 1964, Lofgren Papers.
31. Edwin McMillan letter to Milton White, April 20, 1964, Lofgren Papers.
32. Norman Ramsey letter to Edwin McMillan, May 13, 1964, Green Papers, Box I.
33. William Fry letter to Edwin McMillan, May 26, 1964, McMillan Papers, Box 5.
34. Jack [William] Fry letter to Lee [Teng], n.d., Green papers, Box I.
35. Whereas in the late 1950s basic research had consumed just over 0.1% of federal expenditures, by 1965 it commanded about 0.25%. High-energy physics, which unlike other subfields received almost all its funding (90%) from the AEC, commanded a noticeable portion of the basic research budget for that agency. *Physics in Perspective*, report by Physics Survey Committee,

National Research Council, National Academy of Sciences, Washington, DC, 1972, pp. 645.

36. Also see A. Weinberg 1961, 1967; Price 1961, 1963; Capshew and Rader 1992. For more references and discussion about the evolution and use of the term "big science" see Westfall 2003, 30–56.
37. A. Weinberg 1964. The version that appeared in A. Weinberg 1967, pp. 75–76, contains a detailed explanation of his assessment criteria and other musings on large-scale research.
38. Lyndon Johnson to Hubert Humphrey, January 16, 1964, Secretariat, DOE Archives, Box 1424.
39. During this period an AEC budget typically went through several steps. After BOB input, the budget was sent to the president. Once approved, the AEC budget was incorporated into the presidential budget, which was presented to Congress. The JCAE then held hearings so that the AEC staff and experts could present their views on proposed AEC expenditures. After incorporating revisions, the JCAE members took the AEC budget to separate appropriation hearings in the Senate and House of Representatives. Gerald Tape, private communication to Catherine Westfall, November 4, 1987. For a description of the budget cycle for federal agencies, see Physics Survey Committee, National Research Council 1972, p. 655. Quotes from U.S. Congress, Joint Committee on Atomic Energy 1964, p. 1487.
40. Quotes, respectively, from U.S. Congress, Joint Committee on Atomic Energy 1965, pp. 1508 and 1499.
41. U.S. Congress, Joint Committee on Atomic Energy 1965, p. 1500.
42. As quoted in Kevles 1978, p. 417.
43. Goldwasser 1985.
44. Quotes, respectively, from Donald Hornig letter to Senator John O. Pastore, March 27, 1964, White House Central Files, Box AT 1; and Donald Hornig memorandum to Lyndon Baines Johnson, March 27, 1964, Hornig Papers, Box 1, LBJ Library.
45. Other panel members were Val Fitch, Owen Chamberlain, and Maurice Goldhaber (*Bulletin of the American Physical Society* 1964, p. 496). Quotes, respectively, from U.S. Congress, Subcommittee on Research, Development, and Radiation of the Joint Committee on Atomic Energy 1965, "Appendix 19," pp. 756, 752, and 753.
46. Weisskopf and Weinberg 1964.
47. U.S. Congress, Subcommittee on Research, Development, and Radiation of the Joint Committee on Atomic Energy 1965, "Appendix 19," pp. 744, 754.
48. U.S. Congress, Subcommittee on Research, Development, and Radiation of the Joint Committee on Atomic Energy 1965, "Appendix 18," p. 751, reprinted from *Physics Today*, November 1964, p. 751.
49. Edwin Goldwasser letter to Edwin McMillan, September 9, 1964, Lofgren Papers.

50. Unsigned draft to Edwin McMillan, October 7, 1964, McMillan Papers, Box 5.
51. Quotes from memorandum from Frank T. Cole, Charles G. Dols, Fritz Gruetter, Edward C. Hartwig, Denis Keefe, Quinten Kerns, William A. S. Lamb, Glen R. Lambertson, L. Jackson Laslett, Jack M. Peterson, William M. Salsig, Lloyd Smith, and George H. Trilling to Edward Lofgren, October 26, 1964, McMillan Papers, Box 5; Keefe, Lambertson, and Laslett 1986. Also Edward Lofgren letter to William B. Fretter, October 26, 1964, Lofgren Papers; and William Wenzel letter to M. Stanley Livingston, June 14, 1968, Wenzel Papers, LBL Archives.
52. After a series of sit-ins and negotiations between its leaders, including Mario Savio, and University of California president Clark Kerr, Sproul Hall was occupied in late November. Eight hundred demonstrators were arrested, the largest such arrest to that date in California history. Viorst 1979, pp. 288–296.
53. Quotes, respectively, from Edwin McMillan letter to Wolfgang Panofsky, November 2, 1964, McMillan Papers, Box 5; and William Fry letter to G. Kenneth Green, November 3, 1964, Green Papers, Box I. Also see William Fry letter and enclosures to Edwin McMillan, October 30, 1964; and Edwin McMillan to Paul McDaniel, November 23, 1964, Lofgren Papers.
54. Other panel members included Rodney Cool, Sidney D. Drell, Val Fitch, Roger Hildebrand, F. E. Low, H. K. Ticho, William D. Walker, and William A. Wenzel. U.S. Congress, Joint Committee on Atomic Energy 1965, appendix C, "National Policy for High Energy Physics Program," p. 55.
55. Quotes, respectively, from "Report of the Panel on Elementary Particle Physics to the Physics Survey Committee of the National Academy of Science" draft, November 8, 1964, pp. 5–20 and 5–21. Also see this source, p. 1-1.
56. William Wenzel to Edwin McMillan, November 6, 1964, Wenzel Papers; and McMillan 1984; Lofgren 1984b; Lederman 1984a; and Wenzel 1984.
57. Edwin McMillan to Paul McDaniel, November 23, 1964, McMillan Papers, Box 5.
58. Seitz 1980a; Tape 1986; and Pitzer 1984. Apparently Clark Kerr, the University of California's president, had planned to arrange with other university presidents to form a national corporation, but the AEC stopped him. McMillan to Seitz, December 16, 1964.
59. Seitz 2003.
60. Ramsey 1987a, pp. 157–161.
61. W. B. Fowler, "Meeting at National Academy of Sciences, January 17, 1965, Summary of Notes Taken by Theodore P. Wright," April 13, 1965; and Seitz 1965, pp. 8–9; Seitz 1980b; Seaborg letter to Seitz, March 2, 1965, Secretariat, DOE Archives, Box 1425; Leonard L. Bacon, "Minutes of First Meeting of Board of Trustees of Universities Research Association, Inc.," September 16, 1965, Lofgren Papers; Atomic Energy Commission, "Wide Distribution Shown in AEC List of Proposals for 200 BeV Accelerator," July 9, 1965, press release, Seaborg Papers.
62. Yuan 1965, p. 1. Copy in Fermilab Archives.

63. J. Robert Oppenheimer, "Foreword," in Yuan 1965, p. 5.
64. Julian Schwinger, "The Future of Fundamental Physics," in Yuan 1965, pp. 23 and 51. Also see this source, pp. 33, 55, 59, and 75; and Pickering 1984, pp. 57–60. Also Pais 1986, pp. 552–569; and Riordan 1987.
65. Borrowing a term from Buddhism, Gell-Mann labeled SU(3) "the Eightfold Way," since many strongly interacting particles are grouped into sets of eight. At first scientists did not understand why the classification device worked but found it useful in explaining and predicting results. James S. Trefil notes that SU(3) "bears the same logical relation to elementary particles as the periodic table does to chemical elements." In time the development of SU(3) led to the idea that particles are composed of quarks. Trefil 1980, p. 134.
66. Quotes, respectively, from U.S. Congress, Joint Committee on Atomic Energy 1965, pp. 9 and 13. Also see this source, pp. 10–19.
67. U.S. Congress, Joint Committee on Atomic Energy 1965, p. 20. Also see this source, pp. 41–42, 49.
68. Quote from Lyndon Johnson letter to Chester Holifield, January 26, 1965, in "Correspondence Relating to National Policy on the High Energy Physics Program," in U.S. Congress, Joint Committee on Atomic Energy 1965, pp. 1–2. Also see Orlans 1967, p. 225.
69. Thomas L. Collins, "Long Straight Sections for AG Synchrotrons," CEA-86, July 10, 1961.
70. Quote from U.S. Congress, Subcommittee on Research, Development, and Radiation of the Joint Committee on Atomic Energy 1965, p. 69. Also see this source, pp. 1, 29–40, 43–44, 61–77; and U.S. Congress, Joint Committee on Atomic Energy 1965, p. 48.
71. U.S. Congress, Subcommittee on Research, Development, and Radiation of the Joint Committee on Atomic Energy 1965, pp. 204–237.
72. U.S. Congress, Subcommittee on Research, Development, and Radiation of the Joint Committee on Atomic Energy 1965, p. 12. Also see this source, pp. 5, 6, 11, and 23.
73. U.S. Congress, Subcommittee on Research, Development, and Radiation of the Joint Committee on Atomic Energy 1965, p. 377. Also see this source, pp. 378, 382–383, 388–389.
74. Quotes, respectively, from U.S. Congress, Subcommittee on Research, Development, and Radiation of the Joint Committee on Atomic Energy 1965, pp. 66, 67, and 391. Also see Proton Accelerator Committee of the University of Colorado, "Advantages of the Boulder-Denver Area for the 200–300 BeV Proton Accelerator," unpublished document; and Glenn Seaborg letter to Frederick Seitz, March 2, 1965, Secretariat, DOE Archives, Box 1425.
75. W. B. Fowler, "Meeting at National Academy of Sciences, January 17, 1965, Summary of Notes Taken by Theodore P. Wright," April 19, 1965, unpublished document, p. 4; Glenn Seaborg letter to Frederick Seitz, March 2, 1965, Secretariat, DOE Archives, Box 1425; Seitz 1980b.

76. Bacon, "Minutes of First Meeting of Board of Trustees of Universities Research Association, Inc."
77. Glenn Seaborg letter to Frederick Seitz, March 2, 1965, Secretariat, DOE Archives, Box 1425.
78. Quote from enclosure to Frederick Seitz letter to Glenn Seaborg, April 6, 1965, Secretariat, DOE Archives, Box 1425. Also see Gerald Tape letter to Chester Holifield, April 1, 1976; and Paul McDaniel memorandum to Members of Field Offices and Directors of Divisions, Headquarters, April 13, 1965, Secretariat, DOE Archives, Box 1425; Frederick Seitz letter to Theodore P. Wright, April 13, 1965, Green Papers, Box III; Joseph L. Smith, contract with Frederick Seitz, contract no. AT(49-8)-2783, Secretariat, DOE Archives, Box 1425; and Atomic Energy Commission, "AEC-NAS Enter Agreement on Evaluating Sites for A Proposed New National Accelerator Laboratory," April 28, 1965, press release, Seaborg Papers.
79. Arthur Roberts, "The 200 BeV Accelerator," unpublished song. Also see Web site of Brookhaven National Laboratory. Yaphank was the area near Brookhaven Township, a piece of land that cuts across Long Island. Berlin had been a sergeant in the U.S. Army at Camp Upton, which later was converted to Brookhaven National Laboratory.
80. Quotes, respectively, from LRL 1965a, pp. I-1 and I-5.
81. Donald Edwards, comments to A. Kolb, February 29, 2004.
82. LRL 1965a, p. I-7.
83. LRL 1965b, pp. XIII-1, XIII-13-XIII-33; XIV-1-XIV-11; XVIII-1-XVIII-12.
84. Quotes, respectively, from LRL 1965a, pp. XVI-2 and XVI-3. Also see pp. I-7–I-8; XVI-1, XVI-3, XVI-18, XVI-12, XVII-9, XVI-16.
85. Yuan and Blewett 1961, p. 148; LRL 1965a, pp. XVI-1 to XVI-3; and Edward Lofgren, private communication to Catherine Westfall, February 16, 1988.
86. Hoddeson et al. 1993.
87. Keefe, Lambertson, and Laslett 1986.
88. Lawrence Radiation Laboratory, "Construction Project Data Sheet, Schedule 44," extract from FY1967 budget submission, June 1, 1965, Lofgren Papers; Edward Lofgren, "Proposal for the Construction of a 200 BeV Accelerator," draft, August 10, 1965, Lofgren Papers.
89. Quotes, respectively, from LRL 1965a, pp. I-6, I-7, I-8, and XVI-1. Also I-1, I-5, XVI, XVI-13, and XVI-16.

CHAPTER FOUR

1. Enrico Fermi, speaking at the tenth anniversary of the first controlled chain reaction, University of Chicago, December 2, 1952. Transcription in Fermilab Archives; quote on p. 6 of 11 pages.
2. R. Wilson 1978d.
3. R. Wilson 1978d.
4. Hoddeson et al. 1993.

5. R. Wilson 1978d, pp. 34–38.
6. R. Wilson 1978d, pp. 38–39.
7. R. Wilson 1978d, pp. 39.
8. Jeffrey Appel, "To the History of Physics Library, October 24, 1991, Story from the Origins of Fermilab, Told by Robert R. Wilson at the Board of Overseers Meeting Dinner, October 17, 1991."
9. R. Wilson 1968, quotes on p. 491.
10. R. Wilson 1978d, p. 39.
11. Robert Wilson to Edwin McMillan, September 27, 1965.
12. R. Wilson 1965; also see R. Wilson 1978d.
13. Samuel Devons, "Comments on the High Energy Physics Program in the United States," October 1965, unpublished manuscript, Lofgren Papers, Bancroft Library, University of California, Berkeley.
14. Lofgren 1984a; R. Wilson 1978d; Goldwasser 2004a.
15. Glenn Seaborg, record of conversation, August 27, 1965, Seaborg Papers, Bancroft Library, University of California, Berkeley. Also see Joseph Califano memorandum for Lyndon Johnson, October 22, 1965, Ex FG 202 Box 262, White House Central Files, LBJ Library; and D. Goodwin 1976, p. 296.
16. Some have suggested that the push for reduced scope came, at least in part, from design criticism. Seaborg later told the JCAE, however, that the motivation was strictly budgetary. Melvin Price to John Pastore, April 19, 1967, in U.S. Congress, Subcommittee on Research, Development, and Radiation of the Joint Committee on Atomic Energy 1967, p. v.
17. Edward J. Lofgren, "On the Costs of an Accelerator with Reduced Initial Capabilities," December 13, 1965, unpublished document; and Edward J. Lofgren letter to Paul McDaniel, December 14, 1965, Lofgren Papers.
18. Glenn Seaborg, notes for "Meeting of Board of Trustees, Universities Research Association, Inc.," December 12, 1965, unpublished document, Seaborg Papers.
19. Quotes, respectively, from Seaborg, notes for "Meeting of Board of Trustees, Universities Research Association, Inc.," December 12, 1965; and W. K. H. Panofsky, "SLAC and Big Science: Stanford University," in Galison and Hevly 1992, p. 134. As a November 30, 1965, URA press release explained, management of the organization would "be in the hands of a Board of Trustees elected at a meeting of the Council of Presidents at the [National] Academy [of Sciences] on November 7." Universities Research Association, November 30, 1965, press release, Lofgren Papers.
20. Quotes, respectively, from Seaborg, notes for "Meeting of Board of Trustees, Universities Research Association, Inc.," December 12, 1965; Frederick Seitz letter to Norman Ramsey, December 14, 1965; and Frederick Seitz letter to Norman Ramsey, December 30, 1965.
21. Atomic Energy Commission, "Review of Program Plans for Higher Energy and Higher Intensity in Accelerator Facilities," February 11, 1966, unpub-

lished document, p. 2; Wolfgang Panofsky letter to Paul McDaniel, January 5, 1966; and Edwin Goldwasser letter to Paul McDaniel, January 18, 1966.

22. Quotes, respectively, from R. Wilson 1978d; and "Report to Board of Trustees by the Scientific Subcommittee," January 25, 1966, unpublished document. Also see Atomic Energy Commission, "Review of Program Plans for Higher Energy and Higher Intensity in Accelerator Facilities," p. 2.
23. Quotes, respectively, from R. Wilson 1978d; and "Report to Board of Trustees by the Scientific Subcommittee," January 25, 1966, unpublished document, p. 5. Also see "Meeting on High Energy Physics, Agenda," January 24, 1966, unpublished document, p. 6.
24. Henry D. Smyth to Glenn Seaborg, January 26, 1966; Atomic Energy Commission, "Review of Program Plans for Higher Energy and Higher Intensity in Accelerator Facilities," pp. 6, 8, 6, 7, 24. Wilson was a member of the board of trustees when the board endorsed Smyth's conclusions. He remembers that he again summarized his views for the board at its January 25 meeting but finally agreed that there was no widespread support for alternate design schemes. R. Wilson 1978d.
25. Glenn Seaborg to Chester Holifield, February 16, 1966, LBL Archives.
26. The committee members were Robert Bacher from Caltech; Harvey Brooks from Harvard; Val Fitch from Princeton; William Fretter from the University of California, Berkeley; William Fry from the University of Wisconsin, Madison; John William Gardner from the Carnegie Corporation; Edwin Goldwasser from the University of Illinois; G. Kenneth Green from Brookhaven; Crawford Greenewalt from du Pont; Herbert Longenecker from Tulane University; Emanuel Piore from IBM; and Kenneth Reed from the NAS. Gardner later withdrew from the committee when he was appointed secretary of the Department of Health, Education and Welfare. Site Evaluation Committee, "The Report of the National Academy of Sciences' Site Evaluation Committee," March 1966, unpublished document. Also "Site Selection Committee, Advisory to the Atomic Energy Commission," June 5, 6, 1965, Green Papers, Box V, Brookhaven National Laboratory.
27. W. E. Hughes memorandum to Paul W. McDaniel, June 8, 1965, Secretariat, DOE Archives, Box 1425.
28. Crease 1999a, p. 267.
29. Atomic Energy Commission, "Wide Distribution Shown in AEC List of Proposals for 200 BeV Accelerator," July 9, 1965, press release, Seaborg Papers; Cole 1984; Blewett 1986; Ramey 1986; Paul W. McDaniel letter to Maurice Goldhaber, May 2, 1965, Green Papers, Box III; Theodore Wright letter to Paul McDaniel, June 10, 1965, Green Papers, Box III; and James T. Ramey memorandum to Glenn Seaborg, John Palfrey, Gerald Tape, R. E. Hollingsworth, Spofford English, and Paul McDaniel, July 23, 1965, Seaborg Papers.
30. Quotes, respectively, from Glenn Seaborg, diary, TS, July 28, 1965, and Seaborg letter to Lyndon Baines Johnson, August 28, 1965, Seaborg Papers.

Also see Glenn Seaborg letter to Frederick Seitz, August 24, 1965, Seaborg Papers.

31. Quotes, respectively, from Glenn Seaborg, diary, September 1, 1965, and Seaborg, record of conversation, September 8, 1965, Seaborg Papers. Also see Horace Busby memorandum to Marvin Watson, September 8, 1965, Box Ex FG 262, White House Central Files, LBJ Library; and Lambright 1985, pp. 95, 124.
32. Quotes from Joseph Califano memorandum to Lyndon Johnson, September 10, 1965, Ex FG 202 Box 262, White House Central Files, LBJ Library. Also see Dwight A. Ink letter to William Moyers, September 8, 1965, Ex FG 202 Box 262, White House Central Files, LBJ Library; Glenn Seaborg, diary, TS, September 13, 1965, Seaborg Papers; Glenn Seaborg letter to Frederick Seitz, September 13, 1965, Seaborg Papers; Atomic Energy Commission, "AEC Asks NAS to Evaluate 85 Site Proposals for 200-BeV Accelerator," September 15, 1965, press release, Seaborg Papers; Lambright 1985, p. 13.
33. Quotes as quoted in Duncan Clark letter to Hayes Redmon, September 17, 1965, Ex FG RS PR 8 Box 6, White House Central Files, LBJ Library. Also see Glenn Seaborg, diary, TS, September 15, 1965; and Donald Hornig letter to Joseph Califano, October 19, 1965, FG 202, White House Central Files, LBJ Library.
34. Quotes from Carl M. York memorandum to Professors David Saxon and Harold Ticho, November 8, 1965, Lofgren Papers.
35. Joseph Califano memorandum to Lyndon Johnson, September 10, 1965, FG 202 Box 262, White House Central Files, LBJ Library; interview with Glenn Seaborg by Catherine Westfall, December 16, 1984, personal collection of Catherine Westfall.
36. Quotes, respectively, from Joseph Califano memorandum to Lyndon Johnson, October 22, 1965, Ex FG 202 Box 262, White House Central Files;LBJ Library; and Joseph Califano memorandum for the record, October 25, 1965, Ex FG 202 Box 262, White House Central Files, LBJ Library. Also see Glenn Seaborg letter to Lyndon Johnson, October 12, 1965, agency report, Confidential File; and John V. Vinciguerra letter to Marvin Watson, November 4, 1965, FG 202, White House Central Files, LBJ Library.
37. Site Evaluation Committee, "The Report of the National Academy of Sciences' Site Evaluation Committee," appendixes A and B. Text quotes from Glenn Seaborg letter to Frederick Seitz, September 13, 1965, Seaborg Papers. Also see Site Evaluation Committee, "The Report of the National Academy of Sciences' Site Evaluation Committee," pp. 41–42; G. Kenneth Green letter to Frederick Seitz, draft, n.d., Green Papers, Box V; Tape 1986; and Goldwasser 1985.
38. Quotes from Site Evaluation Committee, "The Report of the National Academy of Sciences' Site Evaluation Committee," p. 7. Also see Panel of Accelerator Scientists, "Report of the Panel of Accelerator Scientists," January 25, 1966, unpublished document, p. 1, Mills Papers; Site Evaluation

Committee, "The Report of the National Academy of Sciences' Site Evaluation Committee," appendix A; Paul McDaniel, "Notes on NAS Site Evaluation Committee on November 22, 1965," unpublished document, Seaborg Papers; and Goldwasser 1985.

39. McDaniel, "Notes on NAS Site Evaluation Committee on November 22, 1965." Also see Goldwasser 1985.
40. Goldwasser letter to Emanuel Piore, October 13, 1965, Green Papers, Box V.
41. Site Evaluation Committee, "The Report of the National Academy of Sciences' Site Evaluation Committee," pp. 7 and 23. Also see this source, pp. 23–39, and appendix A, p. 11.
42. Glenn Seaborg, record of conversation, March 29, 1965, Seaborg Papers; Joseph Califano memorandum for Lyndon Johnson, March 14, 1966, White House Central Files, Ex Fg 202, Box 262, LBJ Library; Paul McDaniel memorandum for Glenn Seaborg, John Palfrey, James Ramey, and Gerald Tape, December 2, 1965, Secretariat, Box 1423, DOE Archives; and Atomic Energy Commission, "AEC to Select 200 BeV Accelerator Site from among Six Locations Recommended by NAS," March 30, 1966, press release, Seaborg Papers.
43. Excerpt of the congressional record, appended to John Burke memorandum to Glenn Seaborg, James Ramey, Gerald Tape, John Palfrey, Seaborg Collection, Box 138, DOE Archives. Also see Glenn Seaborg, record of conversation, March 31, 1966, Seaborg Papers.
44. Quotes, respectively, from Charles Schultze memorandum to Joseph Califano, April 22, 1966, White House Central Files, Ex FG 202, Box 262, LBJ Library; and Charles Schultze letter to Glenn Seaborg, April 23, 1966, Seaborg Collection, Box 140, DOE Archives.
45. Quotes from Glenn Seaborg letter to Charles Schultze, May 12, 1966, Seaborg Collection, Box 138, DOE Archives.
46. Quotes from Glenn Seaborg, record of meeting, June 13, 1966, Seaborg Papers. Also see Tape 1986; and Ramey 1986.
47. AEC n.d., vol. 1, pt. 1, chap. 6, p. 145; Atomic Energy Commission, "Summary Notes of Briefing on Progress Report on 200 BeV Site Analyses," May 10, 1966, unpublished document, Seaborg Papers; and Gerald Tape, private communication to Catherine Westfall, February 4, 1988.
48. D. S. Greenberg, "200-Bev: The Academy Committee Knew Where It Was Going," *Science* 152, no. 3720 (April 15, 1966): 326.
49. Quotes, respectively, from Long Island Association of Commerce and Industry, "Action Program for the 200 BeV Accelerator," 15 April 1966, unpublished document, Green Papers, Box III; and Illinois Information Service, May 31, 1966, press release.
50. Quotes, respectively, from Thomas B. Husband letter to Lyndon Johnson, May 5, 1966, Seaborg Collection, Box 138, DOE Archives; William Wohl letter to the Atomic Energy Commission, April 28, 1966, Seaborg Collection, Box 138, DOE Archives; and Arthur Theriault, draft of an untitled speech given February 16, 1969. Also see Karyl Louwenaar, "The Story of the Village

of Weston: Notes Taken in Interview with Kenneth Reeling," August 18, 1969, unpublished manuscript; and Findlay 1995, p. 36. The plight of those opposing the land acquisition is presented in Lowi and Ginsberg 1976.

51. United States Atomic Energy Commission, December 16, 1966, press release J-282. See also Westfall 1989; and Kolb and Hoddeson 1995.
52. Anton Jachim explains that concerted efforts to lobby for a Midwest site for the 200 GeV accelerator had already begun by the time of the September 1965 Midwest Governors' Conference. Jachim gives further details of Midwest and Illinois efforts; Jachim 1975, pp. 117–129. John Erlewine, "Summary of Proposers' Written and Oral Commitments Re the 200 BeV Accelerator Project," July 25, 1966, unpublished document, Secretariat, Box 7741, DOE Archives.
53. Quotes, respectively, from California Members of the House to Glenn Seaborg, June 24, 1966, Secretariat, Box 7741, DOE Archives; and as quoted by John Erlewine, "Summary of Proposers' Written and Oral Commitments Re the 200 BeV Accelerator Project." Also see Warren Knowles letter to Paul McDaniel, June 10, 1965, Mills Papers; Frank Hausheer letter to Frederick Harrington, June 10, 1965, Mills Papers; Marvin Brickson letter to Frederick Herrington, May 26,1965, Mills Papers; and Proton Accelerator Committee of the University of Colorado, "Advantages of the Boulder-Denver Area for the 200–300 BeV Proton Accelerator," unpublished report, 1966.
54. Tape 1986; and R. Wilson 1978e.
55. Tape 1986.
56. Quote from Glenn Seaborg, record of conversation, July 13, 1966, Seaborg Papers. Also see Henry Traynor memorandum to Glenn Seaborg, James Ramey, and Gerald Tape, July 29, 1966, Secretariat, Box 7741, DOE Archives; Glenn Seaborg, diary, TS, September 15, 1966, Seaborg Papers; and Jachim 1975, p. 127.
57. Quote from Ramsey 1980b. Also see Edwin McMillan memorandum to Emilio Segre, January 19, 1965, Lofgren Papers; Universities Research Association, Inc., October 11, 1966, press release, Lofgren Papers; C. M. York memorandum to F. D. Murphy, November 7, 1966, Lofgren Papers; and Norman Ramsey letter to URA board of trustees, November 18, 1966.
58. Glenn Seaborg, record of conversation, November 28, 1966, Seaborg Papers.
59. John Erlewine memorandum to Glenn Seaborg, Wilfred Johnson, Samuel Nabrit, James Ramey, and Gerald Tape with enclosure "200 BeV Summary"; and Site Evaluation Committee, "The Report of the National Academy of Sciences' Site Evaluation Committee," appendix D1.
60. John Erlewine memorandum to Glenn Seaborg, Wilfred Johnson, Samuel Nabrit, James Ramey, and Gerald Tape with enclosure "200 BeV Summary"; Glenn Seaborg, "Notes to Record," November 29, 1966, Seaborg Papers.
61. Quotes from Harry Traynor memorandum to Glenn Seaborg, Wilfred Johnson, Samuel Nabrit, James Ramey, and Gerald Tape, August 31, 1966.

62. Westfall interviewed four of the five commissioners serving at the time of the site selection. The fifth, John Palfrey, died in 1966. Seaborg mentioned that the commission chose Weston based on the site criteria and received no pressure or interference in final site selection from President Johnson or anyone else. Seaborg 1983, 1984; Tape 1986; Ramey 1986; Seaborg, diary, TS, December 16, 1966, Seaborg Papers.
63. Glenn Seaborg, record of conversation, December 7, 15, 1966, and December 17, 20, 1966, Seaborg Papers; Seaborg, diary, TS, November 29, 1966, and December 16, 1966; Seaborg and Atomic Energy Commission, "AEC Selects Site for 200-BeV Accelerator," December 16, 1966, press release, Seaborg Papers. Kerner was chair of Johnson's Commission on Civil Disorders and prepared the Kerner report in 1968; see Barnhart and Schlickman 1999.
64. McMillan 1984.
65. E.g., see Lambright 1985, p. 62. The counterarguments are given in Westfall 1989; Tape 1986; Ramey 1986; Nabrit 1987; Waldman 1983; Glenn Seaborg, record of conversation, December 7, 1966, Seaborg Papers.
66. Quote in Lowi and Ginsberg 1976, p. 79.
67. Seitz 2003.
68. Those who have studied Johnson's presidency argue that by late 1966 Johnson was so preoccupied with Vietnam that he hardly concerned himself with domestic matters. D. Goodwin 1976, pp. 293–294. As for Dirksen's involvement in the siting issue, Ramey insists that the senator showed little interest in the 200 GeV accelerator. Wilson remembered Dirksen's rudely telling him, "I don't care whether we have that project." Nor did the March 1966 record of congressional interest compiled by the AEC judge Dirksen as having a "strong" interest in the project; Ramey 1986; R. Wilson 1978d; and Atomic Energy Commission, "Congressional Interest in 200 BeV Site Locations," March 18, 1966, unpublished document, Seaborg Collection, Box 138, DOE Archives.
69. Tape 1986.
70. Fred Seitz later reflected on the disparity between his memory and Seaborg's on the issue of the site selection. "I am sure about the fact that I received a call from the White House, perhaps the President, stating that he wanted to select the special site from among the six best as seen by the selection committee. This strongly suggests that he used Illinois to garner an important vote from Senator Dirksen, which would be quite typical of Johnson's tactics when in the rough and tumble of politics. . . . Perhaps the answer to this contradiction lies in the fact that the GAC and the AEC staff decided on their own that a Midwestern location was inevitable in view of the overall national situation and came quite naturally to the conclusion that the accelerator should be either in Wisconsin or Illinois, with Illinois being favored by the presence of O'Hare International Airport and a few other things. . . . This does not mean that President Johnson would not have put pressure upon

Senator Dirksen to gain a favor but it would never have been in the open." F. Seitz to L. Hoddeson, February 12, 2004.

71. Quotes from Tape 1986; and McMillan 1984. Also see Lofgren 1984b; Judd 1984; and Blewett 1986.
72. Ramsey to Kolb and Hoddeson, February 20, 2004.
73. Quotes from David Judd to Glenn Seaborg, February 10, 1967, McMillan Papers, Box 2, Bancroft Library, University of California, Berkeley. (The authors thank Robert Seidel for providing the February 10, 1967, document.) Also see John Conway letter to R. E. Hollingsworth, March 17, 1967, in U.S. Congress, Joint Committee on Atomic Energy 1967, p. 44.
74. W. B. McCool, "Summary of Notes of Meeting with Representatives of Universities Research Association," December 19, 1966, unpublished document, Seaborg Papers. Also see Glenn Seaborg, diary, July 20, 1966; Charles Schultze memorandum to Lyndon Johnson, August 10, 1966, White House Central Files, Box FI4, LBJ Library; URA annual report for 1967, December 5, 1967, p. 3.
75. Atomic Energy Commission, "Basis for the Selection of the Chicago (Weston) Site for Location of the 200 BeV Accelerator Laboratory," January 18, 1967, press release.
76. U.S. Congress, Joint Committee on Atomic Energy 1967, p. 47.
77. McCool, "Summary of Notes of Meeting with Representatives of Universities Research Association."
78. Edward Lofgren letter to Norman Ramsey, January 12, 1967, Lofgren Papers.
79. L. Jackson Laslett, private communication to Catherine Westfall, February 22, 1988.
80. Ramsey 1980b.
81. Ramsey 1980b; and Goldwasser 1985.
82. Quotes from David Judd letter to Glenn Seaborg, February 10, 1967, McMillan Papers, Box 5. Also see Glenn Seaborg, diary, TS, February 11, 1967, Seaborg Papers; and Ramsey 1980b.
83. Glenn Seaborg, record of meeting, February 8, 1967, Seaborg Papers.
84. Quotes from R. Wilson 1967e. Also Keefe, Lambertson, and Laslett 1986; R. Wilson 1978e.
85. Glenn Seaborg, diary, TS accounts, February 9, 11, 14, 1967, Seaborg Papers; D. Keefe, "Report on Meeting between LRL Personnel and the Atomic Energy Commission," February 14, 1967, unpublished document, McMillan Papers, Box 2; and William Salsig, D. G. Eagling, J. A. Burt, E. Eno, R. O. Haglund, F. M. Johnson, W. Popenuck, and H. A. Wollenberg, "Preliminary Estimate of Cost Differentials between the Weston Site and the Reference Site (Sierra) for the Design Study Accelerator," unpublished manuscript, Salsig Papers, LBL Archives.
86. Ramsey to Wilson, February 6, 1967.
87. Glenn Seaborg, diary, February 14, 1967, Seaborg Papers.
88. U.S. Congress, Joint Committee on Atomic Energy 1967, pp. 28–29.

89. Quote from U.S. Congress, Joint Committee on Atomic Energy 1967, pp. 37 and 71. See also 28–29, 35–77, 78–81.
90. U.S. Congress, Joint Committee on Atomic Energy 1967, pp. 22–394.
91. Quotes, respectively, from U.S. Congress, Joint Committee on Atomic Energy 1967, pp. 97 and 108. For a description of Kerner's efforts, see Jachim 1975, p. 128; and Barnhart and Schlickman 1999, pp. 157–188.
92. Quote from U.S. Congress, Joint Committee on Atomic Energy 1967, p. 93. Also pp. v, 31–32.
93. U.S. Congress, Joint Committee on Atomic Energy 1967, p. 24.
94. U.S. Congress, Joint Committee on Atomic Energy 1967, p. 24. Also see this source, pp. 25–28. For Berkeley's expandability scheme, see Garren et al. 1967, p. 223.
95. The URA was not granted a permanent contract until January 22, 1968.
96. Quotes, respectively, from U.S. Congress, Joint Committee on Atomic Energy 1967, p. viii, 57, 58. Also see this document pp. 36, 59. E. J. Bloch letter to John Pastore, January 28, 1968.
97. Quotes from "Vs. Scientific Luxury" 1967. Also see Norton et al. 1982, p. 953; AEC n.d., vol. 1, pt. 1, chap. 6, p. 146; and Lyndon Johnson letter to Glenn Seaborg, July 26, 1967, Seaborg Collection, Box 170, DOE Archives.
98. Wilson to Seaborg, February 28, 1967; Wilson to Ramsey, March 1, 1967.
99. Wilson to Seaborg, February 28, 1967.
100. Minutes of the meeting of the URA executive committee, January 30, 1967.
101. R. Wilson 1979a; Goldwasser 2004b.
102. Twain 1984, p. 416.

CHAPTER FIVE

1. R. Wilson 1967a.
2. Hankerson 1978.
3. The opening section of this chapter, on Wilson's personae, is based on the talk, "Wilson's Way," that Hoddeson presented at Wilson's eightieth birthday celebration, held at Fermilab on March 4, 1994 (Hoddeson 1994).
4. R. Wilson 1968, p. 494.
5. R. Wilson 1968. For more discussion of this frontier imagery in relation to Fermilab see Kolb and Hoddeson 1995, including its references.
6. R. Wilson 1977a; Hilts 1982, pp. 15–99.
7. T. Hughes 1989; Heilbron and Seidel 1989; Hoddeson et al. 1993.
8. E.g., see Debus 1978, p. 59.
9. There is some overlap between this biographical summary of Wilson's life and the account in Hoddeson and Kolb 2008.
10. Goldwasser 2000.
11. R. Wilson 1977a.
12. R. Wilson 1987c; Goldwasser 2000.
13. Weart 2000, pp. 151–154.

14. Weart 2000, p. 157.
15. Weart 2000, pp. 175–185.
16. R. Wilson 1970a.
17. Hoddeson et al. 1993, chap. 1; pp. 63–65.
18. R. Wilson 1970a.
19. Wilson, "Radiological Use of Fast Protons," *Radiology* 47 (November 1946): 487–491.
20. R. Wilson 1966, p. 235. The current "do more with less" and "faster, cheaper, better" goal at NASA is actually a version of Wilson's original frugality tradition.
21. The first, Cornell's 300 MeV electron synchrotron, was in 1954 given as a gift to the University of Jerusalem. R. Wilson 1966.
22. Rose Bethe, private communication by phone with Adrienne Kolb, July 26, 2004.
23. Rose Bethe, private communication by phone with Adrienne Kolb, July 26, 2004.
24. Matyas, phone conversation with Kolb, c. June 2004.
25. Duffield 1978.
26. Sazama et al. 2003.
27. The August J. Mier collection of arrowheads from the site was donated in 1977 and remained on permanent display in Wilson Hall until 2001. Mr. Mier died on February 25, 1986.
28. R. Wilson 1987b.
29. Hilts 1982, p. 18
30. Hilts 1982, p. 26
31. Kolb and Hoddeson 1995.
32. "Robert Rathbun Wilson," in *Current Biography Yearbook* (The Bronx: H. Wilson Co., 1989), vol. 50, no. 8, pp. 625–629.
33. R. Wilson 1968.
34. R. Wilson 1987c.
35. R. Wilson 1979a; Goldwasser 1985; Westfall 1997.
36. Goldwasser 2000.
37. R. Wilson 1979a
38. Goldwasser 1987a.
39. R. Wilson 1987b.
40. "Summer Study Program, 1968," December 8, 1967, unpublished report; Goldwasser 1992, 1985.
41. Barnhart and Schlickman 1999, pp. 157–188.
42. Wilson, telegram to Dr. Martin Luther King, June 22, 1967.
43. "Du Page Shows Little Interest in Protest," *Chicago Tribune*, June 24, 1967.
44. "Thirty Years after Kerner Report, Some Say Racial Divide Wider," CNN.com, March 1, 1998.
45. NAL 1968b; DeMuth 1967, Goldwasser 1985, R. Wilson 1987b.
46. NAL 1968b.

47. AEC n.d., vol. 1, pt. 1, chap. 6, p. 147. For more details on civil-rights efforts at the laboratory, see Goldwasser 1969, pp. 8–9.
48. "22 Trainees Join New Village Job Program," *Village Crier* 1, no. 1 (March 1969): 6.
49. Wilson, "Radiological Use of Fast Protons."
50. For more information on the founding of BNL, see Norman Ramsey, "Early History of Associated Universities Inc.; Brookhaven National Laboratory," BNL 992 (T-421), Upton, March 1966; and Needell 1983, p. 119.
51. Edwin Goldwasser, private communication to Adrienne Kolb, November 14, 1994.
52. This procedure changed when Lederman became director, for Lederman preferred to write his own annual report. Ramsey to Kolb and Hoddeson, February 20, 2004.
53. R. Wilson 1981.
54. R. Wilson 1987b; Kotulak 1968.
55. Edwin Goldwasser, private communication to Adrienne Kolb, November 14, 1994.
56. Ramsey 1992.
57. Robert Wilson form letter, March 20, 1967.
58. R. Wilson 1967a.
59. R. Wilson 1967a.
60. R. Wilson 1967a.
61. Goldwasser 1985; Orr 1990. Among those who left were Frank Shoemaker, who led the Main Ring group before returning to Princeton, and Arie Van Steenbergen, who led the Booster group before returning to Brookhaven. Livdahl 1983.
62. Quotes from Goldwasser 1987a; and Orr 1992.
63. Quote from J. McCarthy 1987. Also see Malamud 1987; Cole 1985; and Goldwasser 1985.
64. Employment figures do not include DUSAF employees. For details on recruitment from 1967 to 1974, see Livdahl 1983.
65. Hankerson 1978.
66. R. Wilson 1969a.
67. R. Wilson 1969a.
68. R. Wilson 1969a.
69. R. Wilson 1969a, 1973. John Krige points out that Wilson's vision of the individualistic researcher was representative of a pervasive view prevalent in the international community of high-energy physics. Such ideas were "part of a constantly regenerated ideology which pivots around images of the scientist as an individual creative genius," which persisted into the late 1980s, although by this time most high-energy physicists considered the ideology to be an outdated cliché. Krige 1991.
70. Robert Wilson letters to Norman Ramsey, March 1, 1967, and March 2, 1967, and to Glenn Seaborg, February 28, 1967. Also see Universities Research

Association, "Board of Trustees, Minutes of Meeting, January 15, 1967"; Ramsey 1980b; and Goldwasser 1985. Also see Norman Ramsey letter to Robert Wilson, February 27, 1967.

71. R. Wilson 1967a.
72. R. Wilson 1967a.
73. "Research at 200 GeV," report of the Universities Research Association, August 1967.
74. NAL 1968a, pp. 1–1 and 1–2.
75. NAL 1968a, p. 2–2.
76. NAL 1968a, pp. 2–2, 2–3.
77. David Young 1968; "Weston Protest Plea Goes to Washington: Village Board Objects to Land Acquisition Methods Used by State," *Wheaton Journal*, May 15, 1968; and Louwenaar 1968. Also see URA annual report for 1967.
78. This outreach effort was undertaken by Karyl Louwenaar, who worked for the laboratory's Public Information Office in 1968–1969. See also Ramsey 1987a; Lowi and Ginsberg 1976; Westfall 1988; Kolb and Hoddeson 1995; Robert R. Wilson letter to Charles Percy, February 3, 1969; Livingston 1969a, p. 7; Nolte 1969.
79. "Fifty Years of Farm Machinery," *FermiNews* 1, no. 13 (September 14, 1978): 3.
80. All of the farms and buildings of the 6,800-acre site had been surveyed, evaluated, and numbered by the architectural and engineering consortium DUSAF, representing the combined efforts of the firms Daniel, Mann, Johnson and Mendenhall (DMJM); Urbahn; Seelye; and Fuller as part of the master plan for NAL.
81. Seaborg groundbreaking remarks, December 1, 1968; "Ground Dug for Powerful A-Plant," *Chicago Sun-Times*, December 2, 1968.
82. Arthur J. Snider, "Scientists Buzzing on Weston's Atomic Future," *Chicago Daily News*, December 2, 1968.
83. Plaque presented at the site transfer ceremony, April 10, 1969.
84. Francis Cole, "Monthly Report of Activities," October 1, 1968, pp. 2, 14, November 1, 1968, and December 1, 1968; Hankerson 1978.
85. Duffield 1978.
86. R. Wilson 1987b; Goldwasser 1987a; Duffield 1978.
87. R. Wilson 1987b; Goldwasser 2004b.
88. Leposky 1960, p. 24.
89. See "DUSAF Observes Third Anniversary," *Village Crier*, June 25, 1970, p. 2, for early NAL history in limerick form.
90. Alan H. Rider, "Design for Fermilab's Central Laboratory—the Architect's Point of View," n.d., sent to Fermilab on March 29, 1976. John Peoples, private communication to Kolb, February 2004; Goldwasser 1987a; and R. Wilson 1987b.
91. Malamud notes to Hoddeson and Kolb, January 2004.
92. Goldwasser 1987a; R. Wilson 1987b; Ramsey 1987b; "Questions Raised on the Design of the 200 BeV Accelerator," n.d., unpublished document.

93. Hilts 1982, p. 86.
94. Peoples and Malamud, private communications to Adrienne Kolb, ca. 2004. Another reason is said to be because the laboratory offices (in the High-Rise) are in Kane County.
95. Leposky 1969, p. 24; and Goldwasser 1987a.
96. Robert Lootens, private communication to Adrienne Kolb, 2003.
97. Gonzales 2004.
98. Gonzales 2004.
99. Chrisman, private communication to Lillian Hoddeson, September 5, 2003.
100. Goldwasser 2000.
101. Duffield 1978.
102. Malamud 1987.
103. Jovanovic 1989.
104. R. Wilson 1970c, p. 1076.
105. R. Wilson 1970c, p. 1076.
106. R. Wilson 1992. Goldwasser 2005.
107. Norman Ramsey letter to Robert Wilson, February 6, 1967.
108. Panel on Elementary Particle Physics, "Report of the Panel on Elementary Particle Physics to the Physics Survey Committee of the National Academy of Science," draft, November 8, 1964, pp. 5–21; and U.S. Congress, Joint Committee on Atomic Energy 1967, p. 291; Ramsey 1980b; Westfall 1988.
109. Don Young, "Comments," spring 2004.
110. R. Wilson 1970c, p. 1083.
111. R. Wilson 1987b. For more discussion of how the pioneer ethic has driven American technology, see Ferguson 1979.
112. Heilbron and Seidel 1989, pp. 283, 284; Collins 1989.
113. R. Wilson 1967c.
114. R. Wilson 1970c.
115. R. Wilson 1977a, p. 29.

CHAPTER SIX

1. R. Wilson 1966, p. 235.
2. R. Wilson 1970c, 1987b.
3. Pestre and Krige 1992, p. 95.
4. Pestre and Krige 1992, p. 90; and Hermann et al. 1990, p. 799.
5. R. Wilson 1987b.
6. Don Young, "Accelerator Beam Quality versus Architectural Expression," spring 2004.
7. R. Wilson 1981.
8. R. Wilson 1987b; Hoddeson 1987.
9. Goldwasser 1987a; Serber 1986; R. Wilson 1979a; Malamud note to Hoddeson and Kolb, January 2004; Don Getz, untitled manuscript, May 1977; Hoddeson 1983, p. 20.

10. Heilbron and Seidel 1989, p. 264; Hoddeson 1989; R. Wilson 1967a, 1967b.
11. Duffield 1978.
12. Goldwasser 1987a; Don Getz, untitled manuscript, May 1977.
13. Kitagaki 1953; and M. White 1953; quote from Arie Van Steenbergen, "200–400 BeV Accelerator Summer Study," July, August 1967, p. 7. See also Keefe, Lambertson, and Laslett 1986; and R. Wilson 1987a.
14. Hinterberger et al. 1971.
15. H. Hinterberger, J. Satti, C. Schmidt, R. Sheldon, and R. Yamada, "Bending Magnets of the NAL Main Accelerator," *IEEE Transactions on Nuclear Science* 18, no. 3 (June 1971): 853–856, 1971 Particle Accelerator Conference, Chicago, March 1–3, 1971. See also Lee Teng's "Comments on the Conference," p. 3 of the same volume.
16. The lattice, refined by Collins when the components were installed, also contained Collins's straight sections for beam handling and radio frequency acceleration.
17. R. Wilson 1987a; NAL 1968a, pp. 4-1, 5-5; Livingston 1968a, p. 21; R. Wilson 1967c, p. 5. The plan was formally presented to the accelerator community by Ernie Malamud in an invited paper presented to the PAC in 1971 titled, following Wilson's suggestion, "Status of the 500 GeV Accelerator."
18. B1 magnets have ½ inch vertical gaps while the B2 magnets have 2 inch gaps. The sizes are different to take into account the changing dimensions of the beam. Livingston 1968a p. 21; R. Wilson 1967c, pp. 4–5; Malamud and Walker 1970, p. 3; Collins 1989.
19. The Berkeley magnets were estimated to weigh 19,415 tons and cost $26.6 million, while the NAL magnets were estimated to weigh 9,750 tons and cost $20.9 million. LRL 1965a, pp. III-9, III-10, XVI-4; NAL 1968a, pp. Ar-4, 16-3.
20. Quote from Ramsey 1980b. Also Don Getz, untitled report, May 1977; "Questions Raised on the Design of the 200 BeV Accelerator," n.d., Mills Papers; LRL 1965a, p. XVI-16; NAL 1968a, p. 16-11.
21. R. Wilson 1969b. Also Malamud 1989. The mysterious number 137, the inverse of the fine structure constant, is associated with the electron, relativity, and quantum theory. Also linked with the Hebrew word *Kabbalah*, it is a symbol that physicists use to indicate "how much we don't know." Lederman 1993, p. 28; Brown and Hoddeson 1983, Weisskopf 1976, p. 78.
22. Goldwasser 2000, 2005; Wilson and Kolb 1997, p. 344.
23. Weart 1979, p. 328; High Energy Physics Advisory Panel, "The Status and Problems of High Energy Physics Today," January 1968, p. 38; Atomic Energy Commission, "Atomic Energy Commission Summary Notes of 200 BeV Accelerator Briefing," September 1, 1967, Seaborg Files.
24. Robert Wilson to Glenn Seaborg, February 28, 1967; Ramsey 1980b.
25. By the late 1960s, some of the modeling of the accelerator—for example, of the magnet lattice—was done with computer programs.
26. Livingston 1969a p. 1; Malamud 1987.
27. Malamud 1987.

28. Livdahl 1983, pp. 9, 15; Francis Cole, "Monthly Report of Activities," February 28, 1969, p. 1.
29. Although the accelerator theory group obtained a PDP-10 computer as a gift when the Princeton-Pennsylvania Accelerator was closed, the group continued to use the more powerful computing facilities at NYU and ANL for many years. MacLachlan 1989a; Francis Cole, "Monthly Report of Activities," April 1, 1968, p. 2; June 1, 1968, pp. 3, 4, 6; R. Wilson 1967c, p. 5.
30. Malamud 1989; Francis Cole, "Main Ring Group Meeting," 27 March 1968; Cole, "Monthly Report of Activities," August 1, 1968, p. 4; September 1, 1968, pp. 4–5; November 1, 1968, p. 3; February 28, 1969, p. 9. See also Malamud's notes to Hoddeson and Kolb, January 2004.
31. Francis Cole, "Monthly Report of Activities," April 30, 1969, p. 6; Livingston 1969a pp. 11, 12.
32. Malamud notes to Hoddeson and Kolb, January 2004.
33. For more information on land acquisition and funding difficulties, see Westfall 1988, chap. 6. Also "Main Accelerator Section Monthly Report," December 1968; Francis Cole, "Monthly Report of Activities," September 1, 1968, p. 6; Yamada 1989; "Monthly Report Main Accelerator Section," November 1968; "Minutes of Staff Meeting Main Ring," August 27, 1969, 28 January 1970.
34. Francis Cole, "Monthly Report of Activities," June 1, 1968, p. 6; October 1, 1968, p. 3; "Main Accelerator Section Monthly Report," 31 March 1969.
35. Wilson, private communication to Kolb, 1992.
36. Malamud 1989. Yamada 1989; Francis Cole, "Monthly Report of Activities," April 30, 1969, pp. 1, 14; May 31, 1969, p. 2.
37. E. Malamud E-mail to Drasco Jovanovic, January 3, 2000; Livdahl 1983, p. 10; R. Wilson 1987b.
38. LRL 1965a; NAL 1968a; Kerns 1987.
39. Don Young 2005.
40. Don Young, "Comments," spring 2004.
41. Don Young 1987; Malamud 1987.
42. Young, "Comments," spring 2004.
43. Francis Cole, "Monthly Report of Activities," April 30, 1970, p. 6; June 30, 1969, p. 1; July 31, 1970, p. 1; and August 31, 1970, p. 9.
44. Francis Cole, "Monthly Report of Activities," October 31, 1969, p. 7; August 1, 1968; and September 30, 1969.
45. Francis Cole, "Monthly Report of Activities," November 30, 1969, p. 8; December 31, 1969, p. 7; February 28, 1970, p. 6; May 31, 1970, p. 2; and August 31, 1970.
46. Sanford 1976, p. 169; Francis Cole, "Monthly Report of Activities," October 31, 1971, p. 2. Also NAL 1968a, p. 5-3; R. Wilson 1987a; 1967b; 1967c, p. 4; Don Edwards, comments to Hoddeson and Kolb, February 2004.
47. Drasko Jovanovic, E-mail to Kolb and Hoddeson, February 6, 2004.
48. "Theoretical Physics Section Is Formed," *Village Crier*, November 1969, p. 1.

49. Clavelli 2001.
50. Clavelli 2001; Jackson 1999.
51. Quote in Clavelli 2001.
52. Clavelli 2001.
53. Sullivan 2005.
54. The Theoretical Physics Steering Committee, formed in December 1972, included Henry Abarbanel, Stephen L. Adler, J. D. Jackson (chair), Martin Einhorn, Ben Lee, Arthur Roberts, Val Telegdi, Sam Trieman, and Jimmy Walker). J. D. Jackson, "Recollections of Edwin L. Goldwasser and the Beginnings of the NAL Theoretical Physics Program," address presented at Goldwasser eightieth birthday symposium, Fermilab, March 10, 1999.
55. Francis Cole, "Monthly Report of Activities," May 31, 1969, p. 5; Cole, "Monthly Progress Report," September 30, 1969, pp. 8, 9; Cole, "Progress Report on the NAL Accelerator," *Particle Accelerators* 2 (1971): 5; "Monthly Report Main Ring Section," October 1969, p. 2.
56. NAL design report, January 1968. pp. 5–19.
57. Malamud, E-mail to Kolb and Hoddeson, January 9, 2005, based on his private documents. Also see J. Schivell and C. Schmidt, "Alignment Techniques for the NAL Main Accelerator," paper L-19 at 1971 Particle Accelerator Conference. According to Malamud, the laser alignment system "shoots a light beam 200 ft. (straight line) from one quadrupole to the quadrupole 200 ft. away. . . . A centering device placed on the quadrupole half way in between detects the beam. All three points (laser, 100 ft. away laser centering device and 200 ft away target) are precisely referenced to notches in the quadrupole laminations." See Ernest Malamud handwritten note; and Malamud, "Status of the 500 GeV Accelerator," *IEEE Transactions on Nuclear Science* 18 (1971): 948–952; Francis Cole, "Monthly Report of Activities," November 30, 1969, p. 9; January 31, 1970, p. 6; "Minutes of the Main Ring Staff Meeting," January 7, 1970.
58. Rich Orr to Adrienne Kolb, February 2, 2004.
59. Quotes, respectively, from R. Wilson 1987b; and tape recording by Henry Hinterberger and Robert Wilson, on June 21, 1982. Also see Livingston 1969a, p. 13; NAL 1969a, p. 8.
60. "Minutes of the Main Ring Staff Meeting," December 3, 1969; "Minutes of the Main Ring Staff Meeting," December 17, 1969; Yamada 1989.
61. R. Wilson 1987a.
62. Quotes, respectively, from "Minutes of the Main Ring Staff Meeting," December 3, 1969; and Francis Cole, "Monthly Progress Report," December 31, 1969, p. 8; and April 30, 1970, pp. 11–12.
63. Weart 1979, p. 328. Also p. 327.
64. Transcript, Second User's Meeting, December 2, 1968; U.S. Congress, Joint Committee on Atomic Energy 1971a, p. 1214; and Francis T. Cole, "Monthly Report of Activities," July 31, 1969, p. 1.
65. Francis Cole, "Monthly Progress Report," October 31, 1969, p. 8.

66. R. Wilson 1987b; "Minutes of the Main Ring Staff Meeting," December 10, 1969. For more information on the further development of mass-produced magnets at Fermilab, see Hoddeson 1987.
67. Robert Wilson, "Notes on Talk to Employees," June 4, 1970. Also Malamud 1989; Francis Cole, "Monthly Progress Report," December 31, 1969, p. 8; Malamud 1987.
68. Malamud notes to Hoddeson and Kolb, January 2004.
69. Ramsey letter to Kolb and Hoddeson, February 20, 2004; Robert Wilson, "Statement Made by R. R. Wilson at the Annual Meeting of the NAL Users Organization on Friday, April 10, 1970."
70. R. Wilson 1969a; MacLachlan 1989a. Also Jovanovic 1989; Malamud 1989.
71. "Minutes of the Main Ring General Meeting," May 28, 1970; Francis Cole, "Monthly Report of Activities, August 31, 1970, p. 12; September 30, 1970, p. 3.
72. Quote from Francis Cole, "Monthly Report of Activities," September 30, 1970, p. 1. Also R. Wilson 1987a.
73. Malamud notes to Hoddeson and Kolb, January 2004.
74. Malamud 1987; D. Edwards 1987; Francis Cole, "Monthly Report of Activities," October 31, 1970, p. 1.
75. "Main Accelerator Monthly Report," October 1970; Francis Cole, "Monthly Report of Activities," 31 March 1971, p. 1; April 30, 1971, p. 1.
76. R. Wilson, "Statement Made by R. R. Wilson at the Annual Meeting of the NAL Users Organization."
77. Quote from "Minutes of Main Ring Section Meeting on the Protomain," 16 March 1970. Also Yamada 1989.
78. Quote from Malamud 1989. Also Ernest Malamud to Mrs. Hanson, March 7, 1980, Malamud Papers.
79. U.S. Congress, Subcommittee on Research and Development and Radiation of the Joint Committee on Atomic Energy 1972, p. 1433.
80. See Francis Cole, "Monthly Report of Activities," June 30, 1971, p. 2, May 31, 1971, p. 2; and October 31, 1971, p. 2; Hoddeson et al. 1993; Willard Hanson to distribution, October 18, 1971; Malamud 1989.
81. Although Edwards had differences with Wilson's accelerator-building approach, she admits that he managed to extract "a sort of beauty." He would "insist that such and such a thing had to have such and such a dimension and that frustrated people, and yet what was produced was beautiful." H. Edwards 1987; Jovanovic 1989; Orr 1987; R. Wilson 1987a; Goldwasser 1987a; and Lederman 1990b.
82. Quotes from Francis Cole, "Monthly Report of Activities," August 1, 1971, p. 2. Also Ernest Malamud to all members of Main Ring Section, August 2, 1971; "Steering Meeting," September 23, 1971; R. Wilson 1987b, p. 14; Francis Cole, "Monthly Report of Activities," October 31, 1971, p. 3. On Felicia's role see "Tiny Ferret Aids Construction of NAL Meson Lab," *Village Crier* 3, no. 35 (September 2, 1971): 1.

83. Quotes from Robert Wilson to Norman Ramsey, October 29, 1971. Also Orr 1989; Robert Wilson, "Formation of the Accelerator Section," memorandum to the staff, October 21, 1971.
84. Orr 1987; Francis Cole, "Monthly Report of Activities," February 30, 1972, p. 2; U.S. Congress, Joint Committee on Atomic Energy, *Hearings*, 92nd Cong., 2nd sess. (Washington DC: GPO, 1972), p. 1434.
85. Goldwasser 1987a; R. Wilson 1987b; Collins 1989.
86. Quote from Collins 1990a; Young, "Comments," spring 2004. Also Yamada 1989; and Malamud 1989; Collins 1989; Jovanovic 1989, MacLachlan 1989a.
87. Ramsey to Hoddeson and Kolb, 2004.
88. Drasko Jovanovic, "What the Operations Report Did Not Say," Fermilab Directorate Personnel Files, January 1977.
89. Jovanovic, "What the Operations Report Did Not Say"; Hilts 1982, p. 97.
90. Orr 1987.
91. Wilson wanted to use the remaining funds to support research on the Energy Doubler. Hoddeson 1987, p. 35. Also Francis Cole, "Monthly Report of Activities," 31 January 1972, p. 1; 1 March 1972; U.S. Congress, Joint Committee on Atomic Energy 1972, appendix 3. p. 1731.
92. Eventually the telescope was funded. See chapters 4 and 5 in R. Smith 1989, quote on p. 87. See also pp. 89, 90, 99, 100, 109, 115.
93. Quote from U S. Congress, Joint Committee on Atomic Energy 1972, p. 1438.
94. Atomic Energy Commission, "AEC Names 200 BeV Accelerator in Honor of Enrico Fermi," press release, April 29, 1969, Seaborg Papers, Bancroft Library, University of California, Berkeley.
95. Proceedings of dedication, Fermi National Accelerator Laboratory, May 11, 1974, Fermilab booklet.

CHAPTER SEVEN

1. Wilson 1978g.
2. R. Wilson 1987b.
3. Robert Wilson to Edwin McMillan, September 27, 1965; R. Wilson 1965.
4. Wilson and Kolb 1997.
5. Wilson and Kolb 1997.
6. Traweek 1988, p. 33, and also pp. 18–45.
7. NAL 1968a, p. 2-4. R. Wilson 1987b, 1990b. An early NAL report reiterated the "policy that members of the regular scientific staff" were not "employed solely for the purpose of pursuing their own physics-research interests." "Physics-Research Section Monthly Report," August 1969.
8. NAL 1968a, p. 2-4; R. Wilson 1987b. An example of attempts to encourage resident users to collaborate with visitors can be found in Goldwasser memorandum to Physics Research Section, May 1, 1970.

9. Lach 1989; and Carrigan 1989.
10. R. Wilson 1987b.
11. Goldwasser 2004a.
12. Goldwasser 1992.
13. Goldwasser 1992.
14. Norman Ramsey letter to members of the high-energy physics community, March 6, 1967; J. R. Sanford to prospective members of the NAL users' organization, April 16, 1968.
15. Goldwasser 1987b; also Goldwasser 1987a.
16. "Summer Study Program, 1968," December 8, 1967; F. T. Cole, "Monthly Report of Activities," September 30, 1969, NAL-31, p. 10; August 31, 1970, NAL-46, pp. 1–2. Also Livingston 1969a, pp. 32–34; Ramsey 1975, pp. 11–12.
17. The *Village Crier* was the internal newsletter for NAL from 1968 to 1979, conveying news of the laboratory, the staff, and related physics information. Press releases were generated by the public information officer, Carl Larson, and later Margaret Pearson.
18. Goldwasser 1992.
19. Other members of the original committee were Thomas Fields, Tom Kirk, Wolfgang Panofsky, D. Reeder, Robert Sachs, Nicholas Samios, and William J. Willis. Francis T. Cole, "Monthly Report of Activities," December 31, 1969, NAL-35, pp. 1–2. Quotes from Goldwasser 1992. Also Francis T. Cole, "Monthly Report of Activities," March 31, 1969, NAL-40, p. 1; Kirk 1990b.
20. Robert R. Wilson, "Notice to NAL Users," March 26, 1970; G. Giacomelli, A. F. Greene, and J. R. Sanford, "A Survey of the Fermilab Research Program," physics reports, 19C/4 July 1975 (Amsterdam: North-Holland Publishing Co., 1975); Sanford 1976: 151–98.
21. Kirk 1990b. Also O'Halloran 1991; and Goldwasser 1992.
22. R. Wilson 1979b, p. 259
23. Giacomelli, Greene, and Sanford, "A Survey of the Fermilab Research Program."
24. Transcription from December 1968 users' meeting, roll 6, p. 18. Also Goldwasser 1992.
25. Quotes, respectively, from Bodnarczuk 1988, p. 68; and Goldwasser 1987a; also John Peoples, private communication to Adrienne Kolb, 2004.
26. A rich source of anecdotal information about the Fermilab 15-foot bubble chamber can be found in Bodnarczuk 1988, see esp. pp. 75–78; Livingston 1969a, p. 10; Fowler 1973, p. 1.
27. The end of Fermilab's bubble chamber era was marked in July 1990 with the completion of experiment E-782, which used the Tohoku bubble chamber.
28. Among them E-2B, 21A, 26, 53A (a "bubbler" experiment), 76, 98, 203, 262, 319, 398, 531, and 632.
29. E. L. Goldwasser, "Vignettes of Fermilab History: Remarks Made at the Robert R. Wilson Celebration," April 27, 1979.

30. Another collaboration, E-21A, led by Barry Barish, vied with E-1A to be first in the new neutrino beamline, but Wilson granted first access to HWPF.
31. Galison 1983; and Perkins 1997.
32. Lederman 1990a. Lederman's collaborator Bruce Brown shared stories of "being a pioneer trudging in the mud." He recalled seeing a picture of the setup of the P-Center area, before the Upsilon experiment was installed there, showing "snow falling into the experimental pit because the pit was still open." Like the other experimental areas where conditions were also spare, the Proton Area had no private toilets, only a community one, which, as Bruce Brown recalls, "in the developing spirit of the times, we called the 'Peoples John'" (a play as well on the name of the head of the Proton Area, John Peoples). B. Brown 1990c; C. Brown 1990.
33. "Experiments and Experimental Facilities," in "1968 Summer Study," 3:45–47. Lederman 1968. NAL 1968a. Lederman 1978, pp. 72–80, on p. 73; Kaplan 1994, pp. 1–2.
34. Drell and Yan 1970; Christenson et al. 1970.
35. T. D. Lee and G. C. Wick, "Negative Metric and the Unitarity of the S-Matrix," *Nuclear Physics B* 9 (1969): 209–243.
36. Jöstlein 1990. See also Kaplan 1994, pp. 4; Lederman 1978, p. 79; and Lee et al. 1970.
37. D. Saxon joined the proposal in December. L. M. Lederman, W. Lee, J. Appel, D. Saxon, M. Tannenbaum, L. Read, J. Sculli, T. White, and T. Yamanouchi, "Study of Lepton Pairs from Proton-Nuclear Interactions: Search for Intermediate Bosons and Lee-Wick Structure," addendum to proposal 70, December 1970.
38. Lederman to Wilson, February 24, 1971, Appel Papers, Fermilab Archives. Also Robert Peters to George MacPherson, January 21, 1971; and Columbia University, "Purchase Order to Bourns, Inc.," Appel Papers, Fermilab Archives.
39. Quotes from C. Brown 1990. Also Lee et al. "Study of Lepton Pairs from Proton-Nuclear Interactions" addendum to proposal 70; "Equipment Loan Agreement," July 1, 1971.
40. Quotes, respectively, from B. Brown 1990a; and Appel 1990.
41. Leon Lederman to Robert Wilson, February 16, 1971, Appel Papers, Fermilab Archives.
42. Saxon to Appel, Gaines, Lederman, Lee, and Read, "Results of Discussions at NAL," May 17, 1971. Also Lederman 1990b.
43. Lederman to Sanford, May 28, 1971.
44. Lederman to Wilson, June 7, 1971.
45. Jeff Appel, private communication to Hoddeson and Kolb, May 26, 2004.
46. Wilson to Lederman, June 28, 1971.
47. URA annual report for 1972, January 17, 1973, p. 2.
48. Orr 1990.
49. Universities Research Association, "Annual Report," 1972, p. 4; 1973, p. 2.

50. Quotes, respectively, from Appel to Lederman, September 25, November 6, and November 21, 1972. Also Francis Cole, "Monthly Report of Activities," October 31, 1972, p. 4. Donald Edwards notes that the extraction problems here resulted from transverse coupling. This problem was corrected by Rae Stiening and Edwards. See Fermilab E-027, November 1972.
51. Francis Cole, "Monthly Report of Activities," December 31, 1972, p. 1; January 31, 1972; and February 28, 1973, pp. 1–2; Roy Rubinstein, "Fermilab Research Program 1989 Workbook," pp. 109–112.
52. Nebeker 1994.
53. Appel to Lederman, September 25, 1972; and November 6, 1972.
54. Appel to Read, January 18, 1973; Francis Cole, "Monthly Report of Activities," February 28, 1973, pp. 1–2.
55. Quotes from Peoples to Appel, January 29, 1973. Also Appel to Peoples, February 6, 1973; Francis Cole, "Monthly Report of Activities," February 28, 1973, pp. 1–2. Appel notes to Hoddeson and Kolb, May 2005.
56. Lederman to Wilson, January 1974.
57. Quotes from Appel to Lederman, November 6, 1972.
58. NAL monthly report of activities, September 30, 1972, p. 4.
59. Jeff Appel, private communication to Hoddeson and Kolb, July 2004.
60. Appel, private communication to Westfall, 2001; Roy Rubinstein, "Fermilab Research Program 1989 Workbook," pp. 116; Appel 1990.
61. Lederman to Appel, November 6, 1972.
62. Francis Cole, "Monthly Report of Activities," April 3, 1973, p. 4; H. L. Allen, "Sequence of Events during Accelerator and Experimental Area Operations from 2105 on 5/23 until 0500 on 5/24 as Pieced Together," May 24, 1973; "Results of Experiment 70 Run of End of May."
63. Appel 1990, p. 28.
64. "Accelerator Utilization Summary—September 1973," *NALREP*, October 1973, p. 21. Also F. T. Cole, "Monthly Activities," June 30, 1973, p. 2; "Accelerator Performance—July 1973," *NALREP*, August 1973, p. 13; "Accelerator Utilization Summary—August 1973," *NALREP*, September 1973, p. 19.
65. H. Paar, H. Snyder, and J. Yoh, "Memo on Comparison of LF and μ ST," October 31, 1973; "Accelerator Utilization Summary—September 1973," p. 21–22; Jeffrey Appel "Appel to Experiment #70," memorandum; Appel, "Phase II Informal Analysis Discussion—December 10, 1973," December 10, 1973, memorandum to E-70; "Accelerator Utilization Summary—October 1973," *NALREP*, November 1973, pp. 9–12; and Appel, "E-70 Phase II Meeting—October 15, 1973," October 22, 1973, memorandum to Experiment #70.
66. David Saxon, "Hodoscope Reconstruction Routines," October 2, 1973; B. Brown 1990a.
67. Jeffrey Appel, "E-70 Phase II Meeting—October 15, 1973," October 22, 1973, memorandum to E-70; Appel, "Post Run Discussions of November 16, 1973," November 29, 1973, memorandum to E-70.

68. Jeffrey Appel, "Phase II Analysis Discussion—December 10, 1973," December 10, 1973, memorandum to E-70; Appel, "Group Meeting of January 8, 1974," January 11, 1974, memorandum to E-70; H. Paar and J. R. Repellin, "Electrons from Uninteresting Sources," February 14, 1974.
69. B. Brown 1990a, pp. 12–14.
70. J. A. Appel, "NAL Chart for NAL E-70," 1971; unsigned notes, November 22; David Saxon, "Phase III Magnet Studies Experiment 70," note, December 16, 1971.
71. Appel et al. 1974b, p. 722; Lederman 1990b, p. 5.
72. B. Brown 1990b; B. Brown to E-70 Experiments, "Phase II—A Minimal and a Maximal System," November 21, 1973; Lee et al. 1970.
73. Quotes, respectively, from B. Brown to E-70 Phase III experiments, "Hodoscopes Plans for Phase III," November 27, 1973; and B. Brown 1990b, pp. 16–17. Also Kaplan 1994, p. 4.
74. B. Brown to E-70 Phase III experiments, "Hodoscopes Plans for Phase III," November 27, 1973; J. Appel and J. Yoh to E-70, "Current Best Guess on Phase III Parameters," December 18, 1978; I. Gaines, J. Yoh, and D. Hom to E-70, "Phase III Electronics," December 25, 1973.
75. B. Brown 1990b, p. 73.
76. D. C. Hom, "Phase III Memo," February 5, 1974; Hom, "Phase III Memo," February 7; J. A. Appel and D. C. Hom, "New Lead Glass Cerenkov Proposal," May 7, 1974; B. Brown 1990b, p. 6; J. A. Appel to E-288, "Basic Options for Two Arm Trigger," May 16, 1974; B. C. Brown to E-70 and E-288, "Current Spectrometer Design E-288," May 21, 1974; Universities Research Association, "Annual Report," 1974, p. 5.
77. Quote from "Agreement," January 31, 1974. Also Lederman 1990b.
78. Appel et al. 1974.
79. Appel et al. 1974; Fermilab, "1989 Fermilab Research Program Workbook," p. 13; Kaplan 1994, p. 5.
80. B. Brown 1990b; H. P. Paar, J. Appel, M. Bourquin, I. Gaines, L. M. Lederman, J.-P. Repellin, D. Saxon, D. Hom, H. Snyder, J. Yoh, B. Brown, J. M. Gaillard, and T. Yamanouchi, "Search for Direct Electron Production in 300 GeV Proton Collisions," abstract submitted to the American Physical Society, April 22–25, 1974; Appel et al. 1974b, p. 722.
81. Appel et al. 1974b, p. 5. Also Lederman 1990a, p. 5.
82. B. Brown 1990b; Lederman 1990b; "Fermi National Accelerator Laboratory 1974 Historical Calendar of Program Accomplishments," Fermilab research program workbook, 1974, p. 16; and Appel 1990, p. 30.
83. C. Brown 1990, p. 13.
84. Appel et al. 1975, p. 9.
85. Riordan 1987, p. 297; Mary K. Gaillard, Benjamin W. Lee, and Jonathan L. Rosner, "Search for Charm," *Reviews of Modern Physics* 47 (1975): 277.
86. Riordan 1987.

87. Yamanouchi 1990, p. 10; Yoh 1990; B. Brown 1990b, p. 42; Appel 1990, p. 29; C. Brown 1990, p. 13.
88. Lederman 1990b, p. 4. Also B. Brown 1990b, p. 42; Weiss 1990a, p. 2.
89. Glashow 1988, p. 236.
90. As quoted in Riordan 1987, p. 306. This source contains a thorough, readable account of the discovery of the J/ψ. Also Glashow 1988, p. 246.
91. Hoddeson et al. 1997.
92. Quotes, respectively, from Lach 1989; and Peoples 1993a.
93. E. L. Goldwasser, "Recent Program Advistory Committee Meetings," *NAL-REP*, January 1975, pp. 1–8, quote on p. 6.
94. Peoples 1993a.
95. D. C. Hom, "E-288 On-Line Discussions," December 5, 1974; J. M. Weiss, "Hadron Rejection for E-288 With and Without Momentum Measurement," December 2, 1974; J. Yoh, "Signal and Background Rates for E-288 Side Lepton/Hadron," December, 1974; Appel 1990, p. 44; party announcement, December 9, 1974.
96. Over the next two years, these efforts to pursue hadron calorimetry would lead to the successful measurement of hadron pairs by E-494, a joint and parallel effort using the E288 apparatus. Bruce Brown, "Muon Identifier and Some Hadron Calorimetry Considerations," April 30, 1974; and Brown to E-288 Experimenters, "Hadron Physics in E-288—Some Possible Goals," January 3, 1975. Also B. Brown to E-70-E-288 Experimenters, "Test of Water Cerenkov Counter Ideas Using Muons," September 18, 1974.
97. B. Brown 1990b, p. 46 and 39. Also Kephart 1990, p. 15; Appel et al. 1974a; Appel 1976; C. Brown 1990, p. 34; Yoh 1990, p. 5; B. Brown 1990b, p. 6 and 42; "Agenda E-288," November 12, 1975.
98. Yoh 1990.
99. The collaboration was joined by a group from Stony Brook led by Good, which also included professors R. J. Engelman, Jöstlein, R. L. McCarthy, and H. Wahl, postdoc Kephart, and graduate student Daniel Kaplan. The newly constituted group was later joined by the following postdocs or graduate students: J. C. Sens, K. Ueno, Steve Herb, A. S. Ito, and R. J. Fisk. Kaplan 1994, p. 5.
100. Appel 1976.
101. B. Brown 1990b; Jöstlein 1990, p. 5, 22.
102. B. Brown 1990b, p. 21, 24.
103. Yoh 1990, p. 14.
104. Quotes from Kaplan 1994, p. 20; Yoh 1990. (Jeff Appel added, "You cannot imagine how luxurious this was compared to the outdoor options of the earlier period." Remark to A. Kolb, September 2003.)
105. Quotes, respectively, from B. Brown 1990c; and Weiss 1990a. See also Lederman 1990b.
106. Quotes respectively, from Jöstlein 1990; and Yoh 1990.

107. Quotes, respectively, from Appel 1976; and Appel 1990.
108. Appel notes to Hoddeson and Kolb, May 2005.
109. Kaplan 1994, p. 7. For an example of Lederman's accounting of the missed J/ψ occurrences, see "Spokesman Leon Lederman (An Unauthorized Autobiography)," *Village Crier*, January 6, 1977.
110. Quotes, from Kaplan 1994, p. 7, 8. Also C. Brown 1990.
111. Quotes, respectively, from Kaplan 1994, p. 8; Yoh 1990; and B. Brown 1990b.
112. Appel 1990.
113. Hom et al. 1976b, pp. 1236, 1239. The term dilepton began to be used to refer to a pair of electrons, or an electron plus antielectron pair; the term dihadron was used to refer to a pair of hadrons.
114. Kaplan 1994, pp. 8–9; and C. Brown 1990. Also B. Brown 1990b; B. C. Brown et al., "A Swimming Pool Hadron Calorimeter," *IEEE Transactions on Nuclear Science* 25, no. 1 (February 1978): 347.
115. Hom et al. 1976c, p. 1; Yoh 1990.
116. Yoh 1976, 1990. See also Hom et al. 1976a, p. 1374; Appel 1990; B. Brown 1990b.
117. Fermilab, "1989 Fermilab Research Program Workbook," May 1989, pp. 13, 23; R. McCarthy 1990; B. Brown 1990b.
118. Kaplan 1994, p. 10.
119. B. Brown 1990b.
120. Quotes, respectively, from R. McCarthy 1990; and C. Brown 1990. The E-494 data analysis led to a series of papers, beginning with Kephart et al. 1977.
121. Kaplan 1994, p. 11.
122. Lederman 1978, p. 76.
123. Lederman to Kolb, August 10, 2004.
124. Kaplan 1994, p. 12.
125. Kaplan 1994, p. 13.
126. Kaplan 1994, p. 13.
127. L. M. Lederman to File, n.d., received by Wilson June 9, 1977. Quotes, respectively, from Kaplan 1994, p. 13; and C. Brown 1990.
128. Quote from C. Brown 1990. Also Kaplan 1994, p. 13.
129. Lederman to Kolb, August 10, 2004; Lederman 1989.
130. Lederman to Kolb and Hoddeson, May 29, 2004.
131. Kaplan 1994, p. 14–15; Herb et al. 1977, p. 252.
132. Lederman to Kolb, August 10, 2004.
133. "The Discovery of Upsilon" 1977, p. 223.
134. As quoted in "!!Extra!! Fermilab Experiment Discovers New Particle 'Upsilon': A New Chapter in High Energy Physics," *Village Crier*, undated special issue, generally assumed to be late June or early July 1977, p. 3; and Kaplan 1994, p. 23.
135. Quotes from B. Brown 1990b. Also C. Brown 1990; Yoh 1990; R. McCarthy 1990.
136. Peoples 1993a.

137. Universities Research Association, "Annual Report," 1973, p. 1; 1974, p. 1; 1976, p. 1.
138. Quote from Lederman and Quigg 1979. Also Universities Research Association, "Annual Report 1975" January 15, 1976; "Annual Report," 1976; and A. F. Green, "What Happened at Fermilab during 1977?" Fermilab report, February 1978.
139. R. Wilson 1992.
140. Richter 1992; Peoples 1993a.
141. R. Wilson 1992.
142. Richter 1992; Peoples 1993a; Carrigan 1989; and Orr 1989.
143. One might profitably ask, in some future historical study, why Lederman's experiment E-288 and the Rubbia, Mann, and Cline study of neutral currents, E-1a, had such different outcomes. Both were approached with Wilson's cost-cutting and flexible style. Can these alternative outcomes be attributed to luck, or was there some important difference in the way the two efforts went about working at NAL?

CHAPTER EIGHT

1. R. Wilson 1977b.
2. R. Wilson 1977b.
3. R. R. Wilson, "Testimony before Representative Mike McCormack's Subcommittee on Energy, Research, Development, and Demonstration of the House Committee on Science and Technology," March 3, 1977.
4. Onnes 1913b, quote on p. 64; Onnes 1913a, on p. 65; Dahl 1984; Hoddeson et al. 1992; also Gorter 1946, 3–7.
5. Pioneering work in Germany in 1941 by G. Aschermann, E. Friedrich, E. Justi, and J. Kramer showed niobium nitride to be superconducting at 16.1 K, about the temperature of pumped liquid hydrogen. Aschermann et al. 1941.
6. Hulm 1983.
7. Hulm et al. 1981; Hulm 1983; An early announcement of the high-field superconductor is Kunzler et al. 1961.
8. Solenoids have superconducting properties different from those of straight wires; W. B. Fowler, "History of the Energy Doubler," unpublished manuscript, December 21, 1984, 8 pages. This brief account gives an excellent summary of the initial technological hurdles in developing viable magnets.
9. John Purcell built the superconducting coils for Argonne's 12-foot and Fermilab's 15-foot bubble chambers, and some half dozen superconducting units in the range of 3 tesla for use as beamline magnets. Fowler 1984. A useful summary of the early history of superconducting magnets for high-energy physics is included in Reardon 1977.
10. Adams 1971. Figure 2 refers explicitly to the superconducting portion at 1000 GeV.

11. P. Smith 1971; P. Turowski, J. H. Coupland, and J. Perot, "Pulsed Superconducting Dipole Magnets of the GESSS Collaboration," in *International Conference on High Energy Accelerators, IX, Proceedings* (Stanford, 1974), pp. 174–178; and McInturff 1984.
12. G. R. Lambertson, W. S. Gilbert, and J. B. Rechen, final report on the Experimental Superconducting Synchrotron (ESCAR), March 1979, LBL-8211.
13. For a history of ISABELLE, see Crease 2005; and Month 2003. The name ISABELLE, an extension of the acronym ISA, stood for intersecting storage accelerator, and was the name of Brookhaven physicist John Blewett's sailboat.
14. Lawrence W. Jones, notes of discussions on superconducting magnets in relation to storage rings and colliding beams, at O'Hare airport, May 21–22, 1967; A. van Steenbergen, "200–400 BeV Accelerator Summer Study, National Accelerator Laboratory, Jul–Aug 1967," notes taken at the Oak Brook meetings; R. Wilson 1978a, p. 7; Wilson to Hoddeson, December 11, 1986; author's conversations with Jones and Francis Cole.
15. R. R. Wilson to NAL users, December 8, 1972.
16. R. R. Wilson, "An Energy Doubler for the 500 BeV Synchrotron," notes, September 1970; Reardon and Strauss 1973.
17. Lundy 1983; and Rode 1984.
18. R. R. Wilson, "1000 GeV in the NAL Synchrotron," April 1, 1970. See figure 5 of H. Edwards 1985. The balls are clearly visible between the top and bottom magnets on the left.
19. U.S. Congress, Joint Committee on Atomic Energy 1971b. See pp. 1191–1247. Quotes are on pages 1206 and 1214. The figure is on p. 1205. The first published mention of the Doubler appears to be in Wilson 1970b.
20. As explained in chapter 10, the Doubler was redesigned to better serve Fermilab's future, and the cost went up. As Lederman later wrote, "The increased costs were totally driven by operational requirements. . . . It was expected to do experiments, which were never in Wilson's dream of a superconducting machine." Lederman to Hoddeson and Kolb, May 29, 2004.
21. The report of the Accelerator Conference of September 1971 contained half a dozen references to NAL's proposed Energy Doubler. McDaniel to Wilson, July 19, 1971; Wilson, "Annotations from the September 1971 Aaccelerator Conference Report with References to NAL Energy Doubler."
22. R. Sheldon and B. Strauss, ".5 Meter Prototype Energy Doubler-Quadrupole Magnet," July 1971, FN 235.
23. W. B. Fowler and P. J. Reardon, "Preliminary Suggestions for Starting the Construction of Prototype Magnets and Refrigeration Systems for the Proposed Energy Doubler," February 24, 1972.
24. Reardon 1984.
25. "Confirmation of Meeting Announcement," August 29, 1972, and Wilson's notes for meeting of September 1, 1972.
26. The term "pancake" refers to a flat magnet winding all in one plane, while "shell" refers to a more complex three-dimensional pattern.

27. Energy Doubler Magnet Evaluation Monthly Progress Reports, November 1972 and January 1973.
28. Reardon and Strauss 1973, I-3; D. Edwards 1984; and Rode 1984.
29. VanderArend was an imaginative outside contractor with considerable experience in cryogenics. He had worked in the 1950s for the air force and NASA on liquid hydrogen as a fuel for airplanes, as well as at the cryogenic engineering laboratory of the National Bureau of Standards in Boulder, Colorado, where he became involved with liquid helium technology. In the 1960s, he had also worked on conceptual issues in the building of the Argonne 12-foot liquid hydrogen bubble chamber. During this time at Argonne, VanderArend formed the engineering and consulting firm Cryogenics Consultants, Inc. (CCI), with which NAL later contracted to build the helium pipeline for the Doubler at low cost. VanderArend 1983.
30. Reardon and Strauss 1973, I-4.
31. Edwin L. Goldwasser to F. C. Mattmueller, December 21, 1972.
32. Mattmueller to Wilson, February 13, 1973.
33. Robert F. Bacher to Wilson, December 14, 1973, FN-263; Universities Research Association, Inc., annual report for 1974, January 1, 1975. See also Livdahl 1984.
34. Richard Lundy to Kolb, January 29, 2004; Strauss 1978.
35. Today variations of both are used in all superconducting accelerator magnets. The design of the Rutherford cable is based on the "Litz wire" phenomenon discussed earlier by Peter Chester, based on the discovery that current densities grow higher as wires are made thinner and that twisting the wires causes canceling of eddy currents and thus achieves greater stability. George Gallagher-Daggit used the principle at the Rutherford Laboratory to make the first "Rutherford cable." Gallagher-Dagitt 1974; M. Wilson 1972. For a discussion of problems of the braided cable, see Palmer 1984; and Tannenbaum 1984.
36. Koepke 1984; Energy Doubler design study, January 31, 1974, pt. 7, pp. 3–5.
37. Energy Doubler design study, January 31, 1974, pt. 7, p. 5; and B. P. Strauss and D. F. Sutter, "Evaluation of Matched Superconducting Dipoles," TM-456, December 11, 1973.
38. Energy Doubler design study, January 31 1974, pt. 6, p. 13; Strauss and Sutter, "Evaluation of Matched Superconducting Dipoles"; Willard Hanson to Henry Hinterberger, March 28, 1975; Wilson to Hoddeson, December 11, 1986.
39. VanderArend and Fowler 1973; W. B. Fowler and P. VanderArend, "Refrigeration System for the NAL Energy Doubler," CCI reports/Energy Doubler, subcontract no. 8073, April 26, 1976; D. Edwards 1984; W. B. Fowler and P. VanderArend, "The Cryogenic System for the Proposed NAL Energy Doubler," TM-421, VII1–VII9.
40. Appel 1984.
41. R. Wilson 1978a; Livdahl 1984.

42. Fowler had been exploring superconducting magnets for bubble chamber wire; Strauss, with an MIT degree in metallurgy and material science, had worked on superconductivity at Avco-Everett Research Laboratory and at Argonne.
43. Biallas 1984.
44. Biallas 1984.
45. Livdahl 1983.
46. Reardon 1984; Livdahl 1984. Also W. B. Fowler, personal communication.
47. Tollestrup, notebook, November 1975.
48. Tollestrup notes, winter 1975–1976, green pages showing concentric circles with 1.5-inch and 3.75-inch radii.
49. Tollestrup, research notebook, December 1985–February 1986. Drawing 1620-MA-96120 in the Fermilab drawing files shows the first collar design, probably by Henry Hinterberger. Hinterberger, Strauss, and Hanson settled the keystoning. Romeo Perin of CERN made the first suggestion of the Roman arch in September 1975. R. Perin to G. Biallis, March 22, 1977. We thank A. Tollestrup for calling our attention to this letter.
50. Construction notes on the Doubler magnets, book 3; and Livdahl 1984.
51. First magnet book, E series, December 31, 1975; Tollestrup 1983; Tannenbaum 1984. ISABELLE researchers later suggested that wrapping their braided wires with Kapton might have eliminated the high-field training of their early magnets.
52. Rolland Johnson and Peter Limon, "Modest Colliding-Beams Meeting," *NALREP*, April 1976, pp. 1–10; "Colliding Beams at Fermilab," *Village Crier* 8, no. 13 (March 25, 1976): 1–2.
53. Tollestrup to Hoddeson, E-mail, February 13, 2005.
54. Tollestrup to Hoddeson, E-mail, February 13, 2005.
55. Tollestrup 1987.
56. Magnet Measuring Group, notebooks; Yamada 1983.
57. Hoddeson et al. 1993, chap. 1.
58. Tollestrup 1983; and Koepke 1984.
59. Koepke 1984
60. Tollestrup 1983.
61. Lundy 1983.
62. Orr 1983; and Lundy 1983.
63. 1 Tesla = 10 kG. "USA Accelerator Conference," *CERN Courier* 17, no. 4 (April 1977): 97–101, on p. 100.
64. Documentation of these developments is contained in the W. B. Fowler Doubler collection.
65. The ring is divided into six sectors, designated A, B, C, D, E, and F. Each sector is controlled and monitored from four service buildings, e.g., for sector A, at stations A1, A2, A3, and A4. In turn, aboveground service buildings separated by 100 feet connect with stations inside the tunnel, A11, A12, A13, etc. The system was invented by Ernie Malamud. Other service buildings are

centered above each long Collins straight section. The Energy Doubler added additional buildings for the liquid helium cooling system.

66. H. Edwards 1983b; "Starting Up the Energy Saver," Fermilab annual report, 1983, pp. 37–57.
67. H. Edwards 1983b, pp. 1–9, on p. 6; also Rode 1984.
68. Orr 1983
69. Limon 1983; Koepke 1984.
70. VanderArend 1983.
71. VanderArend 1983
72. Fowler et al. 1975, pp. 1125–1128; see also C. Rode, D. Richied, S. Stoy, and P. C. VanderArend, "Energy Doubler Refrigeration System," *IEEE Transactions on Nuclear Science* 24 (1977): 1328–1330; Rode 1984; Kuchnir 1985; and W. B. Fowler, personal communications to Lillian Hoddeson, 1983 and 1984. Biallas largely designed the cryostat, based on work by VanderArend, Fowler, and Moyses Kuchnir. M. Kuchnir, "E22–14 Cryostat Boil-Off," Fermilab internal report, TM 740, July 1977.
73. Livdahl 1984.
74. Mravca 1983.
75. Earlier meetings of HEPAP in 1974, 1975, and 1976, had prioritized DOE's high-energy construction projects: SLAC's PEP was the first priority and ISABELLE was second. The Doubler was third.
76. They specified an additional $10 million for "Fermilab to begin the exploitation of the Tevatron for fixed target 1 TeV physics." *High Energy Physics Advisory Panel Report of the Subpanel on New Facilities* (Washington, DC: ERDA, June 1977), pp. 8–12.
77. R. R. Wilson to J. R. Schlesinger, October 22, 1977.
78. R. Wilson 1970c.
79. Limon 1983.
80. R. Wilson 1992.
81. Ramsey letter to Kolb and Hoddeson, February 20, 2004; see "Wilson Submites Resignation," *Village Crier* 10, no. 7 (February 16, 1978): 1–2.
82. Peoples notes to Hoddeson and Kolb, January 16, 2004. According to Peoples, Wilson had written his letter of resignation in September or October of 1977 and showed it to several people, including himself and Rich Orr.
83. Wilson to fellow-workers at Fermilab, February 9, 1978.
84. Wilson to Ramsey, February 9, 1978; and "Wilson Submits Resignation," p. 1. See also "Fermi Lab Director Quits over Funds," *Chicago Tribune*, February 9, 1978.
85. Tollestrup 1983.
86. Harold Ticho to staff of Fermi National Accelerator Laboratory, February 23, 1978.
87. Wilson, handwritten note to colleagues, May 5, 1978.
88. R. Wilson 1978d, p. 50.
89. R. Wilson 1978d.

90. "Wilson Metal Sculpture Erected," *FermiNews* 1, no. 3 (May 25, 1978): 2.
91. Ramsey to Kolb and Hoddeson, February 20, 2004.
92. R. Wilson 1978d, 1978e.
93. Sazama et al. 2003.
94. Robert R. Wilson, private communication to Hoddeson, May 1978. Also Jackie Coleman, private communication to Hoddeson and Kolb, 2003; and Sazama et al. 2003.
95. R. Wilson 1978e.
96. R. R. Wilson to all staff, May 24, 1978.
97. Goldwasser to laboratory, June 6, 1978.
98. E. Goldwasser to laboratory, June 16, 1978.
99. Robert R. Wilson to staff, July 17, 1978.
100. Norman Ramsey to the staff, July 17, 1978.
101. Livdahl to Hoddeson and Kolb, February 13, 2004.
102. Ned Goldwasser, "A Toast to Bob Wilson on his 80th Birthday," March 4, 1994.
103. Hoddeson 1987.
104. Combination of Wilson draft of April 20, 1978, and final publication of foreword to "Fantasies of Future Fermilab Facilities," *Reviews of Modern Physics* 51, no. 2 (April 1979): 259–273. Quote from Robert Browning.
105. Lederman, private communication to Hoddeson and Kolb, June 4, 2004.
106. R. Wilson 1992.
107. Peoples notes to Kolb and Hoddeson, January 16, 2004.
108. Lederman, private communication to Hoddeson, 1999.
109. "Employees Hear Lederman Appointment," *FermiNews* 1, no. 25 (October 26, 1978): 1–3, quote on p. 2.

CHAPTER NINE

1. Thomas 1981, p. 43; included in Lederman 1980.
2. Lederman, private communication to Lillian Hoddeson, July 2001.
3. M. Browne 1981, quote on p. 48.
4. Lederman with Teresi 1993, p. 8.
5. Lederman 1987.
6. The Nobel Lecture of L. M. Lederman (1989b), quote on p. 548.
7. Lederman's simple parity violation experiment, analogous to the historic, independently conceived, experiment of Madame C. S. Wu, studied the asymmetries in weak interactions in low-energy nuclear physics. Like Wu, Lederman verified T. D. Lee and C. N. Yang's startling theoretical prediction of the nonconservation of parity in weak interactions.
8. After Lederman returned to Columbia, Richard Garwin led the g-2 experiment, followed by Emilio Picasso, Frances Farley, and others.
9. Danby et al. 1962, p. 36.

10. He was targeted by demonstrating Columbia University students in 1968–1969 for this association. Many distinguished physicists contributed to Jason's research reports and recommendations, among them Luis Alvarez, James Bjorken, Sidney Drell, Freeman Dyson, Val Fitch, Richard Garwin, Murray Gell-Mann, Marvin Goldberger, George Kistiakowsky, Robert Oppenheimer, Matt Sands, Jeremiah Sullivan, Charles Townes, Sam Trieman, Steven Weinberg, John Wheeler, Eugene Wigner, Fred Zachariasen, and George Zweig.
11. Kolb and Hoddeson 1993.
12. Lederman 1997.
13. Faculty Council profile, Pisa 1988.
14. Lederman 1988, pp. 1–14, on p. 4.
15. C. Brown 1990; Yoh 1990; Jöstlein 1990.
16. Chrisman 2001.
17. "Leon M. Lederman," in *Current Biography Yearbook* (The Bronx: H. Wilson Co., 1989), vol. 50, no. 9, pp. 328–332, quote on p. 331.
18. An often-told Lederman joke, repeated to Hoddeson and Kolb on June 4, 2004.
19. Chrisman 2001.
20. Lederman, private communication to Adrienne Kolb, April 2001. James Leiss, who served as DOE's associate director for energy and nuclear physics in the Office of Energy Research during ISABELLE's final year later noted: "Congressmen don't like to burn their political chips to support a project only to learn that the whole community isn't behind the idea. Leiss 1991. Jefferson Laboratory's later beginning was threatened because of the fallout of ISABELLE's cancellation. J. Bennett Johnson, who chaired the committee that initially approved ISABELLE, angrily asked whether the nuclear physics project would simply turn into another ISABELLE. See Westfall 2001b.
21. R. Wilson 1969b.
22. CDF (the Colliding Detector at Fermilab) was built between 1982 and 1987; the nonmagnetic detector DZero was built between 1984 and 1992 (see chapter 12).
23. Appel 1989, pp. 75–79.
24. Lincoln Read, private communication to Mark Bodnarczuk, 1983.
25. Leon M. Lederman, "International Committee on Future Accelerators," Fermilab report, July 1980, p. 6.
26. Rubinstein, private communication to Bodnarczuk, 1983.
27. K. Brown 1989, pp. 1, 2, 5.
28. *FermiNews* 2, no. 29 (July 19, 1979): 1.
29. The arts program was the idea of Arthur and Janice Roberts; Al Brenner, the head of the Computing Department, served as the first auditorium committee chair. Ruth Ganchiff, Jane Green, Nancy Peoples, Saundra Poces, and Angela Gonzales served as the "hanging committee" that created exhibits in

Fermilab's art gallery; Joanie Bjorken, and later Avril Quarrie, took over the administration of the guest office. See Fermilab annual report, 1988, p. 20.

30. Mravca 1983.
31. Fermilab annual report, 1988.
32. P. Livdahl to Lillian Hoddeson and Adrienne Kolb, February 13, 2004; Teng 2005.

CHAPTER TEN

1. Lederman 1984b.
2. Livdahl announced the establishment, effective October 1, 1978, of the Energy Doubler Magnet Division, placing William Fowler as head and Richard Lundy as deputy head, but the project was by no means secure. Livdahl to Fermilab staff, September 29, 1978.
3. A later meeting of this committee was also held after the decision was made on January 18–19, 1979. Notes by Judy Ward; Richter, Sands, and McDaniel to Lederman, January 22, 1979, and June 18, 19, 1979.
4. L. M. Lederman to scientific staff, October 26, 1978.
5. Livdahl 1984.
6. Lederman to Ryuji Yamada, October 26, 1978, CDF Collection, Fermilab Archives.
7. Hoddeson 1987.
8. Malamud, private communication to Hoddeson and Kolb, 2004.
9. DOE, "General Science and Basic Research-Operating Expenses and Capital Acquisition FY 1979," congressional budget request, construction project data sheet; "DOE Authorizes Fermilab to Build Superconducting Accelerator" 1979, p. 2; James Leiss to John Deutsch, James S. Kane, and William Wallenmeyer, "Energy Saver Recommendations," March 20, 1977.
10. Minutes of the Weekly Magnet Measuring Group, December 11, 1978, no. 4; William Fowler, draft, December 12, 1979, for the 1979 Fermilab annual report; Mravca 1983.
11. The group included such leading accelerator physicists as Frank Cole, Tom Collins, Don and Helen Edwards, David Johnson, Peter Koehler, Sho Ohnuma, Rich Orr, Rae Stiening, Lee Teng, and Alvin Tollestrup.
12. Livdahl 1984. See also Lee Teng, Tom Collins, Helen Edwards, Don Edwards, and Rich Orr to Leon Lederman, September 7, 1978.
13. H. Edwards 1983a.
14. The problem is that the energy of the unextracted portion of the beam can be absorbed by the magnets and cause them to quench.
15. Orr 1983
16. The best overview of the work of the UPC from late 1978 to early 1984 is contained in the "UPC reports," available online and in the Fermilab library. The reports range over topics such as abortable beam, dipole field quality, and

beam extraction. The first crucial year of UPC meetings is, unfortunately, undocumented.

17. Orr 1983.
18. Orr 1983.
19. Orr 1983; Orr to Adrienne Kolb, February 2, 2004; Collins 1979.
20. H. Edwards 1983b.
21. "Summary of Magnet Test Data," June 14, 1979; Also *FermiNews*, July 12, 1979, pp. 1–3.
22. A. V. Tollestrup, "Status of the Fermilab Tevatron Project," April 1979, TM-880.
23. D. Edwards 1984; Tollestrup, "Status of the Fermilab Tevatron Project." Lederman also accepted the UPC's suggestion to separate the trim-coil package and some diagnostic equipment from the quadrupole magnets, which again cost space in the lattice as well as money but promised greater reliability. The wisdom of the decision was not widely appreciated until the machine was turned on four years later.
24. M. Browne 1979; R. R. Wilson to J. R. Schlesinger, October 22, 1977.
25. The laboratory had not requested this increase. It came from the agency, which was not confident that a budget of $38.9 million would suffice; L. Lederman to B. McDaniel, B. Richter, and M. Sands, July 11, 1979; Mravca 1983.
26. Following verbal agreement on July 3, 1979, the news was publicized by the media on July 5, 1979.
27. "DOE Authorizes Fermilab to Build Superconducting Accelerator" 1979, quote on p. 2.
28. "Energy Saver Project Management Plan," July 23, 1979, signed by Leon Lederman, Fred Mattmueller, and James E. Leiss.
29. Mravca 1983; Campbell 1979.
30. Mravca 1983.
31. The change from AEC to ERDA took place in January 1975, and from ERDA to DOE in October 1977.
32. Lehman 2001. Also Galison 1997, p. 606.
33. Mravca 1983.
34. Mravca 1983; Lundy 1983.
35. Hanson, a pioneer of NAL's Booster and Main Ring, and the Tevatron's magnet factory, died in early 1980.
36. Biallas 1984.
37. Lundy 1983.
38. Sho Ohnuma, "Field quality of dipoles in the tunnel," memorandum to the sector test group, September 11, 1978, in R. Yamada, "Energy Doubler Model Magnets," notebook, 1978; A. V. Tollestrup in PMG, minutes of meeting, July 17, 1979; Orr 1983.
39. Limon 1983.

40. Tollestrup in PMG, minutes of meeting of February 28, 1980.
41. Tollestrup in PMG, minutes of meeting of March 18, 1980.
42. It increased the heat leak slightly and necessitated reworking the magnets, which now required more refrigeration and more plumbing. The centering problem proved to be associated with the prestress put on the magnet support structures in order to insure that when the central magnet shrinks (some 20 mils) on cooling, enough force is left to center the coils in the iron.
43. The smart bolts also reduced the quadrupole moment on the dipole magnets almost to zero, because they could easily be moved until the magnet windings induced just enough additional quadrupole moment to cancel the quadrupole moment from the core. Zero quadrupole movement is in practice hard to achieve, because it requires the coils to be wound exactly symmetrically, with no errors in the windings.
44. A. V. Tollestrup in PMG, minutes of meetings of July 30 and September 4, 1980; Mravca 1983.
45. H. Edwards 1983b.
46. Orr 1983.
47. For a historian of technology's analysis of the consequences of reverse salients in the growth of technological systems see T. Hughes 1989, pp. 71–74.
48. Murphy 1984.
49. Murphy 1984.
50. See last magnet production status report attached to P. F. M. Koehler to Saver PMG members, "Minutes of Saver PMG Meeting on March 15, 1983," March 24, 1983. Also H. Edwards 1983b.
51. Toohig 1983.
52. Orr 1983.
53. Limon 1983; and Lundy 1983.
54. Margot Slade and Wayne Riddle, "Fermilab's New Accelerator Is A Smash Hit," *New York Times*, July 10, 1983, p. 4.
55. Program, "Dedication of the Energy Saver," April 28, 1984.

CHAPTER ELEVEN

1. Lederman 1984b.
2. Strings can derive from the investment in a certain kind of physics, a specialized technique, a piece of apparatus. The investment that Fermilab made in its 15-foot bubble chamber gave rise to the string of E-28A, 31A, 45A, 53A, 155, 172, 180, 202, 234, 341, 343, 380, 388, 390, 545, 546, 564, and 632.
3. Bodnarczuk came to this notion in the 1980s but did not publish it until 1997. In the meantime, a few historians had taken up the notion in their work, e.g., Nebeker 1994; and Genuth 1992.
4. Peter Galison has vividly portrayed the increase in size of particle detectors, tracing them from devices of tabletop size to "hangar-size" instruments, like

the absorbers used in 1963 by Lederman, Schwartz, and Steinberger, made of surplus deck plates for naval cruisers and weighing several thousand tons, or the mammoth spark chamber used to seek evidence for neutral currents in Fermilab's E-1A experiment. Galison 1989; also see Galison 1987, pp. 197ff.; and Galison 1983.

5. The SMD was invented between 1978 and 1980 by Robert Klanner, in Europe.
6. Strings similar to those at Fermilab can be identified at other large laboratories, including CERN and Brookhaven—e.g., the photoproduction experiments done at CERN: (NA1, NA14, NA14/2) and (WA4, WA57, WA58, WA69).
7. Fermilab's experimental area offered some flexibility for continuity of research pursuits in alternative areas; e.g., E605 was conducted in the Meson Area, whereas the previous experiments were done in the Proton Area.
8. Charmed particles, in the middle generation of the Standard Model, along with "strange" quarks, are hadrons (e.g., protons and neutrons) that contain at least one "charm" quark, one of the six kinds of known quarks. The other four kinds of quarks are "up" and "down" quarks in the first generation and the "top" and "bottom" quarks in the third generation.
9. A "broad-based" beam containing photons of all energies was also created at Fermilab.
10. The design included the primary and secondary beam transport systems, the photon-tagging system, and the experimental hall where the experiments would be conducted.
11. Nash letter to Wilson, Goldwasser, and the Fermilab PAC, June 6, 1976.
12. For details of the initial beam design see P. Davis, R. Morrison, T. Nash, J. Prentice, J. Cumalat, R. Egloff, G. Luste, F. Murphy, "The Electron Beam Test of October–November, 1974," Fermilab TM-535, December 6, 1974.
13. TPMS design report, May 9, 1977, p. 2.
14. Nash 1990.
15. Goldwasser to Nash on November 19, 1976.
16. C. Heusch to Goldwasser, June 2, 1977.
17. C. Heusch to Goldwasser, June 2, 1977.
18. Nash to Goldwasser, June 15, 1977.
19. Goldwasser to Nash, June 29, 1977.
20. Nash's letter to Goldwasser on October 3, 1977, codified the expected physics results. The approval came in Goldwasser to Nash, November 15, 1977. The experiment was given 5×10^{12} protons incident on the primary production target at a rate of 4 beam spills per minute for 1,000 hours.
21. "The On-Line System for Experiment E-516: How Big Is It?" May 11, 1978.
22. R. Morrison and U. Nauenberg to Lederman, December 26, 1979.
23. Nash to E-516 collaboration, July 21, 1980.
24. Galison 1983.
25. Nash to Morrison, December 9, 1980.

26. Gelfand to Nash, September 25, 1980.
27. Nash to Morrison, December 9, 1980.
28. Morrison 1990.
29. QCD postulates that the quarks within hadrons (e.g., protons) are bound together by the exchange of gauge particles called gluons. When an incident photon beam impinges on the hadron, a "virtual" charmed quark pair can be created. If this pair interacted with a gluon from one of the quarks of a proton or neutron in the experimental target, the imparted energy would knock the virtual quarks apart. As they separated, the effective mass of this high-mass charmed state would increase, causing the virtual quarks to become a "real" pair of charmed D mesons that would decay into the particles that E-516 was attempting to detect with the TPMS. See Nash to Goldwasser, October 3, 1977, p. 4. The diffractive model is based on an earlier framework called "vector dominance," championed in the 1960s by Jun John Sakurai, specifying that the strong interaction between particles could be explained by the exchange of vector mesons. See Nambu 1989, pp. 640ff. The evidence for this diffractive model came from experiments performed at energies below 60 GeV. But the Tevatron provided a proton beam of 800 GeV that produced a photon beam of about 300 GeV. For details see the P-516 proposal, see p. 11 ff.
30. See Appel to Nash on January 30, 1981, for the draft proposal and Nash to Gelfand, February 4, 1981, for the proposal that was submitted to the directorate and designated as P-691.
31. Gelfand to Nash, March 12, 1981.
32. Nash to Nauenberg and Bronstein, September 24, 1981.
33. Steve Bracker, "Bracker's New Track News," November 3, 1981.
34. Nash to Yamanouchi and Gelfand, May 6, 1982.
35. Tom Nash, "A Program for Advanced Electronics Projects at Fermilab," May 11, 1982.
36. Paul Karchin, "Silicon Microstrip Detectors for the TPS," October 18, 1982.
37. Gelfand to Nash, March 3, 1983.
38. For example, problems of the segmented liquid ionization calorimeter (SLIC) fell within the domain of the Santa Barbara contingent, while problems with the Cerenkov counters were considered within the purview of the University of Colorado. Because many of the detector's subsystems were constructed at the home institution of particular groups and then shipped to Fermilab for installation, there had been less social interaction between the members of the collaboration during the earlier phases of the experiment. Interaction increased as the collaboration approached publication. For a description of the computing architecture, see Nash 1983.
39. The first shakedown run of E-516 was in the summer of 1979; the first data run began in late fall 1979. The final data run began early in 1981 and ended at the accelerator shutdown in June 1981, with the data reconstruction processes continuing into the early part of 1983. In addition

to the multiple packages written for the tracking system by Bronstein, Nauenberg, and Bintinger, there were two separate reconstruction packages for the SLIC.

40. See "First Results from the Tagged Photon Spectrometer," photocopies of the overheads from the Wine and Cheese Seminar, by Kris Sliwa, June 10, 1983.
41. Nash to Lederman, Yamanouchi, and Gelfand, June 13, 1983; Nash to Yamanouchi and Gelfand, June 13, 1983. See also Nash, "Fermilab's Advanced Computer Program," Fermilab annual report, 1984, pp. 49–55.
42. Lederman to Nash, June 30, 1983.
43. Witherell to the TPS collaboration, August 24, 1983.
44. Draft of the publication "Inelastic and Elastic Photoproduction of J/psi," September 7, 1983.
45. Handwritten note from Paul Karchin to the collaboration, September 7, 1983.
46. Witherell to entire E-516 collaboration, October 5, 1983.
47. D. J. Summers et al., "Study of the Decay $D^0 \to K^-\pi^+\pi^0$ in High-Energy Photoproduction," *Physical Review Letters* 52, no. 6 (February 6, 1984): 410–413.
48. The TPS collaboration to Yamanouchi, October 15, 1983.
49. Lederman to Nash on November 17, 1983.
50. Nash to the E-691 collaboration, November 28, 1983.
51. "Proposal to Study Photoproduction of Final States of Mass Above 2.5 GeV with a Magnetic Spectrometer in the Tagged Photon Lab," submitted October 1, 1976, by J. Appel, P. Mantsch, and T. Nash (Fermilab), R. J. Morrison (University of California, Santa Barbara), and G. Luste (University of Toronto), pp. 2–3.
52. See "Proposal to Study Photoproduction of Final States of Mass Above 2.5 GeV with a Magnetic Spectrometer in the Tagged Photon Lab."
53. The proposal for E-769 was submitted to the laboratory in November 1985, approved in December 1985, and completed in February 1988, having a substantial time overlap with its predecessor, E-691. The proposal for E-791 was formally submitted to the laboratory in November of 1987, was approved in June 1988, and had data analysis in 1992.
54. Some, like Summers, left and returned; others, like Nash, Morrison, and Nauenberg, left at some point. A few, like Witherell, joined at a later stage.
55. The discontinuities between the two configurations included the addition of two banks of drift chambers to the tracking system (which enabled them to have greater redundancy in resolving the trajectories of particles penetrating the spectrometer) and the significant addition to the online computing capabilities (using a VAX11/780 for data monitoring and the PDP11/55 solely for the acquisition of data from the detector subsystems). For E-516, the PDP11/55 was used for both data acquisition and data monitoring.
56. The first hadron experiment, E-769, had less charm than E-691 but proved techniques that made hadron experiments more tractable.

CHAPTER TWELVE

1. Lederman 1985.
2. For more on multi-institutional, international collaborations, see Warnow-Blewett and Weart 1992, 1995, and 1999.
3. In particular, see Galison's description of the Time Projection Chamber (TPC) and the Superconducting Super Collider in Galison 1997 and Riordan 2001. For an examination of UA1, see Krige 1993. For a comparison of CDF with UA1, the TPC, and the nuclear physics detector CLAS, see Westfall 2001a.
4. The UA1 detector at CERN had about 50,000 channels, the Collider Detector at Fermilab (CDF), the DZero Collider at Fermilab, and the SLD detector at SLAC each had about 100,000 channels, the ALEPH detector at LEP had about 700,000 channels, and the SDC and GEM detectors that were proposed for the now defunct SSCL would have had as many as 50 million channels, depending on the technology that would have been available. For a history of UA1 and UA2, see Krige 1993.
5. Roy Rubinstein, "D0 Workshop," June 1981, Fermilab report, pp. 5–6.
6. Staley 2004.
7. Brown and Hoddeson 1983. Wilson's handwritten notes in September 1948, include a drawing of particle-antiparticle collisions. Cornell University Archives. The concept is referred to on p. 45 of R. Wilson 1978d.
8. Richter 1997, p. 263.
9. Ramsey Panel 1963.
10. Richter 1997; Pellegrini and Sessler 1995.
11. R. Wilson 1978f.
12. NAL 1968a, pp. 18-7–10. "Proton-proton colliding-beam storage rings for the National Accelerator Laboratory design report," design study, 1968, internal document; also see the notes taken by Lawrence W. Jones of the discussions at Oak Brook on superconducting magnets in relation to storage rings and colliding beams, O'Hare airport, May 21–22, 1967; A. van Steenbergen, "200–400 BeV Accelerator Summer Study, National Accelerator Laboratory, Jul–Aug 1967," notes taken at the Oak Brook meetings; R. R. Wilson, "Colliding Beams at Fermilab," workshop on high luminosity, high-energy collisions, March 27–31, 1978, Berkeley, Lawrence Radiation Laboratory report 7574, pp. 7–12, on p. 7; Wilson to Hoddeson, December 11, 1986.
13. "Proton-proton colliding-beam storage rings for the National Accelerator Laboratory design report," design study, 1968; Teng 1993.
14. R. Wilson 1978f.
15. Teng 1993; Proton-proton colliding-beam storage rings design study, 1968, introduction.
16. Carrigan 1971 proposal, NAL-FN 233.
17. They included the electron-positron machine (SPEAR) at SLAC; the electron-positron colliding facility VEPP-2 at Novosibirsk, USSR; ADA, the first

electron-positron collider, built by Bruno Touschek at Frascati in 1960; and ACO, an electron-positron collider which went on line at Orsay in 1965. Those under construction included the electron-positron collider ADONE, which began operating at Frascati in 1969; DORIS, the electron-positron double ring at the Deutsches Elektronen-Synchrotron (DESY); and an electron-positron collider at the Cambridge Electron Accelerator (CEA). For more information see Blewett and Vogt-Nilsen 1971.

18. For information on colliding-beam storage rings and the origins of CEA, see Paris 1999; CERN annual report, 1980, pp. 20–22; Amman 1989; K. Johnson 1997. One of the first confrontations of the ideas of stochastic and electron cooling occurred at a conference in March 1978 at Lawrence Berkeley Laboratory; "Proceedings of the Workshop on Producing High Luminosity High Energy Proton-Antiproton Collisions," March 27–31, 1978, Berkeley, LBL-7574. Note paper by P. McIntyre and A. Ruggiero, "Accommodating Stochastic Cooling at Fermilab," pp. 149–150.
19. URA annual report, 1974, Washington DC, January 24, 1975. R. Wilson 1978f, August 11, 1978.
20. Collins 1990b. The Cambridge Electron Accelerator was closed in 1973.
21. R. Wilson 1978f.
22. Mills 1987.
23. R. Wilson 1978f.
24. The Low panel recommended PEP at SLAC as the number 1 priority, ISABELLE as the second, and R & D for Fermilab's Energy Doubler as the third. John Peoples, note to Hoddeson and Kolb, January 16, 2004. Peoples was on HEPAP and participated in its June 1975 Woods Hole meeting. For the Brookhaven perspective, see also Crease 1999b, pp. 361–362.
25. Richter 1997; R. Wilson 1978f. According to Gary Taubes, Peter McIntyre subsequently convinced Rubbia that proton-antiproton collisions would be better, and less than six months later Rubbia began pushing for proton-antiproton collisions; Taubes 1986.
26. David Cline, Peter McIntyre, D. D. Reeder, L. Sulak, M. A. Green, E. M. Rowe, C. Rubbia, W. S. Tryeciah, and W. Winter (proposal 492); Cline, McIntyre, Reeder, Rubbia, and Sulak (493).
27. Rolland Johnson and Peter Limon, "Modest Colliding-Beams Meeting," *NALREP*, April 1976, pp. 1–10.
28. Johnson and Limon, "Modest Colliding-Beams Meeting," p. 8; CERN annual report, 1976, pp. 72; Taubes 1986, pp. 16–24.
29. Tollestrup 1987, 1990.
30. Goldwasser to A. Tollestrup, May 1976: Tollestrup to N. Goldwasser, May 5, 76.
31. (1) P-478 by James Walker et al., "Proposal to Search for Intermediate Boson Production in Proton Proton Collisions at 200 GeV in the Center of Mass"; (2) P-491 by Gene Fisk, Alvin Tollestrup, Rolland Johnson, Tom Collins, Peter Limon, John Peoples, Robert Walker, Leon Lederman, and Charles

Ankenbrandt, "Clashing Gigantic Synchrotrons"; and (3) P-480 by Luke Mo, Peter Schlein, and Andris Skuja, "Proposal to Search for Heavy Bosons, Heavy Leptons, and Charmed Particles at SSR."

32. Both proposals were from Cline, McIntyre, Reeder, Rubbia, and Sulak: P-492, "Proposal to Construct an Antiproton Source for the Fermilab Accelerators"; and P-493, "Search for New Phenomena Using Very High Energy pp and $\bar{p}p$ Colliding Beam Devices at Fermilab." A. Tollestrup to P. Livdahl, May 26, 1976.
33. Goldwasser 1976. Tollestrup 1990. "Chronology of Collider Detector Facility." When Rubbia's proposal was rejected by Fermilab in 1976, he took his idea to CERN, where it was accepted. CERN then converted its SPS into a Super Proton-Antiproton Synchrotron (S$p\bar{p}$S) colliding-beams facility. In 1981 Rubbia became spokesman for the UA1 experiment, which, with this collider, saw evidence in 1983 of both the $W^{\pm}$ and Z^{0} particles. For these discoveries the 1984 Nobel Prize for Physics was awarded to Rubbia and Simon van der Meer, the inventor of the stochastic cooling method used in Rubbia's experiment, and leadership at the high-energy frontier, measured in terms of the center-of-mass energy of colliding particles, passed briefly to CERN's 540 GeV S$p\bar{p}$S collider (center of mass 270 × 270 GeV).
34. A. Tollestrup to R. Wilson, October 6, 1976.
35. J. Cronin to colleagues, December 30, 1976.
36. Tollestrup 1987.
37. Wilson to laboratory, December 13, 1976.
38. Cronin memo written in pencil, February 7, 1977, CDF Collection, Fermilab Archives.
39. Minutes of Colliding Beams Department Working Group, February 1, 1977.
40. Report on the Subcommittee on Future Accelerators, June 1980, p. 14.
41. Cronin 1987.
42. Lederman 1989 and 1987.
43. Wilson to Cronin, December 13, 1976.
44. Wilson, labwide memo, December 13, 1976.
45. Tollestrup 1987.
46. "Faces around Fermilab," *Village Crier*, January 19, 1978, p. 1; and Fermilab organization chart, May 1, 1978.
47. Robert Wilson to James Walker, December 16, 1977; Wilson, labwide memo, December 19, CDF Collection, Fermilab Archives. Also see Fermilab organization chart, May 1, 1978; and "Faces around Fermilab," p. 1.
48. Peoples 1988.
49. Don Young 2004.
50. Holloway 1988; Tollestrup 1990.
51. Tollestrup 1987. Minutes of the CDF Collaboration Meeting, January 13, 1987, CDF Collection, Fermilab Archives.
52. Minutes of the CDF Collaboration Meeting, January 13, 1978, CDF Collection, Fermilab Archives.

53. Minutes of the CDF Collaboration Meeting, January 29, 1978, February 10, 1978, February 17, 1978, March 3, 1978, CDF Collection, Fermilab Archives.
54. Schochet 1990; and minutes of the CDF Collaboration Meeting, April 28, 1978, CDF Collection, Fermilab Archives.
55. Colliding Detector Facility Collaboration Meeting minutes, May 26, 1978, CDF Collection, Fermilab Archives.
56. Minutes of the CDF Collaboration Meeting, May 19, 1978.
57. Minutes of the CDF Collaboration Meeting, September 8, 1978.
58. Quigg 1993.
59. The Pisa group was supported by the funding agency INFN.
60. Minutes of the CDF Collaboration Meeting, April 28, 1978, and June 9, 1978, CDF Collection, Fermilab Archives. Quote from June meeting.
61. Minutes of the CDF Collaboration Meeting, February 9, 1979; H. Jensen, collaboration-wide memo, February 16, 28, March 30, May 4, July 13, 20, September 7, 21, December 19, 1979, CDF Collection, Fermilab Archives.
62. Minutes of the CDF Collaboration Meeting, June 29, 1979, and July 6, 1979; unnamed author to Berley, December 7, 1979, CDF Collection, Fermilab Archives.
63. Minutes of the CDF Collaboration Meeting, July 27, August 10, 1979, CDF Collection, Fermilab Archives.
64. Minutes of the CDF Collaboration Meeting, July 6, June 29, 1979, August 3, 1979, September 28, 1979, December 14, 1979, CDF Collection, Fermilab Archives.
65. Peoples 1999. Minutes of the CDF Collaboration Meeting, January 11, 1980, May 23, 1980, June 27, 1980, CDF Collection, Fermilab Archives.
66. "What Is a Danny Lehman Review?" *FermiNews*, April 18 1997; Lehman 2001.
67. A. V. Tollestrup to CDF Group, October 20, 1980.
68. A. V. Tollestrup to CDF Group, October 20, 1980.
69. Schwitters 1988b.
70. Schwitters 1988a. For a description of the TPC troubles, see Galison 1997, pp. 619–626. For ISABELLE's difficulties, see Crease 1999b, pp. 405–422.
71. Theriot 1993.
72. Minutes of the CDF Collaboration Meeting, July 17, 1981, May 22, 1981, August 14, 1981, CDF Collection, Fermilab Archives.
73. Design report for the Fermilab Collider Detector Facility (CDF), August 1981, CDF Collection, Fermilab Archives, p. 55.
74. Frisch 1989.
75. Frisch 1989.
76. Schwitters 1990; Frisch 1989. Schwitters explained to collaborators that they needed to reduce the budget to $40 million. The final figures were $37 million in total cost to Fermilab and only $30 million (the figure usually remembered by participants) for detector construction alone. "Cover Agreement for the Collider Detector at Fermilab," February 1983; minutes of the

CDF Collaboration Meeting, February 13, 1981, CDF Collection, Fermilab Archives.
77. Fermilab annual report, 1982, p. 15.
78. Abe et al. 1988.
79. Theriot 1993.
80. Frisch 1999.
81. Frisch 1999.
82. Droege 1990.
83. Minutes of the CDF Collaboration Meeting, November 13, 1981, March 19, 1981, April 25, 1981, September 30, 1983, CDF Collection, Fermilab Archives.
84. Schwitters 1988b; Fermilab annual report, 1983, p. 71–72.
85. Schwitters 1988b; minutes of the CDF Collaboration Meeting, November 21, 1983, January 13, 1984, May 11, 1984, July 20, 1984, October 19, 1984, January 11, 1985, CDF Collection, Fermilab Archives.
86. Schwitters 1988b.
87. R. F. Schwitters to CDF Collaboration, December 20, 1982, CDF Collection, Fermilab Archives
88. Minutes of the CDF Collaboration Meeting, November 4, 1983, 25 March 1983, June 24, 1984, CDF Collection, Fermilab Archives
89. Others on the committee included Tom Collins, Rolland Johnson, and Colin Taylor.
90. Peoples 1988; "The Fermilab Antiproton Source Design Report," June 1981.
91. "Antiproton Source Groundbreaking," *FermiNews* 15, no. 15 (August 21, 1992): 7; Peoples 1988.
92. Carlos Hojvat and George Biallis led the target station work. John Krider was responsible for its instrumentation. Gerry Dugan was the lead physicist of the lithium lens development. Dixon Bogert, Bob Ducar, Al Franck, and Dennis McConnell made critical connections between the target station and the accelerator control system. Steve Holmes, "First Tevatron I Protons: 'Revolutionary' Event," *FermiNews*, May 2, 1985, p. 1; John Krider, "Tev I Target Station Is Ready for Antiproton Production," *FermiNews*, June 27, 1985, pp. 2–3, 5.
93. Peoples 1988.
94. Peoples 1988.
95. Lederman quoted in "Tevatron I," in Fermilab annual report, 1983, pp. 17–23, quote on p. 17.
96. Lederman, p. 12; Fermilab annual report, 1984.
97. *FermiNews*, special edition, October 23, 1985, pp. 1–4. Excavation on DZero's experimental hall started on July 30, 1985. See also "Construction Begins on DØ Experimental Hall," *FermiNews*, August 8, 1985, pp. 3–4.
98. John S. Herrington, "Dedication of the Proton-Antiproton Collider: Tevatron I, Fermilab," October 11, 1985, pp. 25–30, Batavia, Fermi National Accelerator Laboratory.

99. *FermiNews*, special edition, October 23, 1985, pp. 1–4. Excavation on DZero's experimental hall started on July 30, 1985. See quotes esp. in "Construction Begins on DO Experimental Hall," *FermiNews*, August 8, 1985, pp. 3–4.
100. Frisch 1989. Also see minutes of the CDF Collaboration, March 8, 1985, April 8, 1985, June 14, 1985, August 16, 1985, CDF Collection, Fermilab Archives.
101. Frisch 1999. Also "CDF Workshop on 1985 Run," September 23, 1983; "Establishment of a Data Reduction Working Group," November 22, 1985, minutes of the CDF Collaboration Meeting, September 9, 1983, December 6, 1985; "CDF Snapshot," *FermiNews*, October 5, 2000. For details on the detector in 1988, see Abe et al. 1988.
102. Paul Grannis, "Talk on Lapdog," n.d., files of Bernard Pope, Michigan State University, East Lansing.
103. L. Lederman, "Second Colliding Area," *Fermilab Report*, February 1981, p. 2.
104. Pope 2000. Paul Grannis to Lapdoggers, May 28, 1982; Paul Grannis to Norman Gelfand, November 4, 1982; L. Ahrens et al., "Status of P714-LAPDOG," rejected July 1, 1983, February 1, 1983, files of Paul Grannis, State University of New York at Stony Brook.
105. Lederman to P. Grannis, July 1, 1983, files of Paul Grannis.
106. Bernard Pope, notes from conversation with Paul Grannis, June 27, 1983, files of Bernard Pope, Michigan State University, East Lansing.
107. Paul Grannis to Lederman, September 1, 1983, files of Paul Grannis, State University of New York at Stony Brook.
108. Leon Lederman to Paul Grannis, July 1, 1983, Paul Grannis to Lederman, September 1, 1983, files of Paul Grannis, State University of New York at Stony Brook; Grannis 2000.
109. Grannis 2000.
110. Leon Lederman to Paul Grannis, July 1, 1983, Paul Grannis to Lederman, September 1, 1983, files of Paul Grannis, State University of New York at Stony Brook; Grannis 2000.
111. Grannis 2000; Bernard Pope, notes from conversation with Paul Grannis, July 5, 1983; and Grannis to D0 collaborators, July 21, 1983, files of Bernard Pope, Michigan State University, East Lansing.
112. Pope 2000.
113. "Minutes of the D0 Meeting at Stony Brook, 18–19 Jul 1983," files of Bernard Pope, Michigan State University, East Lansing.
114. Pope 2000; also "UA2 Comments by Luigi DiLella," July 1, 1983, files of Bernard Pope, Michigan State University, East Lansing.
115. Grannis 2000.
116. "Design Report: The D0 Experiment at the Formula Antiproton-Proton Collider," November 1984, files of Paul Grannis, State University of New York at Stony Brook.
117. "Design Report: The D0 Experiment at the Formula Antiproton-Proton Collider."

118. Paul Grannis, private communication to Westfall, June 18, 2003.
119. "D0 Detector Review," November 19, 1984, files of Paul Grannis, State University of New York at Stony Brook.
120. "D0 Detector Review."
121. Grannis, "Near Term Needs of E740 from Fermilab," January 10, 1984, files of Paul Grannis, State University of New York at Stony Brook.
122. As quoted in Grannis, "Near Term Needs of E740 from Fermilab"; Grannis to D0 collaborators, July 21, 1983, files of Bernard Pope, Michigan State University, East Lansing.
123. Grannis 2000.
124. Grannis 2000; D0 Detector review no. 2, May 29, 1985.
125. Detector review, 13–14, 1986; Paul Grannis to Andrew Mravca, September 10, 1986; "D0 Atlas," June 1987, files of Paul Grannis.
126. Pope 2000; "D0 DOE Review," October 25–26, 1988, files of Paul Grannis, State University of New York at Stony Brook.
127. Grannis 2000; Fisk 1999; Pope 2000.
128. Grannis 2000. E-mail files recording Grannis's discussions with collaborators in the late 1980s and early 1990s clearly demonstrate Grannis's role and the D0 decision-making style. Fisk 1999; Pope 2000.
129. "Fermilab: D0 Central Calorimeter," *CERN Courier*, March 1990; "Fermilab Experiment Reports Results from World's Highest Energy Accelerator: Background Information," press release, April 1993, files of Paul Grannis, State University of New York at Stony Brook.
130. Paul Grannis to *Science* 253 (August 16, 1991): 719.
131. The final cost in 1997 dollars for CDF and DZero was $43 million and $70 million, respectively, but this CDF cost did not include considerable costs absorbed by Fermilab for "technical manpower" or the contributions of Japanese and Italian contributors. "DOE Detector Contingency Meeting," January 20, 1997, files of Bernard Mecking, Thomas Jefferson National Accelerator Facility, Newport News, VA.
132. *FermiNews*, September 15, 2000, p. 13.
133. This is discussed further in Krige 1993; and Westfall 2001a.
134. Frisch 1999.
135. Twain 1984, p. 416.

CHAPTER THIRTEEN

1. Lederman 1984c.
2. For more information on these developments, see R. Wilson 1978b; W. Owen Lock, "Origins and Early Years of the International Committee for Future Accelerators," draft, December 1982; Marshak 1990; the three volumes of the *History of the Atomic Energy Commission* (Hewlett and Anderson 1962, Hewlett and Duncan 1969, and Hewlett and Holl 1989); and Dickson 1984. Quote is from Seaborg 1987, p. 649.

3. The name was later changed to the Commission on Particles and Fields. See Marshak 1989.
4. R. Wilson 1975; Marshak 1989.
5. "Cooperation between the United States of America and the Union of Soviet Socialist Republics in the Field of the Utilization of Atomic Energy for Peaceful Purposes," memorandum, November 24, 1959.
6. R. Wilson 1978b.
7. Divine 1981.
8. R. Wilson 1978d.
9. R. Wilson 1978b.
10. R. Wilson 1961.
11. R. E. Marshak, "Notes on Adams' Meeting Re: International Accelerator Laboratory," memorandum to Atomic Energy Commission, July 7, 1961, R. E. Marshak papers, AIP. Or as Wilson later expressed it in 1968, a laboratory which "in developing our common culture in physical science," would "provide a force for international harmony." R. Wilson 1968.
12. "Atomic Energy: Scientific and Technical Cooperation in the Field of Peaceful Uses of Atomic Energy," June 21, 1973, Rolland P. Johnson Collection, Fermilab Archives.
13. Lock, "Origins and Early Years of the International Committee for Future Accelerators"; quote in V. Weisskopf letter to R. P. Johnson, December 13, 1974, R. P. Johnson Collection, Fermilab Archives.
14. V. Weisskopf to R. P. Johnson, December 13, 1974, R. P. Johnson Collection, Fermilab Archives.
15. R. Wilson 1984, pp. 9, 112.
16. Leon M. Lederman, "New Orleans—a Proposal," undated document.
17. E. L. Goldwasser, "Normalization of Inter-regional Cooperations and Communications," paper submitted to New Orleans seminar, n.d.
18. R. Wilson 1975, p. 120.
19. J. D. Bjorken, "Physics Issues and the VBA," May 1976.
20. The name was later modified to the International Committee *for* Future Accelerators. Goldwasser 1979.
21. Lederman 1977a.
22. Lock, "Origins and Early Years of the International Committee for Future Accelerators"; and the proceedings of the two ICFA technical workshops from 1978 and 1979; and the proceedings of the 1983 20 TeV Hadron Collider Technical Workshop; see also Diebold 1983; and *Superconducting Super Collider Reference Designs Study* 1984, p. 3.
23. *Report of the Subpanel on Review and Planning* 1980; transmittal letter from Sidney D. Drell to Edward A. Frieman, July 15, 1980, 1–4, DOE/ER-0066.
24. *Report of the Subpanel on Accelerator Research and Development of the High Energy Physics Advisory Panel,* June 1980, DOE/ER-0067; transmittal letter, M. Tigner to S. Drell, August 26, 1980, iii.
25. "CERN Elects Schopper as Director" 1980, pp. 84–85.

26. C. Rubbia, "The Physics Frontier of Elementary Particles and Future Accelerators," *IEEE Transactions on Nuclear Science* 28, no. 3 (June 1981): 3542–3548.
27. Lederman 1999.
28. Keyworth 2000.
29. Keyworth 2000.
30. *Report of the Subpanel on Long Range Planning* 1982, pp. 40–60.
31. *Report of the Subpanel on Long Range Planning* 1982.
32. J. B. Adams, "Framework of the Construction and Use of an International High-Energy Accelerator Complex," February 11, 1981, deliberated at ICFA meeting, October 21, 1981, held at Serpukhov; see minutes of the Sixth ICFA Meeting, draft, November 5, 1981, and minutes of the Seventh ICFA Meeting, held in Paris on July 27–28, 1982, draft, September 2, 1982.
33. Lederman 1982.
34. Lederman 1982.
35. Lederman 1982. Brookhaven historian Robert Crease points out that many at Brookhaven felt that at Snowmass Lederman started a "Fermilab assault on the value and legitimacy of ISABELLE's luminosity claims, and thereby the project itself, convinced that ISABELLE had to be cancelled to make way for plans for the new, grander accelerator." Crease 2005, p. 442.
36. L. M. Lederman to M. Tigner, correspondence, December 7, 1982.
37. James W. Cronin to William Wallenmeyer, January 19, 1983.
38. Krige 2001.
39. "Europe 3, U.S. Not Even Z-Zero," *New York Times*, June 4, 1983, p. A16.; Krige 2001.
40. *Report of the 1983 HEPAP Subpanel on New Facilities for the U.S. High Energy Physics Program* 1983, p. 1; and S. Wojcicki to colleagues, March 11, 1983. Keyworth 2000.
41. Tigner 1983, pp. 1–4 and 52–9.
42. R. R. Wilson to L. M. Lederman, May 20, 1983.
43. For perspective on the Department of Defense's involvement with planning for the SSC in the late 80s see C. Schwartz, "The Department of Defense and the Superconducting Super Collider," unpublished article, April 24, 1990.
44. "Europe 3, U.S. Not Even Z-Zero," p. A16.
45. The Doubler achieved its first 512 GeV beam in July 1983, followed by its first 800 GeV beam in February 1984. Hoddeson 1987. See also Peoples 1984, p. v.
46. D. Jovanovic logbook, April 23, 1983.
47. *Report of the 1983 HEPAP Subpanel on New Facilities for the U.S. High Energy Physics Program* 1983, pp. 1, vii–viii, 5–6.
48. McDonald 1985.
49. Goodwin 1983.
50. Crease considers it ironic "that the U.S. physics community's decision to put all its eggs in the SSC basket, out of a fierce desire to avoid second-rate status with respect to CERN, helped to lodge the U.S. there even more securely after

the SSC's eventual termination." For more on the disadvantages resulting from ISABELLE's cancellation, as well as the story of ISABELLE's demise from the Brookhaven point of view, see Crease 2005, p. 449. Also see Month 2003.

51. Lubkin 1983.
52. B. Richter, memorandum to Ralph DeVries and Wallace Kornack, September 1, 1983, p. 2.
53. See "Future Accelerators Seminar in Japan," *CERN Courier,* October 1984, pp. 319–322; R. Wilson 1980; Yamaguchi 1986; Burton Richter, private communication to Lillian Hoddeson.
54. This section is based in large part on Hoddeson and Kolb 2000.
55. Sources to consult on the frontier outpost include Ambrose 1996; Clifford 1988; Davis 1990: Faragher 1992, 1994; Limerick 1987; and R. White 1991a, 1991b.
56. Quigg 1993, pp. 23–4.
57. SSC newsletters (published by the APS/DPF), February 15, 1984, and March 15, 1984. McIntyre began the TAC under the aegis of the Houston Area Research Center that Mitchell created. Knapp 1997. Before Carlo Rubbia ultimately convinced CERN to build its proton-antiproton collider, he and McIntyre had proposed constructing such a collider at Fermilab, a proposal that Fermilab director Robert Wilson rejected in 1976 in favor of completing the Energy Doubler.
58. Herman Grunder et al. letter to lab directors, November 13, 1983, reprinted in appendix B, "Report of the DOE Review Committee on the Reference Designs Study," May 18, 1984. This funding procedure would later be used in arranging funds for the Central Design Group. Memo from laboratory directors to Grunder et al., December 14, 1983, file labeled "SSC Reference Design Charter."
59. PSSC records in the Fermilab History Room, Bruce Winstein Collection. Quote on p. v of the report *PSSC: Physics at the Superconducting Super Collider Summary Report* 1984.
60. Executive summary of *Superconducting Super Collider Reference Designs Study* 1984, p. iii; "Phase 1 Program Milestones," Tigner Files, Central Design Group, Fermilab Archives.
61. "Phase 1 Program Milestones."
62. "Report of the DOE Review Committee on the Reference Designs Study," May 18, 1984; quote from executive summary. The overall budget increased later as it became possible to estimate the cost of these essential aspects and necessary to include them.
63. *Physics Today,* June 1984, p. 17; August 1984, p. 69.
64. Mantsch 1984, pp. 2–3. Mantsch recalls that in the period just before and after Snowmass, after Lederman presented the idea that would become the SSC, "everyone forgot what else he was doing" and prepared to move forward on the SSC. P. Mantsch, private communication to Hoddeson and Kolb, March 11, 1998.

65. James Leiss to Guy Stever, March 1, 1984, URA Collection, Fermilab Archives.
66. Prior to this point, Fermilab was overseen by the URA board of trustees, but in 1984 a reorganization of URA called for each lab to have its own BOO reporting to the URA president and trustees. This history was later revealed in "Memorandum of Understanding," appendix E of revised unsolicited proposal for URA, February 22, 1988; although prepared in 1988, the memorandum explains the reorganization in 1984. The revised proposal for URA to serve as contractor for the construction and operation of the Superconducting Super Collider Laboratory was drafted on November 9, 1987.
67. Trivelpiece 1996.
68. State-of-the-art computing and communications technology provided by IBM and AT&T were employed to test designs by simulation. Peoples 1984; Donaldson 1984.
69. Gloria B. Lubkin, "UA1 at CERN Says It Has Candidates for Sixth Quark, Top," *Physics Today*, August 1984, pp. 17–18.
70. Trivelpiece to White, November 30, 1984. The National Governors Association gave its endorsement to the SSC in February 1985; *Chronicle of Higher Education*, October 16, 1985.
71. Jackson 1996; M. Tigner to L. Hoddeson, February 10, 1999; "Memorandum of Understanding," appendix E of revised unsolicited proposal for URA, February 22, 1988, to serve as contractor for the construction and operation of the Superconducting Super Collider Laboratory.
72. Choosing the Texas lower-field superferric magnet, judged the CDG, would cause the overall cost of tunneling and radio frequency systems to increase. In addition, these magnets would require a ring about 100 miles in circumference, allowing fewer viable sites than the other designs.
73. "Supercollider: Magnet Decision," *CERN Courier*, November 1985, pp. 383–384; Waldrop 1985, p. 50. J. D. Jackson, notebook B4; Broad 1985.
74. Therese Lloyd, "SSC Faces Uncertain Future," *Scientist* 1, no. 7 (February 23, 1987): 1, 8. J. D. Jackson, notebook B7, pp. 100, 07.29; p. 104, 07.31; p. 106; p. 112; p. 113; and notebook B8, p. 5, 08.19; p. 10, 08.22; p. 29, 09.11; p. 30, private collection of J. D. Jackson.
75. The Balanced Budget and Emergency Deficit Control Act was also known as Gramm-Rudman-Hollings (GRH). Congress and the president had imposed it in December 1985 in an attempt to reduce the budget deficit by 1991; I. Goodwin 1986. Indeed, by the end of 1989, cuts from GRH were imposed on all federally funded programs, including the DOE laboratories.
76. Leon Lederman, unpublished letter to the editor of the *New York Times*, March 10, 1986.
77. Alvin W. Trivelpiece interview in *Westinghouse Advanced Programs Network* newsletter, first quarter 1988, pp. 1, 5–6. See also the correspondence between Lederman and Trivelpiece in this period.

78. Schwarzschild 1986.
79. It was for high-energy physics a time of ongoing financial crunch in the national labs. Leading the argument to reduce the budget was the director of the Office of Management and Budget, Dr. James C. Miller III, who cajoled colleagues to reduce spending on all government programs. Crawford 1987a, p. 625.
80. DOE press release, January 30, 1987. See e.g., I. Goodwin 1987, pp. 47–49; Crawford 1987a, p. 625; Ben E. Franklin, "Reagan to Press for $6 Billion Atom Smasher," *New York Times*, February 2, 1987, pp. 15–16; Malcolm Browne, "Atom Smashing Now, and in the Future: A New Era Begins," *New York Times*, February 3, 1987, pp. 1, 5; Robert E. Taylor, "President Will Request Funds to Build World's Largest Particle Accelerator," *Wall Street Journal*, February 2, 1987, p. 1. The fact that in this month Fermilab reached 1.8 TeV in its Tevatron undoubtedly boosted Trivelpiece's confidence. Kim McDonald, "Reagan Backs Giant $4.4-Billion Particle Accelerator; Scientists Face Major Hurdles in Promoting the Device," *Chronicle of Higher Education*, February 11, 1987, pp. 7, 9.
81. To avoid a conflict of interest, he resigned from the URA board of trustees to become an employee half-time for CDG and half-time directly for URA. Goldwasser 1993a. It was agreed that half of his time would be devoted to writing the proposal and the rest to "odd jobs."
82. "Proposal to Serve as Contractor for the Construction and Operation of the 'Superconducting Super Collider' Laboratory," submitted by Universities Research Association, Inc., March 2, 1987, URA Collection, Fermilab Archives.
83. D. Pewitt, Red Team briefing document, September 30, 1988. Also see Goldwasser 1993a.
84. Goldwasser 1993a.
85. The revision, not written by Goldwasser or his URA committee, consisted essentially of describing the earlier establishment (see n. 66) of a separate board of overseers for Fermilab, an effort to place more distance between Fermilab and URA and to minimize any perception of favoritism for Illinois at a point when site selection was in full swing.
86. Hoddeson 1987.
87. Goldwasser 1993a; Limon, private communication to Lillian Hoddeson, 1993.
88. Hoddeson et. al 1993.
89. Goldwasser 1993a.
90. Ritson 1993.
91. Peoples 1993b.
92. Goldwasser 1993b. The magnet development program was also the context in which the vast new capacity of computer modeling first made substantial impact on the SSC. Peoples 1993b, 4–6; Peoples note to Hoddeson and Kolb, January 16, 2004.

93. The Livermore engineer had earlier been project manager of the Magnetic Fusion Test Facility and had been a major player in the nuclear underground testing project.
94. Tigner to Hoddeson, February 9, 1999.
95. Peoples 1993b. Peoples also noted that an accident at Fermilab about this time worked in his favor. Within a week of starting his new CDG assignment, a superconducting magnet failed at Fermilab in a spectacular fashion: a segment of the coil evaporated, a design glitch capable of destroying the entire magnet. The ensuing investigation led to a much deeper understanding of the magnet's physics.
96. Tigner to Hoddeson, February 9, 1999.
97. Lighter 1994; Hendrickson 2003; Partridge 1961, p. 256. For examples of how the metaphor was used see, e.g., Michener 1974, chap. 6, "The Wagon and the Elephant," pp. 243–349, esp. pp. 295–296, 329, and 332–333; and Gutiérrez 1996, esp. pp. 125–126.
98. Quigg 1993. For a more complete account of this episode see "Panofsky to BOO Files," August 1, 1988, Panofsky Papers, SLAC; also Hoddeson and Kolb 2000.
99. Lederman and Quigg 1988.
100. Quigg 1993.
101. Knapp 1997. Also IISSC newsletter; Appendix E, "Key Personnel," in *Request for Proposals: Number DE-RP02–88ER40486: For the Selection of a Management and Operating Contractor for the Establishment, Management, and Initial Operation of the Superconducting Super Collider Laboratory*, Closing Date: November 4 (Chicago: Chicago Operations Office, 1988). Note: the CDG did not expect to build the SSC without external industrial support. Martin-Marietta never did present a proposal.
102. According to Pewitt, CDG did not want to cooperate. Pewitt 1998. However, according to Tigner, Pewitt never discussed the matter with him. Tigner to Hoddeson, February 9, 1999.
103. Chrisman 1997, 1998.
104. Knapp 1997. J. D. Jackson, notebook B9, pp. 92–93.
105. Knapp 1997.
106. W. K. H. Panofsky to BOO Files, September 6, 1988.
107. Quigg 1993.
108. Schwitters 1997.
109. M. Riordan's notes on conversations with S. Wojcicki, September 1, 1988, and Rene Donaldson, August 24, 1998, personal files of Michael Riordan, Soquel, CA.

EPILOGUE

1. Oppenheimer 1964, p. 3.
2. Personal recollections of Lillian Hoddeson; "Hats Off to Waxahachie" 1988.

3. "Hats Off to Waxahachie" 1988; see also Lederman 1999; and video recording of the event in Fermilab Archives.
4. Lederman, private communication to Hoddeson, November 10, 1988.
5. Those who did included Alex Chao, Tom Elioff, Jim Sanford, Tim Toohig, Tom Kirk, and Roger Coombs. Limon was among those who did not. Wojcicki and J. D. Jackson served later on the SSC's Program Advisory Committee. Those from Fermilab who joined the SSC after URA was awarded the contract to build the SSC included Helen and Don Edwards, Gerry Dugan, William Bardeen, Bruce Chrisman, and Edward West. The CDG physicists' disappointment may be compared with that of Meriwether Lewis and his partner William Clark, at the time James Madison succeeded Thomas Jefferson as president in 1809. Ambrose 1996, p. 452. And just as the different generations of Washington patrons had disparate views of the importance of the Lewis and Clark expedition, so were there different Washington views of the importance of the CDG and the SSC. Letter from Chao to Hoddeson, December 1999.
6. Peoples never intended to stay with CDG; Limon, Kirk, and Quigg prepared to return to the Tevatron.
7. Harry Woolf to Lederman, August 10, 1988.
8. *FermiNews* 12, no. 13 (July 19, 1989): 1–2.
9. Pewitt 1998.
10. "Task Force Sets Funding Priorities for DOE," *FermiNews* 18, no. 1 (October 18, 1991): 1.
11. Later, under George Bush and Admiral Watkins, another substantial weakening of SSC foundations occurred when early in 1989 the SSC left the domain of the Office of Energy Research and was placed under the Office of the SSC (OSSC), overseen by Deputy Secretary of Energy W. Henson Moore, a former Louisiana congressman. See Timothy Toohig's draft, "SSC: the Anatomy of a Failure: A Case Study of Institutional Amnesia," November 3, 1996. By 1990, when DOE reopened the negotiations regarding international funding of the SSC, the political rhetoric of patriotic nationalism impeded negotiations with international partners. Riordan, Hoddeson, and Kolb n.d.
12. Bodnarczuk 1988.
13. *FermiNews*, October 18, 1991, p. 2
14. For example, Philip Livdahl retired in 1987, Richard Lundy left in 1989, and Rich Orr left in 1991.
15. As a cost-saving measure, the laboratory offered attractive voluntary and involuntary retirement packages in 1993–1994, targeted first at the technical support staff. Eventually eighty employees volunteered to retire, and twenty were terminated. In an effort to keep money and talent at Fermilab, some long-term employees remained at Fermilab but worked with SSC contractors. Paula Cashin, private communication to A. Kolb. "Fermilab Says Goodby to Host of Employees," *Ferminews*, October 7, 1994, pp. 5–8.

16. Robert W. Seidel, "From Factory to Farm: Dissemination of Computing in High Energy Physics," to appear in *Historical Studies in the Natural Sciences*.
17. By 1990, the World Wide Web had been launched by CERN physicist Tim Berners-Lee. By 1991 Paul Ginsparg at Los Alamos National Laboratory had set up electronic publishing over the Internet.
18. The software system of LaTex and other desktop publishing tools replaced first-generation word processors and draftspersons. The time-honored catalog of the Fermilab library was contracted out and replaced by an online public access catalog. Elizabeth Anderson, Fermilab librarian, private communication with Kolb, August 24, 2001. The VAX subsequently yielded to networked servers. Segments of the high-energy physics budget were directed to applying computing upgrades into the practice of conducting research.
19. Westfall 1997.
20. *FermiNews* 13, no. 11 (September 21, 1990): p. 1.
21. *FermiNews* 15, no. 8 (May 1, 1992).
22. Pewitt 1998.
23. *FermiNews*, November 15, 1991, pp. 1–2.
24. *FermiNews*, February 7, 1992.
25. Brookhaven came on hard times in 1997 when the lab's director, Nicholas Samios, and its longtime manager, AUI (Associated Universities, Inc.), were both forced out due to bad publicity that stemmed from a tritium leak from a research reactor and associated environmental and safety complaints. See Robert Crease, "Anxious History: The High Flux Beam Reactor and Brookhaven National Laboratory," *Historical Studies in the Physical and Biological Studies* 32 (2001): 41–56.
26. *FermiNews*, May 15, 1992, p. 1.
27. *FermiNews*, September 21, 1990, p. 1; November 15, 1991, p. 1. See Tiger Team Reports in *FermiNews*, February 21, 1992.
28. "The Year of the Tiger," Fermilab annual report, 1992, pp. 50–51.
29. Jeff Appel, private communication, May 25, 2004.
30. John Toll was president of URA from 1989 to 1994.
31. *FermiNews*, April 2, 1993.
32. O'Leary quoted in *FermiNews*, August 20, 1993.
33. Over the next few years of Peoples's tenure, government oversight produced another concern for Fermilab. In February 1995 DOE explored the possibility of adopting a corporate model for managing its multipurpose laboratories. DOE formed a committee chaired by Robert Galvin, chairman of Motorola Corporation, to assess the value of the competence of their contractors. Privatization did not result, but this examination signaled a new trend in the management structures of large national facilities. See the document titled "Secretary of Energy Advisory Board Task Force on Alternative Futures for the National Laboratories."
34. Kevles 1995, esp. pp. xxxii–xxxiv.

35. James Walsh, "Risking Peace," *Time* 142, no. 11 (September 13, 1993): cover story.
36. SSC newsletters, September, October, and November 1993, SSC Collection, Fermilab Archives.
37. We will expore the interweave of many threads which brought about the demise of the SSC in a subsequent book: Michael Riordan, Lillian Hoddeson, and Adrienne Kolb, *Tunnel Visions: The Rise and Fall of the Superconducting Super Collider*.
38. Quotes from *FermiNews*, November 5, 1993.
39. "Vision for the Future of High Energy Physics," Drell report, DOE/ER-0614P, May 1994.
40. *FermiNews*, October 7, 1994.
41. URA had hired Fred Bernthal as the new president of URA.
42. For a full account of the top quark story we refer the reader to Staley 2004.
43. Quote in special edition of *FermiNews*, April 29, 1994.
44. White House, Office of the Press Secretary, press release, August 3, 1994.
45. Fermilab press release, March 2, 1995.
46. Turner 1956, p. 18.
47. Moore 1921, v. 2, p. 147. This quote was assembled by Burnham's partner, Willis Polk, in 1912 (the year Burnham died) from statements Burnham wrote in 1910. Charles Moore's biography of Burnham includes this version of Burnham's famous quote. Thanks to Elizabeth Baughman and Leslie Martin of the Chicago Historical Society Research Center for their assistance.

AUTHORS' STATEMENTS, AND OTHER ACKNOWLEDGEMENTS

1. Robert Wilson, personal communication to Lillian Hoddeson, circa March 1978.
2. R. Wilson 1977a.
3. This anecdote is also included in Hoddeson 2006.
4. See esp. Bodnarczuk 1990, 1997; and Bodnarczuk and Hoddeson 2006.
5. The government has certain rights in this material. Any opinions, findings, and conclusions or recommendations expressed in this material are those of the authors and do not necessarily reflect the views of the National Science Foundation.

Bibliography

Interviews and unpublished material, unless otherwise indicated, are located in the Fermilab Archives.

Abe, F., et al. 1988. "The CDF Detector: An Overview," *The Collider Detector at Fermilab: A Compilation of Articles Reprinted from Nuclear Instruments and Methods in Physics Research-A*. Amsterdam: North-Holland, pp. 388–389.

Adams, J. B. 1971. "The European 300 GeV programme." In *Proceedings of the 8th International Conference on High-Energy Accelerators*, ed. M. H. Blewett and N. Vogt-Nilsen, 25–30. Geneva: European Organization for Nuclear Research.

AEC. 1965. "Policy for National Action in the Field of High Energy Physics." In *High Energy Physics Program: Report on National Policy and Background Information*, by U.S. Congress, Joint Committee on Atomic Energy. Washington, DC: GPO.

AEC. 1967. "Design, Location, and Construction of the 200 BeV Accelerator." Washington, DC: GPO.

AEC. N.d. "The Atomic Energy Commission during the Administration of Lyndon B. Johnson." (See Catherine Westfall, "The First 'Truly National Library': The Birth of Fermilab" [Ph.D. diss., Michigan State University, 1988], vol. 2, chap. 5, n. 70.)

Ambrose, Stephen. 1996. *Undaunted Courage: Meriwether Lewis, Thomas Jefferson, and the Opening of the American West*. New York: Simon and Schuster.

Amman, F. 1989. "The Early Times of Electron Colliders." In *The Restructuring of Physical Sciences in Europe and the United States, 1945–1960*, ed. M. de Maria, M. Grilli, and F. Sebastiani, 449–476. Singapore: World Scientific.

Appel, Jeffrey A. 1976. "Production of High Mass $e^+ e^-$ Pairs at Fermilab." Paper presented at the meeting of the APS, January.

———. 1984 Interview by Lillian Hoddeson, 21 March.

———. 1989. "The Computing Department." In *Fermilab 1988: Annual Report of the Fermi National Accelerator Laboratory*, 75–79. Batavia, IL: Fermilab.

———. 1990. Interview by Frederik Nebeker, 8 May.

Appel, J. A., M. H. Bourquin, I. Gaines, D. C. Hom, L. M. Lederman, H. P. Paar, J. P. Repellin, D. H. Saxon, H. D. Snyder, J. M. Weiss, and J. K. Yoh. 1974a. "Hadron Production at Large Transverse Momentum." *Physical Review Letters* 33, no. 12:719–722.

Appel, J. A., M. H. Bourquin, I. Gaines, D. C. Hom, L. M. Lederman, H. P. Paar, J. P. Repellin, D. H. Saxon, H. D. Snyder, J. M. Weiss, and J. K. Yoh. 1974b. "Observation of Direct Production of Leptons in *p*-Be Collisions at 300 GeV." *Physical Review Letters* 33, no. 12:722–725.

Appel, J. A., M. H. Bourquin, D. C. Hom, L. M. Lederman, J. P. Repellin, H. D. Snyder, J. M. Weiss, J. K. Yoh, B. C. Brown, P. Limon, T. Yamanouchi. 1974c. "A Study of Di-Lepton Production in Proton Collisions at NAL." NAL proposal no. 288 (February). FA.

Appel, J. A., M. H. Bourquin, I. Gaines, D. C. Hom, L. M. Lederman, H. P. Paar, J.-P. Repellin, H. D. Snyder, J. M. Weiss, J. K. Yoh, B. C. Brown, C. N. Brown, J.-M. Gaillard, J. R. Sauer, and T. Yamanouchi. 1975. "Search for *phi* Mesons Produced at High Transverse Momentum in *p*-Be Collisions at 300 GeV." *Physical Review Letters* 35, no. 1:9–12.

Arnison, G., et al. 1983a. "Experimental Observation of Isolated Large Transverse Energy Electrons with Associated Missing Energy at $\sqrt{s} = 540$ GeV." *Physics Letters B* 122, no. 1:103–116.

Arnison, G., et al. 1983b. "$p\bar{p}$ Collisions Yield Intermediate Boson at 80 GeV, as Predicted." *Physics Today*, April, 17–20.

Aschermann, G., et al. 1941. "Supraleitfähige Verbindungen mit extrem hohen Sprungtemperaturen." *Physische Zeitschrift* 42:349–360.

Atomic Energy Commission High Energy Physics Advisory Panel. 1968. *The Status of High Energy Physics Today*. Washington, DC: GPO.

"Attachment C: Summary of Comments Solicited by the Office of Science and Technology from a Group of Users of High Energy Accelerators." 1965. In *High Energy Physics Program: Report on National Policy and Background Information*, by U.S. Congress, Joint Committee on Atomic Energy. Washington, DC: GPO.

Barnhart, William E., and Eugene F. Schlickman. 1999. *Kerner: The Conflict of Intangible Rights*. Urbana: University of Illinois Press.

"Berkeley Study for a Proton Synchrotron in the 150–300 BeV Range." 1963. UCRL-10869, Lawrence Radiation Laboratory, 9 August. FA.

Biallas, George. 1984. Interview by Lillian Hoddeson, 16 February.

"Big Projects in Federal Republic of Germany." 1981. *CERN Courier*, June, 210–211.

Billington, Ray Allen. 1966. *America's Frontier Heritage*. New York: Holt, Rinehart and Winston.

———. 1973. *Frederick Jackson Turner: Historian, Scholar, Teacher*. New York: Oxford University Press.

Binkley, Morris. 1990. Interview by Lillian Hoddeson, 28 September.

Blewett, John P. 1986. Interview by Catherine Westfall, 17 November.

———. 1989. "Accelerator Design and Construction in the 1950s." In *Pions to Quarks: Particle Physics in the 1950s*, ed. Laurie M. Brown, Max Dresden, and Lillian Hoddeson, 162–179. New York: Cambridge University Press.

Blewett, M. H., and N. Vogt-Nilsen, eds. 1971. *Proceedings of the 8th International Conference on High-Energy Accelerators*. Geneva: European Organization for Nuclear Research.

Bodnarczuk, Mark W., ed. 1988. "Reflections on the Fifteen-Foot Bubble Chamber at Fermilab." Batavia, IL: Fermi National Accelerator Laboratory.

———. 1990. "The Social Structure of Experimental Strings at Fermilab: A Physics and Detector Driven Model." Document available as Fermilab-Pub-91/ 63, Batavia, March. FA.

———. 1997. "Some Sociological Consequences of High-Energy Physicists' Development of the Standard Model." In *The Rise of the Standard Model: Particle Physics in the 1960s and 1970s*, ed. Lillian Hoddeson, Laurie Brown, Michael Riordan, and Max Dresden, 384–393. New York: Cambridge University Press.

Bodnarczuk, Mark, and Lillian Hoddeson. 2008 "Megascience in Particle Physics: The Birth of an Experiment String at Fermilab, 1970–90." *Historical Studies in the Natural Sciences* (in press; scheduled November 2008).

Boorstin, Daniel J. 1983. *The Discoverers*. New York: Random House.

Broad, William J. 1985. "Supermagnet Design Chosen for a 60-Mile Atom Smasher." *New York Times*, 19 September. FA.

Brookhaven National Laboratory. 1961. "300–1000 BeV National Accelerator Design Study at the Brookhaven National Laboratory." FA.

Brown, Bruce. 1990a. Interview by Catherine Westfall and Frederik Nebeker, 2 May.

———. 1990b. Interview by Catherine Westfall and Frederik Nebeker, 3 May.

———. 1990c. Interview by Catherine Westfall, 3 September.

Brown, Charles. 1990. Interview by Catherine Westfall and Frederik Nebeker, 2 May.

Brown, Kevin A. 1989. "Feynman Center Opens New Era for Fermilab Computing." *FermiNews* 12, no. 4 (10 March): 1, 2, 5.

Brown, Laurie M. 1986. Interview by Catherine Westfall, 28 August.

Brown, Laurie M., and Lillian Hoddeson, eds. 1983. *The Birth of Particle Physics: Particle Physics in the 1930s and 40s*. New York: Cambridge University Press.

Brown, Laurie M., Max Dresden, and Lillian Hoddeson, eds. 1989. *Pions to Quarks: Particle Physics in the 1950s*. New York: Cambridge University Press.

Browne, E. Janet,1996. *Voyaging*. Princeton: Princeton University Press.

Browne, Malcolm. 1979. "Researchers Race to Find Particle Vital to Atom Theory." *New York Times*, 26 June.

———. 1981. "Leon Lederman: Captain of Science." *Discovery*, October, 45–50.

Buck, Alice L. 1983. *A History of the Atomic Energy Commission*. DOE/ES/003/1. Washington, DC: U.S. Department of Energy.

Bulletin of the American Physical Society. 1964. Ser. 2, vol. 9, no. 4 (April).

Bush, Vannevar. 1945. *Science, the Endless Frontier: A Report to the President on a Program for Postwar Scientific Research*. Washington, DC: National Science Foundation.

———. 1960. *Science, the Endless Frontier: A Report to the President on a Program for Postwar Scientific Research*. Washington, DC: National Science Foundation. (Reprint of 1945. ed.)

———. 1970. *Pieces of the Action*. New York: William Morrow.

Campbell, Roberta. 1979. *Naperville Sun*, 18 July.

Capshew, James H., and Karen H. Rader. 1992. "Big Science: Price to the Present." *Osiris* 7:3–25.

Carrigan, Richard. 1989. Interview by Catherine Westfall, 24 March.

"CDF Snapshot." 2000. *FermiNews* 23, no. 17 (6 October).

"CERN Elects Schopper as Director." 1980. *Physics Today*, June, 84–85.

Chamberlain, Owen. 1984. Interview by Catherine Westfall, 24 April.

Chandler, Alfred D., Jr. 1977. *The Visible Hand: The Managerial Revolution in American Business*. Cambridge, MA: Harvard University Press, Belknap Press.

Charles, Brian. 1992. "Antiproton Source Groundbreaking." *FermiNews* 15, no. 15 (21 August): 7.

Chrisman, Bruce C. 1997. Interview by Adrienne Kolb, 12 December.

———. 1998. Interview by Adrienne Kolb, Lillian Hoddeson, and Steve Weiss, 16 January.

———. 2001. Interview by Lillian Hoddeson and Adrienne Kolb, 22 February.

Christenson, J., et al. 1970. "Observations of Massive Muon Pairs in Hadron Collisions." *Physical Review Letters* 25, no. 21 (23 November): 1523–1526.

Clavelli, Louis, 2001. "Physics at Weston: Life and Physics in the First Fermilab Theory Group (1969–1971)." Unpublished report. 13 September.

Clifford, James. 1988. *The Predicament of Culture: Twentieth-Century Ethnography, Literature, and Art*. Cambridge, MA: Harvard University Press.

Cole, Francis T. 1978. Interview by Lillian Hoddeson, 21 March.

———. 1984. Interview by Catherine Westfall, 13 July.

———. 1985. Interview by Catherine Westfall, 27 February.

———. 1986. Interview by Catherine Westfall, 29 August.

———. 1987. Interview by Catherine Westfall, 13 March.

———. 1994. "O Camelot! A Memoir of the MURA Years." Unpublished manuscript, 17 February. Fermilab Archives.

Coleman, William. 1966. "Science and Symbol in the Turner Frontier Hypothesis." *American Historical Review* 72, no. 1:22–49.

The Collider Detector at Fermilab: A Compilation of Articles Reprinted from Nuclear Instruments and Methods in Physics Research—A. 1988. Amsterdam: North-Holland.

"Colliding Beams at Fermilab." 1976. *Village Crier* 8, no. 13 (25 March): 1–2.

Collins, Thomas. 1979. "The Great Doubler Shift." Fermilab internal report, TM 874. April.

———. 1989. Interview by Catherine Westfall and Lillian Hoddeson, 29 November.

———. 1990a. Interview by Catherine Westfall, 4 June.

———. 1990b. Interview by Adrienne Kolb and Kyoung Paik, 3 October.

Compton, Arthur Holly. 1956. *Atomic Quest: A Personal Narrative*. New York: Oxford University Press.

Corn, Joseph J. 1983. *The Winged Gospel: America's Romance with Aviation, 1900–1950*. New York: Oxford University Press.

Courant, Ernest D., M. Stanley Livingston, and Hartland S. Snyder. 1952. "The Strong-Focusing Synchrotron: A New High-Energy Accelerator." *Physical Review* 88, no. 5:1190–1196.

Crawford, Mark, 1987a. "Reagan Okays the Supercollider." *Science* 235, no. 4789 (6 February): 625.

———. 1987b. "Science Committee Okays Supercollider." *Science* 238, no. 4826 (23 October): 477.

———. 1988a. "CBO Cautions Congress on SSC." *Science* 242, no. 4876 (14 October): 186.

———. 1988b. "SSC Report Attacked." *Science* 242, no. 4882 (25 November): 1243.

Crease, Robert P. 1999a. *Making Physics: A Biography of Brookhaven National Laboratory, 1946–1972*. Chicago: University of Chicago Press.

———. 1999b. "Quenched! The ISABELLE Story." 350th Brookhaven Lecture, 15 December.

———. 2005. "Quenched! The ISABELLE Saga," pt. 1. *Physics in Perspective* 7, no. 3:330–376, 404–452.

Cronin, James W. 1987. Interview by Adrienne Kolb and Lillian Hoddeson, 17 November.

Cronon, William. 1987. "Revisiting the Vanishing Frontier: The Legacy of Frederick Jackson Turner." *Western Historical Quarterly* 18, no. 2 (April): 157–176.

———. 1991a. *Nature's Metropolis: Chicago and the Great West*. New York: W. W. Norton.

———. 1991b. "Turner's First Stand: The Significance of Significance in American History." In *Writing Western History: Essays on Major Western Historians*. ed. Richard W. Etulain, 73–101. Albuquerque: University of New Mexico Press.

Dahl, Per F. 1984. "Kamerlingh Onnes and the Discovery of Superconductivity: The Leyden Years, 1911–1914." *Historical Studies in Physical and Biological Sciences* 15, no. 1:1–37.

Danby, G., J.-M. Gaillard, K. Goulianos, L. M. Lederman, N. Mistry, M. Schwartz, and J. Steinberger. 1962. "Observation of High-Energy Neutrino Reactions and the Existence of Two Kinds of Neutrinos." *Physical Review Letters* 9, no. 1:36–44.

Davis, Mike. 1990. *City of Quartz: Excavating the Future in Los Angeles*. New York: Verso.

Debus, Allen George. 1978. *Man and Nature in the Renaissance*. Cambridge: Cambridge University Press.

DeMuth, Jerry. 1967. "Dr. King Visits Weston Tent-In, Sees Rights Legislation Erosion." *Chicago Sun-Times*, 24 June.

"DESY HERA Ahead." 1981. *CERN Courier*, June, 205–206.

Devlin, T. 1989. Interview by Lillian Hoddeson, 1 February.

Dickson, David. 1983. "New Push for European Science Cooperation." *Science* 220, no. 4602 (10 June): 1134–1136.

———. 1984. *The New Politics of Science*. New York: Pantheon Books.

Diebold, R. 1983. "The Desertron: Colliding Beams at 20 TeV." *Science* 222, no. 4619 (7 October): 13–19.

"The Discovery of Upsilon." 1977. *CERN Courier* 17, no. 7/8 (July/August): 223–224.

Divine, Robert A. 1981. *Eisenhower and the Cold War*. New York: Oxford University Press.

"DOE Authorizes Fermilab to Build Superconducting Accelerator." 1979. *FermiNews* 2, no. 28 (12 July): 1–3.

Donaldson, Rene. 1984. "Preface." In *Proceedings of the 1984 Summer Study on Design and Utilization of the Superconducting Super Collider*, ed. Rene Donaldson and Jorge G. Morfín. Held in Snowmass, CO. Batavia, IL: URA.

Dorner, R. 1970. "Land Management Objectives." Unpublished internal report, 10 February. In Site Planning Files, Fermilab Archives.

Drell, Sidney D., and Tung-Mow Yan. 1970. "Massive Lepton-Pair Production in Hadron-Hadron Collisions at High Energies." *Physical Review Letters* 25, no. 5 (3 August): 316–320.

Droege, Tom. 1990. Interview by Kyoung Paik and Lillian Hoddeson, 26 September.

Duffield, Priscilla. 1978. Interview by Lillian Hoddeson, 16 May.

Dupree, A. Hunter. 1986. *Science in the Federal Government: A History of Policies and Activities*. Baltimore: John Hopkins University Press.

Edwards, Donald. 1987. Interview by Catherine Westfall, 12 March.

———. 1984. Interview by Lillian Hoddeson, 7 July.

Edwards, Helen. 1983a. Interview by Lillian Hoddeson, 16 December.

———. 1983b. "The Energy Saver Test and Commissioning History." In *Proceedings of the 12th International Conference on High-Energy Accelerators*, 1–9. Batavia, IL: Fermi National Accelerator Laboratory.

———. 1985. "The Tevatron Energy Doubler: A Superconducting Accelerator." *Annual Review of Nuclear and Particle Science* 35:605–660.

———. 1987. Interview by Catherine Westfall, 13 March.

ERDA. 1977. *High Energy Physics Advisory Panel Report of the Subpanel on New Facilities*. Washington, DC: ERDA (June).

"Europe 3, U.S. Not Even Z-Zero." 1983. Editorial, *New York Times*, 4 June.

Faragher, John Mack. 1992. *Daniel Boone: The Life and Legend of an American Pioneer*. New York: Henry Holt and Co.

———, ed. 1994. *Rereading Frederick Jackson Turner: The Significance of the Frontier in American History, and Other Essays*. New York: Henry Holt and Co.

Ferguson, Eugene. 1979. "The American-ness of American Technology." *Technology and Culture* 20:3–24.

"Fermi National Accelerator Laboratory 1974 Historical Calendar of Program Accomplishments." 1974. Fermilab Research Program workbook. FA.

Findlay, John W., 1995. "Atomic Frontier Days: Richland, Washington, and the Modern American West." *Journal of the West* 34, no. 3 (July): 32–41.

Fisk, Gene. 1999. Interview by Catherine Westfall, 28 October.

Fowler, W. 1973. "The 15-Foot Bubble Chamber." *NALREP*, October, 1–2.

———. 1984. "History of the Energy Doubler." Unpublished document, 21 December.

Fowler, W. B., D. Drickey, P. J. Reardon, B. P. Strauss, and D. F. Sutter. 1975. "The Fermilab Energy Doubler: A Two-Year Progress Report." *IEEE Transactions on Nuclear Science*, n.s., 22, no. 3 (June): 1125–1128.

Frisch, Henry. 1989. Interview by Lillian Hoddeson, Adrienne Kolb, and Kyoung Paik, 29–30 November.

———. 1999. Interview by Catherine Westfall, 29 October.

"Future Accelerators Seminar in Japan." 1984. *CERN Courier*, October, 319–322.

Galison, Peter. 1983. "How the First Neutral Current Experiments Ended." *Review of Modern Physics* 55:477–509.

———. 1987. *How Experiments End*. Chicago: University of Chicago Press.

———. 1989. "Bubbles, Sparks, and the Postwar Laboratory." In *Pions to Quarks: Particle Physics in the 1950s*, ed. Laurie M. Brown, Max Dresden, and Lillian Hoddeson, 213–251. New York: Cambridge University Press.

———. 1997. *Image and Logic: A Material Culture of Microphysics*. Chicago: University of Chicago Press.

Galison, Peter, and Bruce Hevly, eds. 1992. *Big Science: The Growth of Large-Scale Research*. Stanford: Stanford University Press.

Galison, Peter, Bruce Hevly, and Rebecca Lowen. 1992. "Controlling the Monster: Stanford and the Growth of Physics Research, 1935–1962." In *Big Science: The Growth of Large-Scale Research*, ed. Peter Galison and Bruce Hevly, 46–77. Stanford: Stanford University Press.

Gallagher-Dagitt, G. E. 1974. "Superconductor Cables for Pulsed Dipole Magnets." Rutherford High Energy Laboratory report M/A25, February.

Garren, Al A., Glen Lambertson, Edward Lofgren, and Lloyd Smith. 1967. "Extendible-Energy Synchrotrons." *Nuclear Instruments and Methods* 54:223–228.

Genuth, Joel. 1992. "Historical Analysis of Selected Experiments at US Sites." In *AIP Study of Multi-institutional Collaborations: Phase I: High Energy Physics*, report 4. New York: American Institute of Physics.

Giacomelli, G., A. F. Greene, and J. R. Sanford. 1975. "A Survey of the Fermilab Research Program." *Physics Reports* 19C, no. 4 (July): 171–232.

Gladding, G. 1990. Interview by Lillian Hoddeson, 4 September.

Glashow, Sheldon. 1988. *Interactions: A Journey through the Mind of a Particle Physicist and the Matter of This World*. New York: Warner Books.

Golden, William T. 1988. *Science and Technology Advice to the President, Congress, and the Judiciary*. New York: Pergamon.

Goldwasser, Edwin L. 1969. "Science and Man: Breaking New Ground at Batavia." *Bulletin of the Atomic Scientists*, October, 7–10.

———. 1976. "Highlights of the Summer PAC Meeting." *NALREP*, July, 1–11.

———. 1979. "Report on the Status and Plans of the International Committee on Future Accelerators." *Proceedings of the 19th International Conference on High Energy Physics*, 961–968. Tokyo: Physical Society of Japan.

———. 1985. Interview by Catherine Westfall, 10 July.

———. 1987a. Interview by Catherine Westfall, 15 May.

———. 1987b. Interview by Catherine Westfall, 10 July.

———. 1992. Interview by Catherine Westfall, 23 March.

———. 1993a. Interview by Adrienne Kolb and Lillian Hoddeson, 8 May.

———. 1993b. Interview by Adrienne Kolb and Lillian Hoddeson, 23 October.

———. 2000. "Robert R. Wilson: A Man for All Seasons." Talk delivered to American Physical Society, Long Beach, CA, 1 May.

———. 2004a. Interview by Lillian Hoddeson, 22 April.

———. 2004b. Interview by Lillian Hoddeson, 20 July.

———. 2005. Interview by Lillian Hoddeson, 23 February.

Gonzales, Angela. 2004. Interview by Lillian Hoddeson and Adrienne Kolb, 3 September.

Goodwin, Doris Kearns. 1976. *Lyndon Johnson and the American Dream*. New York: Harper and Row.

Goodwin, Irwin. 1983. "DOE Answers to Congress as It Officially Kills Brookhaven CBA." *Physics Today*, December, 41–43.

———. 1984. "Tigner Named to Direct R & D Program for SSC." *Physics Today*, August, 69.

———. 1986. "R & D Budget for Fiscal 1987: Life at the Threshold of Pain." *Physics Today*, May, 55–60.

———. 1987. "Reagan Endorses the SSC, a Colossus among Colliders." *Physics Today*, March, 47–49.

Gorter, C. J. 1946. "Superconductivity until 1940 in Leiden as seen from there." *Reviews of Modern Physics* 36:3–7.

Graham, Daniel Orrin. 1983. *The High Nuclear Defense of Cities: The High Frontier Space-Based Defense against ICBM Attack*. Cambridge, MA: Abt Books.

Grannis, Paul. 1999. "The D0 Syndrome." *Science* 253 (16 August): 719.

———. 2000. Interview by Catherine Westfall, 5 December.

Green, A. F. 1978. "What Happened at Fermilab During 1977?" Fermilab Report, February.

Green, Daniel R., and Leon M. Lederman. 1989. *Fermilab Research Results 1978–1988*. July. Batavia, IL: Fermilab Publication.

Greenbaum, Leonard. 1971. *A Special Interest: The Atomic Energy Commission, Argonne National Laboratory, and the Midwestern Universities*. Ann Arbor: University of Michigan Press.

Greenberg, Daniel. 1967. *The Politics of Pure Science: An Inquiry into the Relationship between Science and Government in the United States*. New York: New American Library.

———. 1971. *The Politics of Pure Science: An Inquiry into the Relationship between Science and Government in the United States*. New York: New American Library. (Reprint of 1967. ed.)

———. 1999. *The Politics of Pure Science: An Inquiry into the Relationship between Science and Government in the United States*. 2nd ed. Chicago: University of Chicago Press.

Grossman, James. 1994. Introduction to "The Frontier and the American Mind," talk given by Richard White at the Newberry Seminar in American Social History, Chicago, 18 January.

Gutiérrez, David G. 1996. "Seeing the Elephant: Myth and Myopia: Hispanic People and Western History." In *The West: An Illustrated History*, ed. Geoffrey C. Ward, 118–171. Boston: Little, Brown.

Hankerson, Mack. 1978. Interview by Lillian Hoddeson, 14 September.

Hart, David, 1998. *Forged Consensus: Science, Technology and Economic Policy in the United States, 1921–1953*. Princeton: Princeton University Press.

"Hats Off to Waxahachie." 1988. *FermiNews* 11, no. 21 (18 November): 1, 3, 6–7.

Heilbron, John L., and Robert W. Seidel. 1989. *Lawrence and His Laboratory: A History of the Lawrence Berkeley Laboratory*. Vol. 1. Berkeley: University of California Press.

Heilbron, John L., Robert W. Seidel, and Bruce R. Wheaton. 1981. *Lawrence and His Laboratory: Nuclear Science at Berkeley, 1931–1961*. Berkeley: Office for Science and Technology.

Hendrickson, Robert. 2003. *The Facts on File Encyclopedia of Words and Phrase Origins: Definitions and Origins of over 12,500 Words and Expressions*. 3rd ed. New York: Eurospan.

"HERA: Proton-Electron Colliding Beam Project at DESY." 1980. *CERN Courier* 20, no. 3 (May): 99–104.

Herb, S. W., D. C. Hom, L. M. Lederman, J. C. Sens, H. D. Snyder, J. K. Yoh, J. A. Appel, B. C. Brown, C. N. Brown, W. R. Innes, K. Ueno, T. Yamanouchi, A. S. Ito, H. Jöstlein, D. M. Kaplan, and R. D. Kephart. 1977. "Observation of a Dimuon Resonance at 9.5 GeV in 400 GeV Proton-Nucleus Collisions." *Physical Review Letters* 39, no. 5:252–255.

Heren, Louis. 1970. *No Hail, No Farewell*. New York: Harper and Row Publishers.

Hermann, A., J. Krige, L. Belloni, and D. Pestre. *History of CERN*. Vol. 1. Amsterdam: North Holland.

Hermann, A., J. Krige, U. Mersits, and Dominique Pestre, with L. Weiss. 1990. *History of CERN: Building and Running the Laboratory*. Vol. 2. Amsterdam: North Holland.

Hewlett, Richard G., and Oscar E. Anderson, Jr. 1962. *A History of the United States Atomic Energy Commission.* Vol. 1, *The New World, 1939–1946.* University Park: Penn State University Press.

Hewlett, Richard G., and Francis Duncan. 1969. *A History of the United States Atomic Energy Commission.* Vol. 2, *Atomic Shield, 1947–1952.* University Park: Penn State University Press.

Hewlett, Richard G., and Jack M. Holl. 1989. *A History of the United States Atomic Energy Commission.* Vol. 3, *Atoms for Peace and War, 1953–1961.* Berkeley: University of California Press.

Hilts, Philip. 1982. *Scientific Temperaments: Three Lives in Contemporary Science.* New York: Simon and Schuster.

Hinterberger, Henry, and Robert R. Wilson. 1982. Tape recording of a conversation about Fermilab history, 21 June.

Hinterberger, H., J. Satti, C. Schmidt, R. Sheldon, and R. Yamada,1971. "Bending Magnets of the NAL Accelerators." 1971 Particle Accelerator Conference. *IEEE Transactions on Nuclear Science* 18, no. 3 (June 1971): 853–856.

Hirsch, Susan E., Robert I. Goler, and Sam Bass Warner. 1990. *A City Comes of Age.* Chicago: Chicago Historical Society.

Hoddeson, Lillian. 1983. "Establishing KEK in Japan and Fermilab in the US: Internationalism, Nationalism and High Energy Accelerator Physics during the 1960s." *Social Studies of Science* 13:1–48.

———. 1987. "The First Large-Scale Application of Superconductivity: The Fermilab Energy Doubler, 1972–1983." *Historical Studies in Physical and Biological Sciences* 18, no. 1:25–54.

———. 1989. "The Los Alamos Implosion Program in World War II: A Model for Postwar American Research." In *The Restructuring of Physical Sciences in Europe and the United States, 1945–1960,* ed. M. de Maria, M. Grilli, and F. Sebastiani, 31–34. Singapore: World Scientific.

———. 1994. "Wilson's Way." Talk delivered for Robert R. Wilson's 80th birthday celebration at Fermilab, 4 March.

———. 2006. "The Conflict of Memories and Documents: Dilemmas and Pragmatics of Oral History." In *The Historiography of Contemporary Science, Technology and Medicine: Writing Recent Science,* ed. Ronald Edmund Doel and Thomas Söderqvist, Routledge Studies in the History of Science, Technology and Medicine, 187–200. New York: Routledge.

Hoddeson, Lillian, and Adrienne Kolb. 2000. "The Superconducting Super Collider's Frontier Outpost, 1983–1988." *Minerva* 38, no. 3:271–310.

———. 2003. "Vision to Reality: From Robert R. Wilson's Frontier to Leon M. Lederman's Fermilab." *Physics in Perspective,*vol. 5, no. 1:67–86.

———. 2008. "Robert Rathbun Wilson." In *New Dictionary of Scientific Biography,* 7:326–332. Detroit: Charles Scribner's Sons, Thomson Gale.

Hoddeson, Lillian, Helmut Schubert, Steve J. Heims, and Gordon Baym. 1992. "Collective Phenomena." In *Out of the Crystal Maze: Chapters from the His-*

tory of Solid State Physics, ed. Lillian Hoddeson et al., 489–616. New York: Oxford.

Hoddeson, Lillian, Paul W. Henriksen, Roger A. Meade, and Catherine L. Westfall. 1993. *Critical Assembly: A Technical History of Los Alamos during the Oppenheimer Years, 1943–1945*. New York: Cambridge University Press.

Hoddeson, Lillian, Laurie Brown, Michael Riordan, and Max Dresden. 1997. *The Rise of the Standard Model: Particle Physics in the 1960s and 1970s*. New York: Cambridge University Press.

Holden, Constance. 1981. "Former Carolina Governor to Head DOE." *Science* 211 (6 February): 555.

Holl, Jack M. 1997. *Argonne National Laboratory, 1946–1996*. Urbana: University of Illinois Press.

Holloway, Lee. 1988. Interview by Lillian Hoddeson, 28 February.

Hom, D. C., L. M. Lederman, H. P. Paar, H. D. Snyder, J. M. Weiss, J. K. Yoh. 1976a. "Production of High-Mass Muon Pairs in Proton-Nucleus Collisions at 400 GeV." *Physical Review Letters* 37, no. 21:1374–1377.

Hom, D. C., L. M. Lederman, H. P. Paar, H. D. Snyder, J. M. Weiss, J. K. Yoh, J. A. Appel, B. C. Brown, C. N. Brown, W. R. Innes, T. Yamanouchi, and D. M. Kaplan. 1976b. "Observation of High-Mass Dilepton Pairs in Hadron Collisions at 400 GeV." *Physical Review Letters* 36, no. 21:1236–1239.

Hom, D. C., et. al. 1976c. "Production of High Mass Muon Pairs in Hadron Collisions at 400 GeV." Preprint, Fermilab-Pub-76-075, Batavia (September).

Hughes, Jeff. 2002. *The Manhattan Project: Big Science and the Atom Bomb*. New York: Columbia University Press.

Hughes, Thomas P. 1983. *Networks of Power: Electrification in Western Society, 1800–1930*. Baltimore: Johns Hopkins University Press.

———. 1989. *American Genesis: A Century of Invention and Technological Enthusiasm, 1870–1970*. New York: Viking Penguin Press.

Hulm, J. K.1983. "Superconductivity Research in the Good Old Days." *IEEE Transactions on Magnetics* 19:161–166.

Hulm, J. K., J. E. Kunzler, and B. J. Matthias. 1981. "The Road to Superconducting Materials." *Physics Today* 34, no. 1:34–43.

Jachim, Anton J. 1975. *Science Policy Making in the United States and the Batavia Accelerator*. Carbondale: Southern Illinois University Press.

Jackson, J. David. 1996. Interview by Adrienne Kolb and Lillian Hoddeson, 3 May.

———. 1999. "Recollections of ELG and the Beginnings of the NAL Theoretical Physics Program." Paper presented at the "The Early Days of Fermilab: A Symposium in Honor of Edwin Goldwasser's Eightieth Birthday," 10 March, Fermilab. Manuscript in Fermilab Archives.

Jacobs, Michael E. 1980. "CERN Elects Schopper as Director." *Physics Today* 33, no. 6 (June): 84–85.

Johnson, Kjell. 1997. "The CERN Intersecting Storage Rings: The Leap into the Hadron Collider Era." In *The Rise of the Standard Model: Particle Physics in the*

1960s and 1970s, ed. Lillian Hoddeson, Laurie Brown, Michael Riordan, and Max Dresden, 261–284. New York: Cambridge University Press.

Johnson, Lyndon B. 1966. "The Strengthening of Academic Capacities for Science Throughout the Country." In *Equitable Distribution of R & D Funds by Government Agencies*, by U.S. Congress, Subcommittee on Government Research of the Committee on Government Operations, 89th Cong., 2nd sess. Washington, DC: GPO.

Jöstlein, Hans. 1990. Interview by Catherine Westfall and Frederik Nebeker, 3 May.

Jovanovic, Drasko. 1989. Interview by Catherine Westfall and Lillian Hoddeson, 29 November.

Judd, Dave. 1984. Interview by Catherine Westfall, 1 May.

Kaiser, David 2004. "The Postwar Suburbanization of American Physics." *American Quarterly* 56, no. 4 (December): 851–888.

Kaplan, Daniel M. 1994. "The Discovery of the Upsilon Family." IIT-HEP-94/1. Physics Department, Illinois Institute of Technology.

Katz, James E. 1978. *Presidential Politics and Science*. New York: Praeger.

Kay, Lily E. 1993. *The Molecular Vision of Life: Caltech, the Rockefeller Foundation, and the Rise of the New Biology*. New York: Oxford University Press.

Kerns, Quentin. 1987. Interview by Catherine Westfall, 12 March.

Keefe, Denis, Glen Lambertson, and L. Jackson Laslett. 1986. Interview by Catherine Westfall, 22 December.

"Kennedy and Johnson Open the Campaign: Acceptance Speeches of the Democratic Nominees." 1960. *U.S. News and World Report*, 25 July, 100–102.

Kephart, Robert. 1990. Interview by Frederik Nebeker, 13 June.

Kephart, Robert, et al. 1977. "Measurement of the Dihadron Mass Continuum in p-Be Collisions and a Search for Narrow Resonances." *Physics Review Letters* 39:1440–1443.

Kevles, Daniel. 1978. *The Physicists: The History of a Scientific Community in Modern America*. New York: Alfred A. Knopf.

———. 1995. *The Physicists: The History of a Scientific Community in Modern America*. Cambridge, MA: Harvard University Press.

Keyworth, George. 2000. Interview by Lillian Hoddeson, 12 March.

"Keyworth Says U.S. Can't Be Tops in All R & D." 1981. *Science and Government Report* 11, no. 12 (15 July): 8.

Killian, James Rhyne. 1977. *Sputnik, Scientists, and Eisenhower: A Memoir to the First Special Assistant to the President for Science and Technology*. Cambridge, MA: MIT Press.

Kirk, Thomas. 1990a. Interview by Mark Bodnarczuk and Lillian Hoddeson, 10 August.

———. 1990b. Interview by Catherine Westfall, 10 August.

———. 1991. Interview by Catherine Westfall, 11 November.

Kitagaki, T. 1953. "A Focusing Method for Large Accelerators." *Physical Review* 89, no. 5:1161–1162.

Knapp, Edward, 1997. Interview by Lillian Hoddeson, 14 July.
Koepke, Karl. 1984. Interview by Lillian Hoddeson, 14 February.
Kolb, Adrienne, and Lillian Hoddeson. 1993. "The Mirage of the World Accelerator for World Peace and the Origins of the SSC, 1953–1983." *Historical Studies in the Physical and Biological Sciences* 24, no. 1:101–124.
———. 1995. "A New Frontier in the Chicago Suburbs: Settling Fermi National Accelerator Laboratory, 1963–1972." *Journal of the Illinois State Historical Society* 88, no. 1:2–18.
Kotulak, Ronald.1968. "Head of Atom Smasher May Resign." *Chicago Tribune*, 15 June.
Krige, John. 1991. "Some Socio-historical Aspects of the Multi-institutional Collaborations in High-Energy Physics of CERN between 1975 and 1987." CHS-32. CERN, Geneva.
———. 1993. "Some Socio-historical Aspects of Multinational Collaborations in High-Energy Physics at CERN Between 1975 and 1985." In *Denationalizing Science*, ed. E. Crawford et. al., 233–262. The Netherlands: Kluwer Academic Publishers.
———, ed. 1996. *History of CERN*. Vol. 3. Amsterdam: North-Holland.
———. 2001. "Distrust and Discovery: The Case of the Heavy Bosons at CERN." *Isis* 92, no. 3 (March): 517–540.
Kuchnir, Moyses. 1985. Interview by Lillian Hoddeson and Adrienne Kolb, 1 May.
Kunzler, J. E., E. Buehler, F. S. L. Hsu, and J. H. Wernick. 1961. "Superconductivity in Nb_3Sn at High Current Density in a Magnetic Field of 88 kgauss." *Physics Review Letters* 6, no. 3:89–91.
Lach, Joseph. 1989. Interview by Catherine Westfall, 23 March.
Lambright, Henry. 1985. *Presidential Management of Science and Technology: The Johnson Presidency*. Austin: University of Texas Press.
Lawrence, Ernest O. 1941. "The New Frontiers in the Atom." *Science* 94:221–225.
Lederman, Leon M. 1963. "The Truly National Laboratory." In 1963 Super-High Energy Summer Study, report, Brookhaven National Laboratory, AADD-6.
———. 1968. "Beam Dump Experiment: Dimuons and Neutrinos." Preprint report.
———. 1977a. "VBA." *IEEE Transactions on Nuclear Science*, n.s., 24, no. 3 (June): 1903–1908. (Also published in *Physics Today*, May 1977, 19–20.)
———. 1977b. "Spokesman Leon Lederman (an Unauthorized Autobiography)." *Village Crier*, 6 January, 6.
———. 1978. "The Upsilon Particle." *Scientific American* 239, no. 4 (October): 72–80.
———. 1979. *Aesthetics and Science*. Proceedings of the International Symposium in Honor of Robert R. Wilson, 27 April, Batavia, Fermi National Accelerator Laboratory, 1–23. Bellwood, IL: Sleepeck Printing Co.
———. 1980. "The State of the Laboratory." In *Fermilab 1980: Annual Report of the Fermilab National Accelerator Laboratory*, 1–8. FA.

———. 1982. "Fermilab and the Future of HEP." In *Proceedings of the 1982 DPF Summer Study on Elementary Particle Physics and Future Facilities*, ed. Rene Donaldson, Richard Gustafson, and Frank Paige (DPF/APS), 125–127. DOE/NSF. Batavia: Fermi National Accelerator Laboratory.

———. 1984a. Interview by Catherine Westfall, 20 July.

———. 1984b. Talk presented at the dedication of the Energy Saver, April 28.

———. 1984c. "The Value of Fundamental Science." *Scientific American* 251, no. 5 (November): 40–47.

———. 1985. Talk presented at the dedication of the proton-antiproton collider, Tevatron I, October 11.

———. 1987. "Biography." Unpublished manuscript.

———. 1988. "The State of the Laboratory." In *Fermilab 1988: Annual Report of the Fermi National Accelerator Laboratory*, 1–14.

———. 1989. "Observations in Particle Physics from Two Neutrinos to the Standard Model," *Reviews of Modern Physics* 61, no. 3 (3 July): 547–560.

———. 1990a. Interview by Frederik Nebeker, 7 May.

———. 1990b. Interview by Catherine Westfall, 26 November.

———. 1991. *Science: The End of the Frontier?* Washington, DC: AAAS.

———. 1993. "What Can We Learn from the Super Collider's Demise?" *Scientist*, 29 November, 12.

———. 1997. "The Discovery of the Upsilon, Bottom Quark, and B Mesons." In *The Rise of the Standard Model: Particle Physics in the 1960s and 1970s*, ed. Lillian Hoddeson, Laurie Brown, Michael Riordan, and Max Dresden, 101–113. New York: Cambridge University Press.

———. 1999. Interview by Lillian Hoddeson and Adrienne Kolb, 19 November.

Lederman, Leon M., and Chris Quigg. 1979. *Fermilab Research Results*. Washington, DC: GPO.

———. 1988. *Appraising the Ring: Statements in Support of the Superconducting Super Collider*. Washington, DC: Universities Research Association.

Lederman, Leon M., with Dick Teresi. 1993. *The God Particle: If the Universe Is the Answer, What Is the Question?* New York: Houghton Mifflin.

Lee, W., L. M. Lederman, J. Appel, M. Tannenbaum, L. Read, J. Sculli, T. White, and T. Yamanouchi. 1970. "Study of Lepton Pairs from Proton-Nuclear Interactions: Search for the Immediate Bosons and Lee-Wick Structure." NAL-Proposal-70, 17 June.

Lehman, Daniel. 2001. Interview by Catherine Westfall, 24 January.

Leiss, James E. 1981. "Relevance of Accelerator/Storage Ring Technological Developments to U.S.A. Science and Technology—Past, Present, and Future." *IEEE Transactions on Nuclear Science*, n.s., 28, no. 3 (June): 3553–3555.

———. 1982. "Impact of Accelerators on Science and Technology." *Physics Today* 34 (July): 70–72.

———. 1991. Interview by Catherine Westfall, 9 December. Jefferson Laboratory, Newport News, VA.

Leposky, George. 1960. "What's the Best Way to Build a $250 Million Atom Smasher?" *Inland Architect*, August/ September, 24.

Leslie, Stuart W. 1993. *The Cold War and American Science: The Military-Industrial-Academic Complex at MIT and Stanford.* New York: Columbia University Press.

"Lewis and Clark: How an Amazing Adventure 200 Years Ago Continues to Shape How America Sees Itself." 2002. *Time* 160, no. 2 (8 July): 6, 36–76.

Lighter, J. E., ed. 1994. *Random House Historical Dictionary of American Slang.* Vol. 1. New York: Random House.

Limerick, Patricia. 1987. *The Legacy of Conquest: The Unbroken Past of the American West.* New York: W. W. Norton.

Limon, Peter. 1983. Interview by Lillian Hoddeson, 7 November.

Livdahl, Philip V. 1983. "A Brief Summary of Fermilab during Initial Construction Years." TM-1223, Fermilab, November.

———. 1984. Interview by Lillian Hoddeson, 26–27 September.

Livingston, M. Stanley. 1968a. "Early History of the 200-GeV Accelerator." NAL-12, National Accelerator Laboratory, Batavia, June.

———. 1968b. *Particle Physics: The High Energy Frontier.* New York: McGraw-Hill.

———. 1969a. "Design Progress at the National Accelerator Laboratory: 1968–1969."

———. 1969b. *Particle Accelerators: A Brief History.* Cambridge, MA: Harvard University Press.

Livingston, M. Stanley, and John Blewett. 1962. *Particle Accelerators.* New York: McGraw-Hill.

Lofgren, Edward. 1984a. Interview by Catherine Westfall, 24 April.

———. 1984b. Interview by Catherine Westfall, 3 May.

———. 1984c. Interview by Catherine Westfall, 16 May.

———. 1987. Interview by Catherine Westfall, 9 October.

Louwenaar, Karyl Louwenaar. 1968. "The Story of the Village of Weston." Unpublished manuscript, 24 June.

Lowi, Theodore J., and Benjamin Ginsberg. 1976. *Poliscide.* New York: McMillan.

LRL. 1965a. "200 BeV Accelerator Design Study." Vol. 1. June.

LRL. 1965b. "200 BeV Accelerator Design Study." Vol. 2. June.

Lubkin, Gloria B. 1983. "Panel Says: Go for a Multi-TeV Collider and Stop Isabelle." *Physics Today*, September, 17–20.

———. 1984a. "R & D Funding for the Supercollider." *Physics Today*, October, 21.

———. 1984b. "SSC Design Goes to DOE: ICFA Discusses CERN Hadron Collider." *Physics Today*, June, 17–19.

———. 1984c. "UAI at CERN Says It Has Candidates for Sixth Quark, Top." *Physics Today*, August, 17–18.

Lundy, Richard. 1983. Interview by Lillian Hoddeson, 11 November.

MacLachlan, James. 1989a. Interview by Catherine Westfall, 24 October, 27–28 November.

Malamud, Ernest. 1983. "Early History of the Fermilab Main Ring." 20 October. FA.

———. 1987. Interview by Catherine Westfall, 12 March.

———. 1989. Interview by Catherine Westfall and Lillian Hoddeson, 24 October.

Malamud, Ernest, and James K. Walker. 1970. "Progress and Prospects at the National Accelerator Laboratory."

Mantsch, Paul. 1984. "Spirit of Snowmass Spreads across Land." *FermiNews*, 14 June, 2–3.

Margolis, Jon. 1992. "The Frontier Never Was as Pretty as the Pictures." *Chicago Tribune*, 8 July.

Marshak, Robert E. 1989. "Scientific and Sociological Contributions of the First Decade of the Rochester Conferences to the Restructuring of Particle Physics (1950–60)." In *The Restructuring of the Physical Sciences in Europe and the United States, 1945–60*, ed. M. de Maria, M. Grilli, and F. Sebastiani, 745–786. Singapore: World Scientific.

———. 1990. "The Khrushchev Détente and Emerging Internationalism in Particle Physics." *Physics Today* 43, no. 1 (January): 34–42.

Marshall, Elliot. 1987. "Big versus Little Science in the Federal Budget." *Science* 236 (17 April): 249.

McCarthy, Jack. 1987. Interview by Catherine Westfall, 31 March.

McCarthy, R. 1990. Interview by Frederik Nebeker, 16 April.

McDonald, K. 1985. "Gigantic Particle Accelerator Will Have No Modern Rival—If It's Built." *Chronicle of Higher Education* 31, no. 7 (16 October): 1, 10–11.

McDougall, Walter A. 1985. *The Heavens and the Earth: A Political History of the Space Age*. New York: Basic Books.

McInturff, A. 1984. Interview by Lillian Hoddeson, 14 February.

McMillan, Edwin. 1984. Interview by Catherine Westfall, 16 May.

Mervis, Jeffrey. 1987. "New Research Chief Sees Foreign Cooperation on SSC." *Scientist*, 10 May, 6.

Michener, James. 1974. "The Wagon and the Elephant." In *Centennial*, 243–349. New York: Random House.

Mills, Frederick. 1983. Interview by Lillian Hoddeson, 15 December.

———. 1984. Interview by Catherine Westfall, 13 July.

———. 1987. Interview Lillian Hoddeson and Adrienne Kolb, 16 November.

Month, Melvin. 2003. *Weep for Isabelle: A Rhapsody in a Minor Key: The Rise and Decline of American Big Physics*. New York: AvantGarde Press.

Moore, Charles. 1921. *Daniel H. Burnham*. Boston: Houghton Mifflin Co.

Morrison, Rollin. 1990. Interview by Mark Bodnarczuk, July 3.

Mravca, Andrew. 1983. Interview by Lillian Hoddeson, 14 December.

Murphy, C. Thornton. 1984. Interview by Lillian Hoddeson, 22–23 March and 6 April.

Nabrit, Samuel. 1987. Interview by Catherine Westfall, 6 October.

NAL. 1968a. "Design Report." National Accelerator Laboratory, January.

———. 1968b. "Policy Statement on Human Rights." National Accelerator Laboratory, 15 March.

Nambu, Yoichiro. 1989. "Gauge Principle, Vector-Meson Dominance, and Spontaneous Symmetry Breaking." In *Pions to Quarks: Particle Physics in the 1950s*. ed. Laurie M. Brown, Max Dresden, and Lillian Hoddeson, 638–642. New York: Cambridge University Press.

Nash, Thomas. 1983. "Specialized Computers for High-Energy Experiments." *Physics Today* 36, no. 5 (May): 36–37.

———. 1990. Interview by Mark Bodnarczuk, 18 July.

"National Policy for High Energy Physics Program." 1965. In *High Energy Physics Program: Report on National Policy and Background Information*, by U.S. Congress, Joint Committee on Atomic Energy. Washington, DC: GPO.

Nebeker, Frederik, 1994. "Strings of Experiments in High-Energy Physics: The Upsilon Experiment." *Historical Studies in the Physical and Biological Sciences* 25, no. 1:137–164.

Needell, Allan. 1983. "Nuclear Reactors and the Founding of Brookhaven National Laboratory." *Historical Studies in the Physical Sciences* 14, no. 1:93–122.

Nolte, Robert. 1969. "Last Weston Folks Prepared to 'Make Room for Progress.'" *Chicago Tribune*, 24 August.

Norton, Mary Beth, David Katzman, Paul Escot, Howard Chudacoff, Thomas Patterson, and William Tuttle, eds. 1982. *A People and a Nation: A History of the United States*. Vol. 2. Dallas: Houghton Mifflin Co.

O'Halloran, Thomas. 1991. Interview by Catherine Westfall, 4 November.

Onnes, H. Kamerlingh. 1913a. "Report on Research Made in the Leiden Cryogenic Laboratory between the Second and Third International Congress of Refrigeration." *Communications from the Physical Laboratory of the University of Leiden* 134d:35–70.

———. 1913b. "The Sudden Disappearance of the Ordinary Resistance of Tin, and the Supraconductive State of Tin." *Communications from the Physical Laboratory of the University of Leiden* 133d:51–68.

Oppenheimer, J. Robert. 1964. "Space and Time." In *The Flying Trapeze: Three Crises for Physicists*. London: Oxford University Press.

Orlans, H. 1967. *Contracting for Atoms*. Washington, DC: Brookings Institution.

Orr, J. Richie. 1983. Interview by Lillian Hoddeson, 9 November.

———. 1987. Interview by Catherine Westfall, 12 March.

———. 1989. Interview by Catherine Westfall, 24 March.

———. 1990. Interview by Catherine Westfall, 23 May.

———. 1992. Interview by Catherine Westfall, 27 May.

Paik, Kyoung. 1983 "The Origin of CDF and the Fermilab Collider Project." Unpublished manuscript, Department of History, University of Illinois at Urbana-Champaign.

Pais, Abraham. 1986. *Inward Bound: Of Matter and Forces in the Physical World*. New York: Oxford University Press.

Palmer, Robert. 1984. Interview by Lillian Hoddeson, 4 May.

Paris, Elizabeth. 1999. "Ringing in the New Physics: The Politics and Technology of Electron Colliders in the United States, 1956–1972." Ph.D. diss., University of Pittsburgh.

———. 2003. "Do You Want to Build Such a Machine? Designing a High Energy Proton Accelerator for Argonne National Laboratory." Argonne History Group, Argonne manuscript, 20 November.

Partridge, Eric. 1961. *A Dictionary of Slang and Unconventional English from the Fifteenth Century to the Present Day*. New York: Macmillan Co.

Pellegrini, Claudio, and Andrew Sessler. 1995. "Chapter 1: Introduction." In *The Development of Colliders*, ed. Claudio Pellegrini and Andrew Sessler, 1–6. New York: American Institute of Physics.

Peoples, John. 1984. "Introduction." In *Proceedings of the 1984 Summer Study on Design and Utilization of the Superconducting Super Collider*, ed. Rene Donaldson and Jorge G. Morfín. Batavia, IL: Fermi National Accelerator Laboratory.

———. 1988. Interview by Adrienne Kolb and Lillian Hoddeson, 8 March.

———. 1993a. Interview by Catherine Westfall, 19 March.

———. 1993b. Interview by Lillian Hoddeson and Adrienne Kolb, 23 October.

———. 1999. Interview by Catherine Westfall, 27 October.

———. 2004. Interview by Lillian Hoddeson and Adrienne Kolb, 16 January.

"Perfect VBA Site Found?" 1977. *Physics Today*, May, 19–20.

Perkins, Donald. 1997. "Gargamelle and the Discovery of Neutral Currents." In *The Rise of the Standard Model*, ed. Lillian Hoddeson, Laurie Brown, Michael Riordan, and Max Dresden, 428–446. New York, Cambridge University Press.

Pestre, Dominique. 1990. "La Seconde Génération d'Accelerateurs Pour Le CERN, 1956–1965: Étude Historique d'un Processus de Décision de Gros Équipement en Science Fondamentale." CHS-19.

Pestre, Dominique, and John Krige. 1992. "Some Thoughts on the Early History of CERN." In *Big Science: The Growth of Large-Scale Research*, ed. Peter Galison and Bruce Hevly. Stanford: Stanford University Press.

Pewitt, D. 1998. Interview by M. Riordan, 3 May.

Physics Survey Committee, National Research Council. 1972. *Physics in Perspective*. Washington, DC: National Academy of Sciences.

Pickering, Andrew. 1984. *Constructing Quarks: A Sociological History of Particle Physics*. Chicago: University of Chicago Press.

Pinch, Trevor. 1986. *Confronting Nature: The Sociology of Solar-Neutrino Detection*. Boston: D. Reidel Publishing Co.

"Piore Panel Report—1960." 1965. In *High Energy Physics Program: Report on National and Background Information*, by U.S. Congress, Joint Commission on Atomic Energy. Washington, DC: GPO.

Pitzer, Kenneth. 1984. Interview by Catherine Westfall, 17 May.
"Policy Makers Invoke New Paradigm for Science and Technology." 1994. *APS News*, June, 2.
Pope, Bernard. 2000. Interview by Catherine Westfall, 17 October.
"The President: Something Borrowed." 1967. *Newsweek*, 27 February, 28.
President's Science Advisory Committee. 1960. *Scientific Progress, the Universities, and the Federal Government.* Washington, DC: GPO.
Price, Derek J. de Solla, 1961. *Science since Babylon.* New Haven: Yale University Press.
———. 1963. *Little Science, Big Science—and Beyond.* New York: Oxford University Press.
Proceedings of the 1984 ICFA Seminar on "Future Perspectives in High Energy Physics" May 14–20, 1984 KEK. 1984. Ed. S. Ozaki, S. Kurokawa, and Y. Unno. KEK Report 84-14, September.
PSSC: Physics at the Superconducting Super Collider: Summary Report. 1984. Batavia, IL: Fermilab.
Quigg, Chris. 1993. Interview by Adrienne Kolb and Lillian Hoddeson, 5–7 May.
Ramey, James T. 1986. Interview by Catherine Westfall, 21 November.
Ramsey, Norman. 1966. "Early History of Associated Universities, Inc. and Brookhaven National Laboratory." BNL 992 (T-421) Upton. March.
———. 1975. "History of the Fermilab Accelerator and URA." 2 May.
———. 1978. Interview by Lillian Hoddeson, 19 December.
———. 1980a. Interview by Lillian Hoddeson, 22 January.
———. 1980b. Interview by Lillian Hoddeson, 26 and 27 February.
———. 1987a. "The Early History of URA and Fermilab." In *Annual Report*, 157–161. Batavia, IL: Fermilab.
———. 1987b. Interview by Lillian Hoddeson, 26 and 27 February.
———. 1992. Interview by Catherine Westfall, 29 June.
Ramsey Panel. 1963. *Report of the Panel on High Energy Accelerator Physics of the General Advisory Committee to the Atomic Energy Commission and the President's Science Advisory Committee, April 26, 1963, Transmitted on May 20, 1963.* U.S. AEC TID-18636. Washington, DC: GPO.
Reagan, Michael D. 1969. *Science and the Federal Patron.* New York: Oxford University Press.
Reardon, Paul J. 1977. "High Energy Physics and Applied Superconductivity." *IEEE Transactions on Magnetics* 13, no. 1:704–718.
———. 1984. Interview by Lillian Hoddeson, 4 May.
Reardon, P. J., and B. P. Strauss, eds. 1973. "Some Preliminary Concepts about the Proposed Energy Doubler Device for the 200/500 GeV Proton Accelerator at the United States National Accelerator Laboratory, Batavia, Illinois." TM-421, May.
Reingold, N., 1987 "Vannevar Bush's New Deal for Research: Or the Triumph of the Old Order." *Historical Studies in the Physical and Biological Sciences* 17:300–344.

Report of the Subpanel on Review and Planning for the U.S. High Energy Physics Program. 1980. July.

Report of the 1983 HEPAP Subpanel on New Facilities for the U.S. High Energy Physics Program. 1983. U.S. Department of Energy, Office of Energy Research Division of High Energy Physics, July.

Report of the Subpanel on Long Range Planning for the U.S. High Energy Physics Program of the High Energy Physics Advisory Panel. 1982. DOE/ER-0128. January.

Report of the DOE Review Committee on the Reference Designs Study, May 18. 1984.

Rhodes, Richard. 1986. *The Making of the Atomic Bomb*. New York: Simon and Schuster, Touchstone.

Richter, Burton. 1992. Interview by Catherine Westfall, 24 June.

———. 1997. "The Rise of Colliding Beams." In *The Rise of the Standard Model: Particle Physics in the 1960s and 1970s*, ed. Lillian Hoddeson, Laurie Brown, Michael Riordan, and Max Dresden, 261–284. New York: Cambridge University Press.

Riordan, Michael. 1987. *The Hunting of the Quark: A True Story of Modern Physics*. New York: Simon and Schuster.

———. 2001. "A Tale of Two Cultures: Building the Super Collider, 1988–1993." *Historical Studies in the Physical and Biological Sciences* 32:125–144.

Riordan, Michael, Lillian Hoddeson, and Adrienne Kolb. N.d. "Tunnel Visions: The Rise and Fall of the Superconducting Super Collider." Unpublished manuscript (in progress).

Ritson, David. 1993. "Demise of the Texas Supercollider." *Nature* 366 (16 December): 607–610.

Rode, Claus. 1984. Interview by Lillian Hoddeson, 28 September.

Sands, Matthew. 1959. "Ultra High Energy Synchrotrons." In "1959 MURA Summer Study on Design and Utilization of High-Energy Accelerator," MURA Report 465, by Midwestern Universities Research Association, 10 June 1959.

———. 1960. "A Proton Synchrotron for 300 GeV." Internal Caltech report, Caltech Synchrotron Lab.

Sanford, James R. 1976. "The Fermi National Accelerator Laboratory." *Annual Review of Nuclear Science* 26:151–198.

Sazama, Cynthia, et al. 2003. Interview of Cynthia Sazama, Jean Lemke, Jackie Coleman, Marilyn Paul, and Barb Kristen by Adrienne Kolb, 11 and 22 September.

Schlesinger, Arthur, Jr. 1983. *A Thousand Days: John F. Kennedy in the White House*. New York: Greenwich House.

Schwarzschild, Bertram. 1986. "Panel Reaffirms High-Field Magnet Choice for Supercollider." *Physics Today*, July, 21–23.

Schweber, Silvan. 1997. "A Historical Perspective on the Rise of the Standard Model." In *The Rise of the Standard Model: Particle Physics in the 1960s and 1970s*, ed. Lillian Hoddeson, Laurie Brown, Michael Riordan, and Max Dresden, 645–684. New York: Cambridge University Press.

Schwitters, Roy. 1986. "SSC Briefing Materials" (March). Unpublished document. SSC Collection.

———. 1988a. Interview by Lillian Hoddeson, 3 March.

———. 1988b. Interview by Lillian Hoddeson, 10 March. CDF Collection.

———. 1990. Interview by Lillian Hoddeson and Kyoung Paik, 15 November.

———. 1997. Interview by M. Riordan, 22 March.

Seaborg, Glenn. 1983. Interview by Catherine Westfall, 4 February.

———. 1984. Interview by Catherine Westfall, 16 February.

———. 1987. "Atoms for Peace Program." In *Encyclopedia Americana* 2:649. Danbury, Conn.: Grolier.

Seidel, Robert W. 1983. "Accelerating Science: The Postwar Transformation of the Lawrence Radiation Laboratory." *Historical Studies in the Physical Sciences* 13, no. 2:375–400.

Seitz, Frederick. 1965. "National Academy of Sciences Meeting of University Presidents, January 17, 1965." In *Hearings*, by U.S. Congress, Subcommittee on Research, Development, and Radiation of the Joint Committee on Atomic Energy, 89th Cong., 1st sess. Washington, DC: GPO.

———. 1980a. Interview by Lillian Hoddeson, 2 February.

———. 1980b. Interview by Lillian Hoddeson, 7 February.

———. 2003. Interview by Adrienne Kolb and Lillian Hoddeson, 20 June.

Serber, Robert. 1986. Interview by Catherine Westfall, 24 February.

Shochet, Melvyn J. 1990. Interview by Lillian Hoddeson and Adrienne Kolb, 27 September.

Shoemaker, Nancy. 1993. "Teaching the Truth about the History of the American West." *Chronicle of Higher Education*, 27 October, A48.

Slotkin, Richard. 1992. *Gunfighter Nation: The Myth of the Frontier in Twentieth-Century America*. New York: Atheneum.

Smith, Merrit Roe. 1993. "Frontiers of Change." *STS News*, September, 1–2. MIT, Program in Science, Technology and Society.

Smith, Peter. 1971. "Superconducting Synchrotron Magnets: Present Status." In *Proceedings of the 8th International Conference on High-Energy Accelerators*, ed. M. H. Blewett and N. Vogt-Nilsen, 35–46. Geneva: European Organization for Nuclear Research.

Smith, Robert. 1989. *The Space Telescope: A Study of NASA, Technology, and Politics*. Cambridge: Cambridge University Press.

Sokolov, Raymond. 1983. "Fermilab: Utopia on the Prairie." *Wall Street Journal*, 11 February.

Staley, Kent W. 2004. *The Evidence for the Top Quark: Objectivity and Bias in Collaborative Experimentation*. Cambridge: Cambridge University Press.

Star Trek 20th Anniversary Special. 1991. Video. Irvine, CA: Interplay Productions.

Strauss, Bruce. 1978. Interview by Lillian Hoddeson, 14 September.

Sullivan, Jeremiah. 2005. Interview by Lillian Hoddeson, 22 February.

Superconducting Super Collider Reference Designs Study. 1984. Unpublished document, Central Design Group, Berkeley, 8 May.

"A Switch in Time." 1968. *Newsweek*, 11 March, 38.

Tannenbaum, Michael. 1984. Interview by Lillian Hoddeson, 19 June.

Tape, Gerald. 1986. Interview by Catherine Westfall, 21 November.

Taubes, Gary. 1986. *Nobel Dreams: An Experiment at the Edge of the Universe*. New York: Random House.

Taylor, George Rogers, ed. 1956. *The Turner Thesis: Concerning the Role of the Frontier in American History*. Boston: D. C. Heath and Co.

Teng, Lee. 1993. Interview by Kyoung Paik, 16 April.

———. 2006. "Accelerators and I." Manuscript, 13 May. Argonne National Laboratory.

Theriot, Dennis. 1993. Interview by Adrienne Kolb, 22 November.

Thomas, Lewis. 1981. "Notes of a Biology Watcher." *Discover*, January.

Tigner, M. 1981. "Accelerator R/D in the U.S. High Energy Physics Program: Past, Present and Future." *IEEE Transactions on Nuclear Science*, n.s., 28 (June): 3549–3552.

———. 1983. "Report of the 20 TeV Hadron Collider Workshop." Unpublished report, Cornell University, Ithaca.

Tollestrup, Alvin V. 1978. Interview by Lillian Hoddeson, 16 November.

———. 1983. Interview by Lillian Hoddeson, 9 November.

———. 1987. Interview by Lillian Hoddeson and Adrienne Kolb, 24 November.

———. 1990. Interview by Lillian Hoddeson and Kyoung Paik, 5 July.

Tollestrup, Alvin V., and Robert Walker. 1985. Interview by Catherine Westfall, 4 May.

Toohig, Timothy. 1983. Interview by Lillian Hoddeson, 10 November.

"To Trivelpiece, a Superconducting Super Collider is Exhilarating." 1987. *Chronicle of Higher Education*, 11 February, 7, 9.

Traweek, Sharon. 1988. *Beamtimes and Lifetimes: The World of High Energy Physicists*. Cambridge, MA: Harvard University Press.

Trefil, James S. 1980. *From Atoms to Quarks: An Introduction to the Strange World of Particle Physics*. New York: Charles Scribner's Sons.

Trenn, Thaddeus J. 1983. *America's Golden Bough: The Science Advisory Intertwist*. Cambridge: Oelschlager, Gunn, and Hain.

Trivelpiece, Alvin V. 1996. Interview by Steve Weiss, 6 November.

Turner, Frederick Jackson. 1956. "The Significance of the Frontier in American History." Speech delivered at the Chicago World's Columbian Exposition, 12 July 1893. In *The Turner Thesis: Concerning the Role of the Frontier in American History*, ed. George Rogers Taylor, 1–18. Boston: D. C. Heath and Co.

———. 1994. "The Significance of the Frontier in American History." Lecture to the American Historical Association, meeting at the World's Columbian Exposition, Chicago, 1893. In *Rereading Frederick Jackson Turner*, ed. John Mack Faragher, 31–60. New York: Henry Holt and Co.

Turowski, P., J. H. Coupland, and J.Perot. 1974. "Pulsed Superconducting Dipole Magnets of the GESSS Collaboration." In *Proceedings of the 9th International*

Conference on High Energy Accelerators, 175–178. Menlo Park, CA: Stanford Linear Accelerator Laboratory.

Twain, Mark. 1984. *Life on the Mississippi*. New York: Viking Penguin; originally published in 1883.

U.S. Atomic Energy Commission. High Energy Physics Advisory Panel. 1968. "The Status and Problems of High Energy Physics Today." January. FA.

U.S. Congress. 1964. *Government and Science: Hearings before the Subcommittee on Science, Research, and Development of Committee on Science and Astronautics*. Washington, DC: GPO.

U.S. Congress, Joint Committee on Atomic Energy. 1964. *Atomic Energy Commission Authorization, FY 1965*. 88th Cong., 2nd sess. Washington, DC: GPO.

———. 1965. Appendix C, "National Policy for High Energy Physics Program." In *High Energy Physics Program: Report on National Policy and Background Information*. Washington, DC: GPO.

———. 1967. *Atomic Energy Commission Authorizing Appropriations, FY 1968*. 90th Cong., 1st sess. Washington, DC: GPO.

———. 1969. *Atomic Energy Commission Authorizing Legislation, Fiscal Year 1970*, pt. 1, *Hearings before the Joint Committee on Atomic Energy*. 91st Cong., 1st sess., 17–18 April. Washington DC: GPO.

———. 1971a. *Hearings*. 91st Cong., 1st sess. Washington, DC: GPO.

———. 1971b. *AEC Authorizing Legislation, Fiscal Year 1972: Hearings before the Joint Committee on Atomic Energy*. 92nd Cong., 1st sess.: Physical Space Research, Nuclear, and Nuclear Waste Management Programs, 9, 16, 17 March. Pt. 3. Washington DC: GPO.

———. 1972. *Hearings*. 92nd Cong., 2nd sess. Washington, DC: GPO.

U.S. Congress, Subcommittee of the Select Committee on Small Business,1963. *The Role and Effect of Technology in the Nation's Economy*. 88th Cong., 1st sess. Washington, DC: GPO.

U.S. Congress, Subcommittee on Government Research of the Committee on Government Operations. 1966. *Hearings, "Equitable Distribution of R & D Funds by Government Agencies."* 89th Cong., 2nd sess. Washington, DC: GPO.

U.S. Congress, Subcommittee on Research, Development, and Radiation of the Joint Committee on Atomic Energy. 1965. *Hearings*. 89th Cong., 1st sess., Washington, DC: GPO.

———. 1967. *Hearings*. 90th Cong., 1st sess., Washington, DC: GPO.

———. 1972. *Hearings*. 92nd Cong., 2nd sess. Washington, DC: GPO.

U.S. Congress, Subcommittee on Science, Research and Development of the Committee on Science and Astronautics. 1964. *Government and Science: Distribution of Federal Research Funds*. 88th Cong., 2nd sess., Washington, DC: GPO.

VanderArend, Peter. 1983. Interview by Lillian Hoddeson, 15 December.

VanderArend, P., and W. B. Fowler. 1973. "Superconducting Accelerator Magnet Cooling Systems." *IEEE Transactions on Nuclear Science*, n.s., 20, no. 3 (June): 119–121.

Viorst, Milton. 1979. *Fire in the Streets: America in the 1960's*. New York: Simon and Schuster.

"Vs. Scientific Luxury." 1967. *New York Times*, 16 July, editorial.

Waldman, Bernard. 1983. Interview by Catherine Westfall, 4 February.

Waldrop, M. Mitchell. 1984. "The Supercollider, 1 Year Later." *Science* 225 (3 August): 490–491.

———. 1985. "Magnets Chosen for Supercollider." *Science* 230 (4 October): 50.

Walker, Robert. 1960. "The Future of High Energy Physics at Caltech." Caltech Report, document. 17 May.

Ward, Geoffrey C. 1996. *The West: An Illustrated History*. Boston: Little, Brown.

Warnow-Blewett, Joan, and Spencer R. Weart, eds. 1992. *AIP Study of Multi-institutional Collaborations Phase I: High-Energy Physics*. New York: American Institute of Physics.

———. 1995. *AIP Study of Multi-institutional Collaborations Phase II: Space Sciences and Geophysics*. College Park, MD: American Institute of Physics.

———. 1999. *AIP Study of Multi-institutional Collaborations Phase III: Ground-Based Astronomy, Materials Science, Heavy Ion and Nuclear Physics and Computer-Mediated Collaborations*. College Park, MD: American Institute of Physics.

Wattenberg, Albert. 1988. "The Fermi School in the United States." *European Journal of Physics* 9:88–93.

Weart, Spencer. 1979. "The Physics Business in America, 1919–1940: A Statistical Reconnaissance." In *The Sciences in the American Context: New Perspectives*, ed. Nathan Reingold. Washington, DC: Smithsonian Institute Press.

———. 2000. "From Frontiersman to Fermilab: Robert R. Wilson." *Physics in Perspective* 2, no. 2 (June): 141–203. (Based on interview of Wilson by Weart, 19 May 1977.)

Weinberg, Alvin M. 1961. "Impact of Large-Scale Science." *Science* 134 (July 21): 161–164.

———. 1964. "Criteria for Scientific Choice." *Physics Today* 17, no. 3 (March): 42–48.

———. 1967. *Reflections on Big Science*. Cambridge, MA: MIT Press.

Weinberg, Steven. 1993. *Dreams of a Final Theory*. New York: Pantheon Books.

Weiss, Jeffrey. 1990a. Interview by Frederik Nebeker, 1 May.

———. 1990b. Interview by R. Nilan and Rick Nebeker, 3 May.

Weisskopf, V. 1976. "Group meets in U.S.S.R. on Very Big Accelerator." *Physics Today* 29, no. 5:19.

Weisskopf, Victor, and Alvin Weinberg. 1964. "Two Open Letters." *Physics Today* 17, no. 6 (June): 46–47.

Wenzel, William. 1984. Interview by Catherine Westfall, 2 May.

Westfall, Catherine. 1988. "The First 'Truly National Library': The Birth of Fermilab." Ph.D. diss.: Michigan State University.

———. 1989. "The Site Contest for Fermilab." *Physics Today* 42, no. 1 (January): 44–52.

———. 1997. "Science Policy and the Social Structure of Big Laboratories, 1964–1979." In *The Rise of the Standard Model: Particle Physics in the 1960s and 1970s*, ed. Lillian Hoddeson, Laurie Brown, Michael Riordan, and Max Dresden, 364–383. New York: Cambridge University Press.

———. 2001a. "Collaborating Together: The Stories of the TPC, UAI, CDF, and CLAS." *Historical Studies in the Physical and Biological Sciences* 32:163–178.

———. 2001b. "A Tale of Two Laboratories: Readying for Research at Fermilab and Jefferson Laboratory." *Historical Studies in the Physical and Biological Sciences* 32, no. 1:369–407.

———. 2003. "Rethinking Big Science: Modest, Mezzo, Ground Science and the Development of the Bevalac, 1971–1993." *ISIS* 94:30–56.

———. 2004. "Vision and Reality: The EBR-II Story." *Nuclear News* 47, no. 2 (February): 25–32.

Westfall, Catherine, and Lillian Hoddeson. 1996. "Thinking Small in Big Science: The Founding of Fermilab, 1960–1972." *Technology and Culture* 37, no. 3:457–492.

Westwick, Peter. J. 2003. *The National Labs: Science in an American System, 1947–1974*. Cambridge, MA: Harvard University Press.

White, M. G.,1953. "Preliminary Design Parameters for a Separated-Function Machine." Unpublished manuscript, Princeton, NJ, 3 March.

White, Richard. 1988. "Frederick Jackson Turner." In *Historians of the American Frontier*, ed. John R. Wunder, 660–681. New York: Greenwood.

———. 1991a. *"It's Your Misfortune and None of My Own": A History of the American West*. Norman: University of Oklahoma Press.

———. 1991b. *The Middle Ground: Indians, Empires, and Republics in the Great Lakes Region*, 1650–1815. New York: Cambridge University Press.

———. 1994. "The Frontier and the American Mind." Talk given at the Newberry Seminar in American Social History, Chicago, 18 January.

White, Richard, and Patricia Nelson Limerick. 1994. *The Frontier in American Culture: An Exhibition at the Newberry Library, August 26, 1994–January 7, 1995*. Ed. James R. Grossman. Berkeley: University of California Press.

Wilmington, Michael. 1994. "Westerns, Well Listen Up, Pilgrim." *Chicago Tribune*, 22 May.

Wilson, M. N.,1972. "Rate Dependent Magnetization in Flat Twisted Superconducting Cables." Unpublished document, Rutherford High Energy Laboratory Report M/A26, September.

Wilson, Robert R. 1960. "Ultrahigh Energy Accelerators: Summary of a Discussion Held at Rochester, NY, August 28, 1960." Unpublished document.

———. 1961. "Ultrahigh-Energy Accelerators." *Science* 133 (May): 1602–1607.

———. 1965. "Some Proton Synchrotrons, 100–1000 GeV." Unpublished document, 22 September.

———. 1966. "An Anecdotal Account of Accelerators at Cornell." In *Perspectives in Modern Physics: Essays in Honor of Hans A. Bethe on the Occasion of His 60th Birthday*, 225–244. New York: John Wiley, Interscience Publishers.

———. 1967a. "Edited Version of Talk at 1st NAL Users Meeting." Unpublished manuscript, 7 April.

———. 1967b. "National Accelerator Laboratory Synchrotron." Unpublished manuscript, 23 July.

———. 1967c. "Some Aspects of the 200 GeV Accelerator." Paper presented at the Sixth International Conference on High Energy Accelerators, Cambridge, MA, 12 September. Cambridge, MA: Cambridge Electron Accelerator.

———. 1967d. "Wilson Talks about Weston." Unpublished document. FA.

———. 1967e. Tape recording of "1967 Berkeley Meeting."

———. 1968. "Particles, Accelerators and Society." 1968 Richtmyer Lecture. *American Journal of Physics* 36 (June): 490–495.

———. 1969a "Wilson to Whole Staff." Unpublished address, 1 October.

———. 1969b. "Sanctimonious Memo #137." Unpublished document, 11 July. FA.

———. 1970a. "Conscience of a Physicist." *Bulletin of the Atomic Scientists* 26, no. 6 (June): 30–34.

———. 1970b. "Future Options at NAL Batavia." In *The International Conference on High Energy Accelerators 8/27–9/21968, Yerevan, SSR*, ed. A. I. Alikhanian., 103–105. Yerevan: Publishing House of the Academy of Sciences of the Armenian SSR.

———. 1970c. "My Fight against Team Research." *Daedalus* 99, no. 4 (Fall): 1076–1087.

———. 1973. "Initial Experiments at NAL." Unpublished manuscript for talk presented at Coral Gables Conference on High Energy Physics and Cosmology, Coral Gables, FL, 25 January.

———. 1975. "A World Laboratory and World Peace." *Physics Today* 28, no. 11 (November): 120.

———. 1977a. Interview by Spencer Weart, 19 May.

———. 1977b. "The Tevatron." *Physics Today* 30, no. 10 (October): 23–30.

———. 1978a. "Colliding Beams at Fermilab." Proceedings of Workshop on High Luminosity, High-Energy Proton-Antiproton Collisions, 27–31 March 1978 Berkeley, CA, Lawrence Berkeley Laboratory Report 7574, 7–12.

———. 1978b. "Toward a World Accelerator Laboratory." Fermilab TM-811, 16 August.

———. 1978c. Interview by Lillian Hoddeson, 12 January.

———. 1978d. Interview by Lillian Hoddeson, 8 and 10 May.

———. 1978e. Interview by Lillian Hoddeson, 12 May.

———. 1978f. "Colliding Beams at Fermilab." Fermilab TM-807, 11 August.

———. 1978g "The Humanness of Physics." Unpublished talk delivered at Fermilab in 1978. Reprinted as informal Golden Book (Batavia, IL: Fermi National Accelerator Laboratory, 1992).

———. 1979a. Interview by Lillian Hoddeson, 12 January.
———. 1979b. "Fantasies of Future Fermilab facilities." *Reviews of Modern Physics*, 51, no. 2:259–273.
———. 1980. "The Next Generation of Particle Accelerators." *Scientific American* 242, no. 1 (January): 42–57.
———. 1981. Interview by Lillian Hoddeson, 23 January.
———. 1984. "A World Organization for the Future of High-Energy Physics." *Physics Today* 37, no. 9 (September): 9, 112.
———. 1987a. Interview by Catherine Westfall, 1 April.
———. 1987b. Interview by Catherine Westfall, 25 May.
———. 1987c. "Starting Fermilab: Some Personal Viewpoints of a Laboratory Director (1967–1978)." *Annual Report of the Fermi National Accelerator Laboratory*, 163–185. Batavia, IL: Fermilab.
———. 1990a. Interview by Lillian Hoddeson, 13 February.
———. 1990b. Interview by Adrienne Kolb and Catherine Westfall, 15 March.
———. 1992. Interview by Mark Bodnarczuk, 24 September.
Wilson, Robert R., and Adrienne W. Kolb. 1997. "Building Fermilab: A User's Paradise." In *The Rise of the Standard Model: Particle Physics in the 1960s and 1970s*, ed. Lillian Hoddeson, Laurie Brown, Michael Riordan, and Max Dresden, 338–363. New York: Cambridge University Press.
Yamada, Ryuji, 1983. Interview by Lillian Hoddeson, 14 December.
———. 1989. Interview by Lillian Hoddeson and Catherine Westfall, 25 October.
Yamaguchi, Y. 1986. "ICFA: Its History and Current Activities." In *Proceedings of the 1985 International Symposium on Lepton and Photon Interactions at High Energies*, 826–847. Kyoto: Nissha.
Yamanouchi, T. 1990. Interview by Frederik Nebeker, 13 June.
Yoh, J. 1976 "From the People Who Brought You the Upsilon, a Bigger (but Not Necessarily Better) Resonance." Internal experiment note, 17 November.
———. Interview by Catherine Westfall and Frederik Nebeker, 4 May.
Young, David. 1968. "Bid to Annex Ruled Illegal in Weston Survival Fight." *Chicago Tribune*, 1 May.
Young, Don. 1987. Interview by Catherine Westfall, 13 March.
———. 2004. Comments on "The Ring of the Frontier: Big Science at Fermilab, 1967–1989." Unpublished document.
———. 2005. Interview by Adrienne Kolb, 21 March.
Yuan, Luke C. L., ed. 1965. *Nature of Matter: Purposes of High Energy Physics*. Washington, DC: United States Clearinghouse for Federal Scientific and Technical Information.
Yuan, Luke C. L., and J. P. Blewett. 1961. "Experimental Program for Requirements for a 300 to 1000-BeV Accelerator." Unpublished document printed by Brookhaven National Laboratory, Brookhaven, NY.

Index

Page references in *italics* denote figures.